P9-CKY-133

Cisco IP Telephony

David Lovell

Cisco Press

Cisco Press
201 West 103rd Street
Indianapolis, IN 46290 USA

Cisco IP Telephony

David Lovell

Published by:
Cisco Press
201 West 103rd Street
Indianapolis, IN 46290 USA

Printed in the United States of America 2 3 4 5 6 7 8 9 0

Library of Congress Cataloging-in-Publication Number: 2001090432

ISBN: 1-58705-050-1

Second Printing June 2002

Warning and Disclaimer

This book is designed to provide information about configuring a Cisco IP Telephony network. Every effort has been made to make this book as complete and as accurate as possible, but no warranty or fitness is implied.

The information is provided on an "as is" basis. The authors, Cisco Press, and Cisco Systems, Inc. shall have neither liability nor responsibility to any person or entity with respect to any loss or damages arising from the information contained in this book or from the use of the discs or programs that may accompany it.

The opinions expressed in this book belong to the author and are not necessarily those of Cisco Systems, Inc.

Feedback Information

At Cisco Press, our goal is to create in-depth technical books of the highest quality and value. Each book is crafted with care and precision, undergoing rigorous development that involves the unique expertise of members from the professional technical community.

Readers' feedback is a natural continuation of this process. If you have any comments regarding how we could improve the quality of this book, or otherwise alter it to better suit your needs, you can contact us through e-mail at feedback@ciscopress.com. Please make sure to include the book title and ISBN in your message.

We greatly appreciate your assistance.

Trademark Acknowledgments

All terms mentioned in this book that are known to be trademarks or service marks have been appropriately capitalized. Cisco Press or Cisco Systems, Inc., cannot attest to the accuracy of this information. Use of a term in this book should not be regarded as affecting the validity of any trademark or service mark.

Publisher	John Wait
Editor-In-Chief	John Kane
Executive Editor	Brett Bartow
Course Developer	David Lovell
Cisco Systems Management	Michael Hakkert
	Tom Geitner
Acquisitions Editor	Amy Lewis
Production Manager	Patrick Kanouse
Development Editor	Christopher Cleveland
Copy Editor	Chuck Gose
Technical Editors	Mick Buchanan
	Paul Giralt
	Ketil Johansen
	Chris Pearce
	John Livengood
	Anne Smith
Team Coordinator	Tammi Ross
Book Designer	Gina Rexrode
Cover Designer	Louisa Klucznik
Production Team	Octal Publishing, Inc.
Indexer	Tim Wright

Corporate Headquarters
Cisco Systems, Inc.
170 West Tasman Drive
San Jose, CA 95134-1706
USA
http://www.cisco.com
Tel: 408 526-4000
800 553-NETS (6387)
Fax: 408 526-4100

European Headquarters
Cisco Systems Europe
11 Rue Camille Desmoulins
92782 Issy-les-Moulineaux
Cedex 9
France
http://www-europe.cisco.com
Tel: 33 1 58 04 60 00
Fax: 33 1 58 04 61 00

Americas Headquarters
Cisco Systems, Inc.
170 West Tasman Drive
San Jose, CA 95134-1706
USA
http://www.cisco.com
Tel: 408 526-7660
Fax: 408 527-0883

Asia Pacific Headquarters
Cisco Systems Australia,
Pty., Ltd
Level 17, 99 Walker Street
North Sydney
NSW 2059 Australia
http://www.cisco.com
Tel: +61 2 8448 7100
Fax: +61 2 9957 4350

Cisco Systems has more than 200 offices in the following countries. Addresses, phone numbers, and fax numbers are listed on the Cisco Web site at www.cisco.com/go/offices

Argentina • Australia • Austria • Belgium • Brazil • Bulgaria • Canada • Chile • China • Colombia • Costa Rica • Croatia • Czech Republic • Denmark • Dubai, UAE • Finland • France • Germany • Greece • Hong Kong • Hungary • India • Indonesia • Ireland • Israel • Italy • Japan • Korea • Luxembourg • Malaysia Mexico • The Netherlands • New Zealand • Norway • Peru • Philippines • Poland • Portugal • Puerto Rico Romania • Russia • Saudi Arabia • Scotland • Singapore • Slovakia • Slovenia • South Africa • Spain Sweden • Switzerland • Taiwan • Thailand • Turkey • Ukraine • United Kingdom • United States • Venezuela Vietnam • Zimbabwe

Printed in the USA on recycled paper containing 10% postconsumer waste.

About the Author

David Lovell is an Educational Specialist at Cisco Systems, Inc., where he designs, develops, and delivers training on Cisco IP Telephony networks that focus on Cisco CallManager.

David has chosen a career in training and instruction to help people improve their knowledge and skills. While at Pepperdine University he played baseball and studied coaching and teaching. He graduated from Pepperdine University with a BA degree and is in the Pepperdine University Hall of Fame for being on the 1992 National Championship baseball team.

David resides in Dallas, Texas, with his wife and two children.

About the Technical Reviewers

Mick Buchanan, CCNA, CIPT, MCP, CNE, began working with CallManager in 1998 as one of the original Selsius Systems Customer Support Engineers. He is currently responsible for maintaining Cisco's Dallas Technical Briefing Center for product demonstrations on the entire line of Cisco AVVID IP Telephony products.

Paul Giralt, CCIE #4793, is a Customer Support Engineer in the Cisco Systems Technical Assistence Center in Research Triangle Park, North Carolina, and has been with the company since 1998. He is currently an escalation resource for all the Cisco AVVID IP Telephony support teams around the world. He also works closely with EVVBU developers on performing real-world testing on Cisco CallManager prior to customer shipment to ensure software quality. He holds a bachelor's degree in computer engineering from the University of Miami.

Ketil Johansen, CCIE #1145, is a Systems Engineer with Cisco Systems in Denmark. Ketil has worked with networking technologies since 1984, and since 1998, he's been focusing on IP telephony technologies.

Chris Pearce is a technical leader in the Cisco CallManager software group at Cisco Systems, Inc. He has 10 years of experience in telecommunications. His primary areas of expertise include call routing, call control, and telephone features. Chris was a member of the team that developed and implemented the Cisco CallManager software from its early stages, and he was directly involved in developing the system architecure and design.

John Livengood is a General Manager with Advanced Network Information (ANI), a Cisco Learning Solutions Partner. ANI is focused on AVVID training. John has 30 years experience in the computer/network industry, 22 years in aviation for Learjet, Beech, and Boeing. John has spent the last two and a half years teaching Cisco IP Telephony courses. John helps EVVBU with course development and beta testing of those courses.

Anne Smith is a technical writer in the CallManager engineering group at Cisco Systems, Inc. She has written comprehensive technical documentation for Cisco Systems since the Selsius Systems acquisition in 1998. She is the author of *Cisco CallManager Fundamentals: A Cisco AVVID Solution* (ISBN: 1-58705-008-0).

Dedications

This book is dedicated to my family. I would not have had the strength to finish this book without the support of my wife, Wendy. I also dedicate this book to my son, Jaime, and daughter, Keely, for reminding me that there are precious things in life that I should stop and take notice of. I also dedicate this book to my brother, Charles, for opening the opportunity for me to work for Cisco Systems, Inc. And finally, I dedicate this book to my mom and dad, Bennett and Toni, for always believing in me that I could do anything.

Acknowledgments

The CIPT course came out of an acquisition of Selsius Systems, Inc., who developed the Cisco CallManager product. Although I am the primary course developer for CIPT, I would like to acknowledge the many teams that contribute to the development of this product from it's early acquisition stages to where the product is now with Cisco Systems: The CAP, Solutions, Integration, technical marketing engineers (TME), and documentation and development engineers that have answered numerous questions about the technology to help bring this course and book together.

I'd like to acknowledge my managers, Wade Hamblin and Scott Veibell, for allowing me to pursue this endeavor and clear the way when tough questions needed answers.

Thanks to Addis Hallmark for his patience and clarity when discussing complex concepts that are included in this book. Addis has been a "go-to" guy for over a year for answers and clarity of how this works in the "real-world."

I'd like to give special recognition to Paul Giralt for providing his expert technical knowledge in editing the book. Paul's expertise is well known within the support ranks of Cisco IP Telephony.

I would like to thank Mick Buchanan for being a great sounding board and for helping work through numerous equipment issues.

Special acknowledgements go to John Livengood, who provided an up close view of what customers want, and also to Ketil Johansen for broadening my scope of global awareness. Also in this group is Graham Gudgin, who provided extra information for Chapter 13 of this book.

I would also like to thank Chris Pearce for his help with the route plan section and for always having time to complement complex dial plan issues with great analogies.

I would like to thank Anne Smith, who toiled with my work and pushed me when I thought I couldn't go on, even while she was recovering from knee surgery.

Thanks to the team at Advanced Network Information (ANI) for the review sessions and countless additions to the course and book.

A big 'thank you' goes out to the production team for this book: John Kane, Amy Lewis, and Christopher Cleveland have been incredibly professional and a pleasure to work with. I couldn't have asked for a finer team.

Contents at a Glance

Table of Contents

Foreword

With the acquisition of Selsius Systems in 1998, Cisco Systems, acquired the heart to their converged strategy for voice, video, and data communications, known as Cisco AVVID (Architecture for Voice, Video, and Integrated Data). Cisco CallManager provides the core call processing capability for Cisco AVVID, enabling enterprises and organizations to deploy voice communications over the same network on which they already manage data communications.

Early in 1999, the first Cisco CallManager training class was developed with the help of development engineers from the Selsius acquisition and veteran training professionals from Cisco and training partners. One of the more impressive training professionals that participated on that team was David Lovell, the author of this book and also the author of all Cisco IP Telephony (CIPT) courses for the last two years. When given the opportunity to hire an education specialist who would be dedicated to Cisco CallManager development, I immediately called David (who lived in the San Jose, California, area at the time) and asked if he would be interested in moving to Dallas to join our Cisco CallManager engineering team. To my pleasant surprise the answer was "Yes!" I believe the quality training material developed and delivered over the last two years speaks for itself, and many thousands have since received Cisco CallManager training from the various Cisco training partners using David's courses.

This book represents the first time information from the CIPT course has been made available outside the classroom. Targeting Cisco CallManager release 3.1(1), the information in this book represents a valuable resource for those preparing to take the CIPT Certification Exam in the near future and also serves to fill in the gaps for those who have become familiar with Cisco CallManager but want to learn more about concepts or procedures that are new to them. Although a publication such as this cannot replace an instructor-led training environment, I believe that the CIPT material presented in this book will be of significant value to many readers, including those who have (or will soon) attend the CIPT course, as well as those who have not. Finally, I know it will be a great help as you prepare for and take the Cisco CallManager certification test.

Scott Veibell
Manager, Software Development Group
Enterprise Voice, Video Business Unit
Cisco Systems, Inc.

Introduction

Professional certifications have been an important part of the computing industry for many years and will continue to become more important. Many reasons exist for these certifications, but the most popularly cited reason is that of credibility. All other considerations held equal, the certified employee/consultant/job candidate is considered more valuable than one who is not.

Goals and Methods

The most important and somewhat obvious goal of this book is to help you pass the Cisco IP Telephony exam. In fact, if the primary objective of this book were different, then the book's title would be misleading; however, the methods used in this book to help you pass the CIPT exam are designed to also make you much more knowledgeable about how to do your job.

One key methodology used in this book is to help you discover the exam topics that you need to review in more depth, to help you fully understand and remember those details, and to help you prove to yourself that you have retained your knowledge of those topics. So this book does not try to help you pass by memorization; rather, it helps you truly learn and understand the topics.

Who Should Read This Book?

This book is designed to provide you with a foundation for working with Cisco IP Telephony products, specifically Cisco CallManager. If your task is to install, support and maintain a CIPT network, this is the book for you. In your support room for Cisco IP Telephony network, you should have this book and the Cisco Press book *Cisco CallManager Fundamentals: A Cisco AVVID Solution* (ISBN:1-58705-008-0). With these two books and the online documentation in Cisco CallManager, you will be able to install, support, and maintain a CIPT network.

How This Book Is Organized

Although this book could be read cover-to-cover, it is designed to be flexible and allow you to easily move between chapters and sections of chapters to cover just the material that you need more work with. Chapter 1 provides an overview of Cisco AVVID and CIPT. Chapters 2 through 8 discuss some basic configuration and Chapters 9 through 13 discuss advanced configuration concepts. Chapter 14 lightly discusses the applications that are in a CIPT network. If you do intend to read them all, the order in the book is an excellent sequence to use.

Chapters 1 through 14 cover the following topics:

- **Chapter 1: Introduction to Cisco IP Telephony (CIPT) Components**—This chapter introduces you to Cisco AVVID and the CIPT components. A light introduction to VoIP and how CIPT is designed.

- **Chapter 2: Navigation and System Setup**—This chapter covers basic administration and configuration in Cisco CallManager Administration. It provides a map of Cisco CallManager Administration and describes the basic steps you can use to configure a Cisco CallManager cluster.
- **Chapter 3: Cisco CallManager Administration Route Plan Menu**—The Route Plan menu in Cisco CallManager Administration is used to configure route patterns, groups and lists. This chapter discusses configuration of a route plan and the route plan flow, using route groups and lists.
  - The major topics in this chapter are the following:
    - Route Plan Flow
    - Digit Manipulation (translation patterns and transformation masks)
    - Providing a Class of Service for Devices (partitions and calling search spaces)
- **Chapter 4: Cisco CallManager Administration Service Menu**—This chapter focuses on the Media Resources menu in Cisco CallManager Administration and how to configure those resources using media resource groups and lists. The main topics in this chapter are: how to group media resources and how they are applied to devices based on their grouping.
- **Chapter 5: Cisco CallManager Administration Feature and User Menus**—This short chapter discusses the Feature and User menu. These two menu items of Cisco CallManager Administration are together as one menu item because the user will usually subscribe to the features in Cisco IP Phone services. The main topics in this chapter are call park, call pickup, Cisco IP Phone services, and adding a user.
- **Chapter 6: Cisco IP Telephony Devices**—This chapter focuses on how to add and configure devices in Cisco CallManager Administration to handle added devices. You need to know what type of device is being added and what information is critical for that device to be added. Phone configuration is a crucial topic which administrators need to be familiar. Knowing what parts of the configuration pages affect what the user will see is critical in providing a good end-user experience. Gateway information can be critical when troubleshooting. Using the description field for devices, especially gateways, can save time during troubleshooting and working with the service provider.
- **Chapter 7: Understanding and Using the Bulk Administration Tool (BAT)**—BAT is a great tool for large deployments for adding, updating, or deleting large numbers of phones, users, and gateways. This is a tool that all administrators should know how to use. Adding thousands of phones associated with users within minutes is a big plus. This chapter is intentionally placed after all the manual configuration chapters to highlight the usefulness of this tool.
- **Chapter 8: Installation, Backups, and Upgrades**—Installing a Cisco CallManager server does not affect call processing, registration, or other features of Cisco CallManager unless the installation is not done correctly. Proper design is critical for a good Cisco CallManager cluster to be built. The focus of this chapter is the installation of a Cisco CallManager server, the post-installation tasks, and the upgrading process for a Cisco CallManager cluster.

- **Chapter 9: LAN Infrastructure for Cisco IP Telephony**—A good infrastructure is critical to the success of a CIPT network solution. The end user does not distinguish between the network going down nor Cisco CallManager failing when they cannot get a dial tone from their phone. Cisco CallManager is an application running on the network, meaning that the network infrastructure needs to made voice-ready.
- **Chapter 10: Call Preservation**—Which calls stay active? These questions are answered in this chapter. This chapter will teach you how to ensure that calls stay active even when a Cisco CallManager fails.
- **Chapter 11: Media Resources**—Which server provides the media resources? This question is answered in this chapter. This chapter will help you in designing your network for allocating resources. The MOH section of this chapter discusses the process of the audio sources and provides the rules for how the MOH server is determined, based on where it is configured in Cisco CallManager Administration.
- **Chapter 12: WAN Design Considerations for Cisco IP Telephony Networks**—Gatekeeper and survivable remote site telephony (SRST) are the major topics in this chapter. SRST has been available since 3.0(5) but will get its real exposure with the Catalyst 4224 Access Gateway Switch.
- **Chapter 13: Best Practices for Cisco IP Telephony Deployment**—This chapter provides tips and techniques recommended by Cisco Systems, Inc., CallManager engineering team to install and maintain a CIPT network.
- **Chapter 14: Applications**—This chapter discusses the applications that are part of the CIPT network.

Icons Used in This Book

Throughout this book, you will see a number of icons used to designate Cisco and general networking devices, peripherals, and other items. The icon legend that follows explains what these icons represent.

Cisco CallManager

Cisco IP Phone 7960 or Generic Cisco IP Phone

Cisco IP Phone 7940

Cisco IP Phone 7910

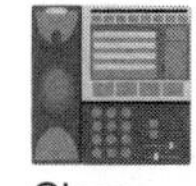

Cisco IP SoftPhone

Stations

Tower
Used for:
Voice mail server
MOH server
SW conference bridge
MTP

Network Device Icons

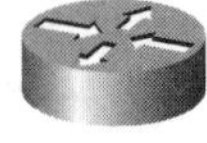
Router

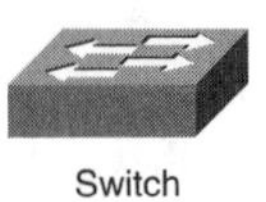
Switch

Layer 3 switch

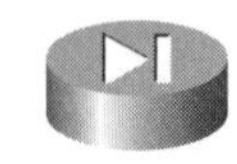
PIX Firewall

Gateway or
3rd-party
H.323
server

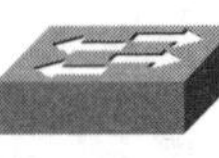
Voice-enabled
switch

Access Server

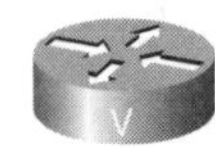
Used for:
Voice-enabled router
Gateway
Gatekeeper
HW Conference bridge
Transcoder
Analog gateway
H.323 gateway
DHCP
DNS

PBX Switch

PBX (small)

Cisco
Directory
Server

LAN/WAN Device Icons

PC

Laptop

Server

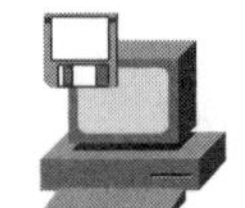
PC w/software

Modem

POTS Phone

Relational
Database

Fax machine

Media/Building Icons

Network Cloud

Ethernet connection

Serial connection

Telecommuter

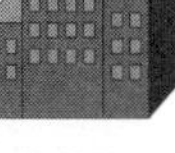

Building

Branch Office

Command Syntax Conventions

The conventions used to present command syntax in this book are the same conventions used in the IOS Command Reference. The Command Reference describes these conventions as follows:

- Vertical bars (|) separate alternative, mutually exclusive elements.
- Square brackets [ ] indicate optional elements.
- Braces { } indicate a required choice.
- Braces within brackets [{ }] indicate a required choice within an optional element.
- **Boldface** indicates commands and keywords that are entered literally as shown. In actual configuration examples and output (not general command syntax), boldface indicates commands that are manually input by the user (such as a **show** command).
- *Italics* indicate arguments for which you supply actual values.

Comments to the Author

The author is interested in your comments and suggestions about this book. Please send feedback to the following address:

ciptbook@cisco.com

PART I

Getting Started with Cisco IP Telephony

Upon completing this chapter you will be able to do the following tasks:

- Given a four-layer box, label each box with the correct layer from the Cisco Architecture for Voice, Video, and Integrated Data (Cisco AVVID) architecture.
- Given some pictures or definitions, identify the core Cisco IP Telephony (CIPT) components.
- Given a list of tasks performed by the Cisco CallManager, identify and describe the primary tasks performed by Cisco CallManager.
- Given photos of the different Cisco IP Phones, identify the model number of those phones.

CHAPTER 1

Introduction to Cisco IP Telephony (CIPT) Components

This chapter discusses the Cisco IP Telephony (CIPT) components. CIPT is part of Cisco's Architecture for Voice, Video and Integrated Data (Cisco AVVID). The Cisco AVVID focus is for a converged network that runs voice, video, and data. CIPT, centered on Cisco CallManager, converges voice and data networks. This chapter introduces Cisco AVVID and the core components of CIPT. The following topics are discussed in this chapter:

- Abbreviations
- Audience
- Prerequisites
- Sources of Information
- Cisco AVVID
- CIPT Components

Abbreviations

This section defines the abbreviations used in this chapter. For more information about terms and abbreviations used in this chapter refer to the *IP Telephony Network Glossary* at the following URL:

www.cisco.com/univercd/cc/td/doc/product/voice/evbugl4.htm

Table 1-1 provides the abbreviations and complete term for abbreviations used frequently in this chapter.

Table 1-1 *Abbreviation with Definition*

Abbreviations	Definitions
AA	Auto Attendant
API	application programming interface
ATM	Asynchronous Transfer Mode
AVVID	Architecture for Voice, Video, and Integrated Data
CIPT	Cisco IP Telephony
CRA	Customer Response Applications
DHCP	Dynamic Host Configuration Protocol
DSP	digital signal processor
DTMF	dual tone multifrequency
GW	gateway
HTML	Hypertext Markup Language
ICD	Integrated Contact Distribution
IOS	Internetworking Operating System
IPCC	IP Contact Center
IVR	Interactive Voice Response
JTAPI	Java TAPI
MCS	Media Convergence Server
MGCP	Media Gateway Control Protocol
OSI	Open System Interconnection
PA	Personal Assistant
PBX	private branch exchange
POP	Post Office Protocol
PPS	Phone Productivity Services
PSTN	public switched telephone network
PTT	Post, Telephone, and Telegraph
QoS	Quality of Service
SMDI	simple message desktop interface
SMTP	Simple Mail Transfer Protocol
TAPI	Telephony Application Programming Interface
TDM	time-division multiplexing

Table 1-1 *Abbreviation with Definition (Continued)*

Abbreviations	Definitions
UM	Unified Messaging
UPS	uninterruptible power supply
UTP	unshielded twisted pair
VG200	Voice Gateway 200
VM	voice messaging
VoIP	Voice over IP
WAN	wide-area network

Audience

This book is written for the individual responsible for installing, configuring, and maintaining a CIPT solution. CIPT spans a variety of disciplines, so you may have a strong networking background or a strong telephony background or a strong applications background or a strong server background. Regardless of your background, this book will try to address all aspects of CIPT. Will the book cover every concept in which you are interested to the depth that you would like? Probably not, but if more information is needed, you can search on www.cisco.com to find numerous links that will assist in your learning.

Because CIPT spans a variety of disciplines, your skills, with the help of this book, will enable you to become a valuable asset to your company. It is very rare to find an individual that is knowledgeable of all the disciplines in a CIPT network. By reading through this book, you will become one of the rare individuals with knowledge of all four disciplines.

Prerequisite Knowledge

To fully benefit from reading this book, you should already possess certain skills. These skills can be gained from completing the *Internetworking Technology Multimedia (ITM)* CD-ROM or through work experience. You should have the following prerequisite experience:

- Cisco Certified Networking Associate (CCNA).
- Building VoIP networks. Gained from the Cisco course, *Cisco Voice Over Frame Relay, ATM, and IP v2.0 (CVOICE).*
- Building a switched LAN environment. Gained from the Cisco course, *Building Cisco Multilayer Switched Networks (BCMSN).*
- Work experience and knowledge of Windows 2000 Server and SQL server.

For more practice to help reinforce the concepts discussed in this book, it will benefit you to have access to Cisco CallManager, Cisco IP Phones, and other equipment discussed in this book.

Sources of Information

Most of the information presented in this book can be found on the Cisco Systems Web site or on CD-ROM. These supporting materials are available in HTML format and as manuals and release notes.

To learn more about the subjects covered in this book, feel free to access the following sources of information:

- *Cisco Documentation* CD-ROM or www.cisco.com.
- ITM CD-ROM or www.cisco.com.
- *Cisco IOS 12.0 Configuration Guide and Command Reference Guide*: www.cisco.com/univercd/cc/td/doc/product/software/ios120/12cgcr/
- *Voice over IP Fundamentals*, by Jonathan Davidson, CCIE. ISBN: 1-57870-168-6.
- *Cisco CallManager Fundamentals: A Cisco AVVID Solution*, by John Alexander, Chris Pearce, Anne Smith, and Delon Whetten. ISBN: 1-58705-008-0.
- *Routing TCP/IP*, Volume I, by Jeff Doyle, CCIE. ISBN: 1-57870-041-8.
- *Internetworking Technologies Handbook*, Third Edition, by Cisco Systems, Inc. ISBN: 1-58705-001-3.
- *Integrating Voice and Data Networks*, by Scott Keagy, CCIE. ISBN: 1-57870-196-1
- *Cisco IP Telephony Solution Guide*: www.cisco.com/warp/public/788/solution_guide/.
- *Cisco IP Telephony Network Design Guides*: www.cisco.com/univercd/cc/td/doc/product/voice/ip_tele/index.htm.

All online documents can all be found at www.cisco.com.

Cisco AVVID

Why is there a movement to converged networks and why use IP? As shown in Figure 1-1, the Web, or Internet/intranet, has the most successful application architecture around. Application servers, either private or public, reside on the Internet or intranet serving applications, such as HTML, Java, RealAudio, MP3, POP/SMTP mail, and others.

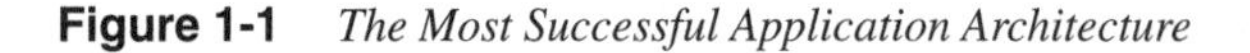

Figure 1-1 *The Most Successful Application Architecture*

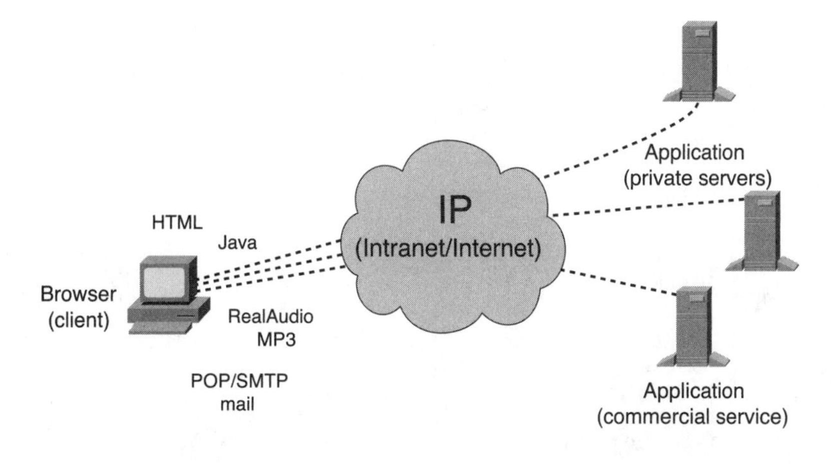

Today's multiple communication networks are entirely separate, each serving a specific application. The traditional PSTN (TDM) network serves the voice application; the Internet and intranets serve data communications; and multiple private and public H.320 networks exist for the purpose of video conferencing.

Often, everyday business requirements force these networks to interoperate. (Have you ever been on a videoconference for video while using the PSTN and a Polycom for audio or used NetMeeting?) As a result, deploying multiservice (data, voice, and video) applications, such as UM or Web-based customer contact centers, is a daunting task because it requires expensive and complex links between proprietary systems (such as PBXs) and standards-based data networks.

Cisco AVVID is a significant milestone in the evolution of IP networking for the enterprise. It enables customers to move from maintaining a separate data network and a closed, proprietary voice PBX system to maintaining one open and standards-based converged network for all their data, voice, and video needs. Also, Cisco AVVID provides customers a path to the New World today, removing their dependency on Old World proprietary suppliers who have yet to migrate from traditional PBXs.

Figure 1-2 displays the classic four layers of Cisco AVVID in the infrastructure layer, the Call Processing layer denoting Cisco CallManager, the Application layer where Contact Center and UM reside, and the Client layer where end devices such as Cisco IP Phones reside. The key concepts are as follows:

- The Client layer brings the applications to the user regardless if the end device is a Cisco IP Phone, a PC using Cisco IP SoftPhone, or a PC delivering UM.
- Applications are physically independent from call processing and the physical voice processing infrastructure; they may reside anywhere within your network.

- Call processing is physically independent from the infrastructure. Thus, you can have a Cisco CallManager in Chicago processing call control for a bearer channel in San Francisco.
- The infrastructure can support multiple client types, hard phones (physical devices such as Cisco IP Phones), soft phones (virtual devices such as Cisco IP SoftPhone), and video phones.

Figure 1-2 *Cisco AVVID Layers*

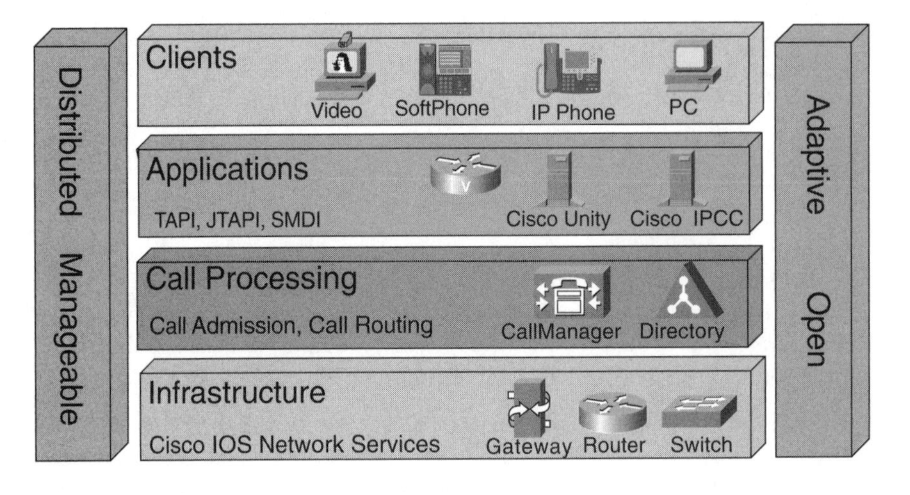

Applications

IP-based voice and video applications increase competitive advantages by improving productivity and enabling exceptional customer care. The following are some of the voice and video applications currently available:

- Cisco Personal Assistant (PA) is an IP-based telephony application that delivers personalization, ease of use, and enhances productivity in the workplace by streamlining voice communications with personal call rules and speech recognition. As an integral part of the Personal Productivity Suite, PA interoperates with Cisco CallManager and scales to meet the present and future needs of enterprises. Users browse voice mail, dial by name, and conference from any telephone using verbal commands instead of the telephone keypad. The administration interface enables users to forward and screen calls in advance or in real time.
- Cisco Integrated Contact Distribution (ICD) provides automated voice call distribution within the enterprise and supports custom contact interaction management.

- Cisco IP PPS is a suite of personal productivity applications for Cisco IP Phones that provides services that allow users to check e-mail, voice mail, calendar, and personal contact information using the large LCD display and interactive soft keys on the Cisco IP Phone.
- Cisco Unity messaging application, a key component of the CIPT solution, provides VM functionality to enterprise communications. Cisco Unity is also able to deliver a unified communications solution for the CIPT solution. A unified communications solution provides access to voice, e-mail, and fax messaging from a single source using various access points.
- Cisco IP Contact Center (IPCC) combines CIPT and contact center solutions. The IPCC delivers an integrated suite of proven products that enable agents using Cisco IP Phones to receive both TDM and VoIP calls. Because the IPCC was intended for integration with legacy call center platforms and networks, it provides a migration path to IP-based customer contact while taking advantage of previous technology investments.
- IP/TV and IP/Video Conferencing products enable distance learning and workgroup collaboration.
- An IP-powered Interactive Voice Response (IVR) solution, Cisco IP IVR combined with Cisco IP AA provides an open, extensible, and feature-rich foundation for delivering IVR solutions over an IP network.
- Cisco WebAttendant replaces the traditional PBX manual attendant console; WebAttendant provides a flexible and scalable IP-based solution.
- Cisco IP SoftPhone provides transportable communication capabilities that increase personal efficiency and promote collaboration.

For more information about the Cisco AVVID application layer go to

www.cisco.com/warp/public/779/largeent/avvid/products/iptel_apps.html.

Call Processing

Cisco CallManager is a software-based call processing application that is the foundation of an end-to-end VoIP network solution. Cisco CallManager can be distributed and clustered over an IP network allowing it to scale to 10,000 users with triple call processing redundancy per device. Multiple clusters can be tied together to enable expansion of up to hundreds of thousands of users. Cisco CallManager provides signaling and call control services to Cisco integrated applications as well as third-party applications. Cisco CallManager is central to the distributed architecture of any converged network.

For more information about the Cisco AVVID call processing layer go to

www.cisco.com/warp/public/779/largeent/avvid/products/call_process.html.

Infrastructure

The following is a breakdown of components of the infrastructure layer of Cisco AVVID:

- **Cisco Certified Servers**—High-availability server platforms for Cisco AVVID from Compaq (such as the ProLiant DL320 and DL380), the MCS-7800 series (such as the MCS-7825 and 7835), and IBM (such as the xSeries 330 and 340). The Cisco certified server platforms are an integral part of a complete, scalable architecture for a new generation of high-quality IP voice solutions that run on enterprise data networks. The servers deliver the high performance and availability demanded by today's enterprise networks and represent a turnkey solution that is easy to deploy and highly cost effective.
- **Switches**—The entire line of Catalyst chassis and stackable switches support a variety of voice features, including inline power for Cisco IP Phones and VoIP gateway cards. Some examples include the Cisco Catalyst 6000, Cisco Catalyst 4000, Cisco Catalyst 3500, and Cisco Catalyst 2900.
- **Integrated IP telephony solution**—An ideal communications solution for branch office and mid-market businesses wanting to deploy New World applications. Some examples are the Cisco Integrated Communications System 7750 (ICS-7750) and the Cisco Catalyst 4200 product line.
- **Voice gateways**—Cisco voice gateways are a reliable, manageable way to connect IP telephony systems to existing switches or analog devices and to provide trunk capabilities that can scale to accommodate IP telephony networks for small, medium, and large offices. The Cisco voice gateway products extend enterprise-class versatility, integration, and cost-savings to remote sites and branch offices by enabling toll-bypass. Some examples are the Cisco Catalyst 6000 T/E1 module, Cisco 2600, Cisco 3600, Cisco AS5300, and Cisco Catalyst 4000.

For more information about the Cisco AVVID infrastructure layer go to

www.cisco.com/warp/public/779/largeent/avvid/products/infrastructure.html.

Clients

Cisco is now delivering second-generation IP-enabled communication devices:

- **Cisco IP Phone 7960**—The Cisco IP Phone 7960 is a second generation full-featured, IP telephone primarily for executives and managers providing six line or speed dial buttons and many more features.
- **Cisco IP Phone 7940**—The Cisco IP Phone is similar to the 7960 with two lines or speed dial buttons.
- **Cisco IP Phone 7910**—The Cisco IP Phone 7910 is a second generation, basic-feature phone mainly for use in lobby areas, break rooms, and hallways. This phone features a single line without the large LCD found on the 7960 and 7940 models.

- **Cisco IP Conference Station 7935**—The Conference Station 7935 couples state-of-the-art conference room speakerphone technologies from Polycom with the Cisco award-winning AVVID technologies.
- **Cisco IP SoftPhone**—The Cisco IP SoftPhone is a PC-based phone that shares characteristics with the Cisco IP Phone 7960. The SoftPhone is a virtual device, allowing the user to effectively bring his or her phone along (on a laptop) during travel and still have the same phone functionality available as if the user were still at his or her desk in the office.
- **Cisco IP Phone 30VIP**—The Cisco IP Phone 30VIP is a first-generation (legacy) Cisco IP Phone (replaced by the 7960). It provides 30 programmable line or feature buttons, high-quality speakerphone/microphone, mute, and a two-line LCD for call status and ID.
- **Cisco IP Phone 12SP+**—The Cisco IP Phone 12SP+ is a first-generation (legacy) Cisoc IP Phone (replaced by the 7910). It provides 12 programmable line or feature buttons, high-quality speakerphone, and a two-line LCD for call status and ID.

NOTE Although the Cisco IP Phone 30VIP and 12 SP+ can be found in some networks, these models can no longer be purchased from Cisco.

For more information about the client layer of Cisco AVVID, go to

www.cisco.com/warp/public/779/largeent/avvid/products/clients.html

VoIP Flavors

There are many flavors of VoIP. Figure 1-3 shows how VoIP was first introduced. Over time, the Internet (and data networking technology in general) has been slowly subsuming these other traffic types. (Desktop conferencing is becoming more ubiquitous, and the mainframe has evolved from being a separate network to being a server on the IP data network.) This convergence has recently—within the past four years—begun absorbing voice and video as applications into the data network. Several large PTT carriers have been using packet switching or Voice over ATM as their backbone technology for years, and enterprise customers have accepted virtual trunking or connecting their disparate PBXs via their wide-area data network to avoid long distance toll charges. Toll-bypass configuration details are discussed in the Cisco Voice over Frame Relay, ATM, and IP (CVOICE) course.

Some examples of Internet telephony solutions include NetMeeting-to-NetMeeting and NetMeeting-to-phone via the Internet.

Figure 1-3 *The Many Flavors of VoIP*

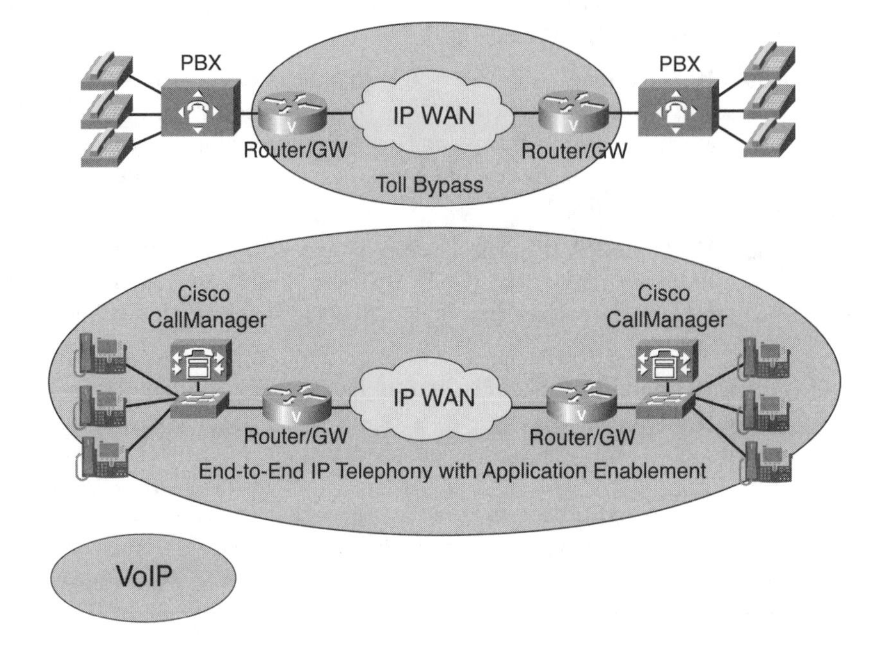

With the introduction of Cisco CallManager, companies can incorporate the toll-bypass solution and provide an end-to-end solution using VoIP. Cisco CallManager provides call processing for Cisco IP Phones and signaling support for end-to-end VoIP. This book discusses the end-to-end solution centered on Cisco CallManager.

The end-to-end solution is built using the following three levels:

- Applications
- Scalable call processing
- QoS-enabled infrastructure

NOTE There are four layers of Cisco AVVID, however, this section focuses on the three mentioned in the preceding list. The fourth, Clients, will be discussed in the "CIPT Components" section in this chapter.

As in building a house, you want to start with the foundation. An infrastructure without QoS will not provide a good VoIP experience for you or the end user. End users expect quality voice and a reliability of five-nines (0.99999) from their voice applications. A QoS-enabled infrastructure is the first step to building an end-to-end VoIP solution.

Scalable call processing can be compared with the electricity delivered to all rooms of a house and the circuit box being able to supply electricity to more rooms when you build extensions to the home. Each time you walk into a room and turn on a light, you want the electrical current available to flow to that room and illuminate the lights. Cisco CallManager provides scalability using a clustered environment and available call processing. When the user makes a call, Cisco CallManager is available to handle the call processing for that call.

Applications are like the electrical appliances we plug into electrical outlets in our home. A variety of appliances (television, radio, refrigerator, computer, microwave, and so on) are plugged in, all working together and providing services that we need. Applications in an end-to-end VoIP solution can provide services such as directory services to the phone, Web content, music, IVR, and much more.

The New World, open packet telephony architecture as shown in Figure 1-4 provides a standards-based packet infrastructure, open Call Control layer, and an open service Application layer. The TDM or circuit-switched technology can be aligned with any of the three layers in an end-to-end VoIP solution.

Figure 1-4 *Open Packet Telephony Architecture*

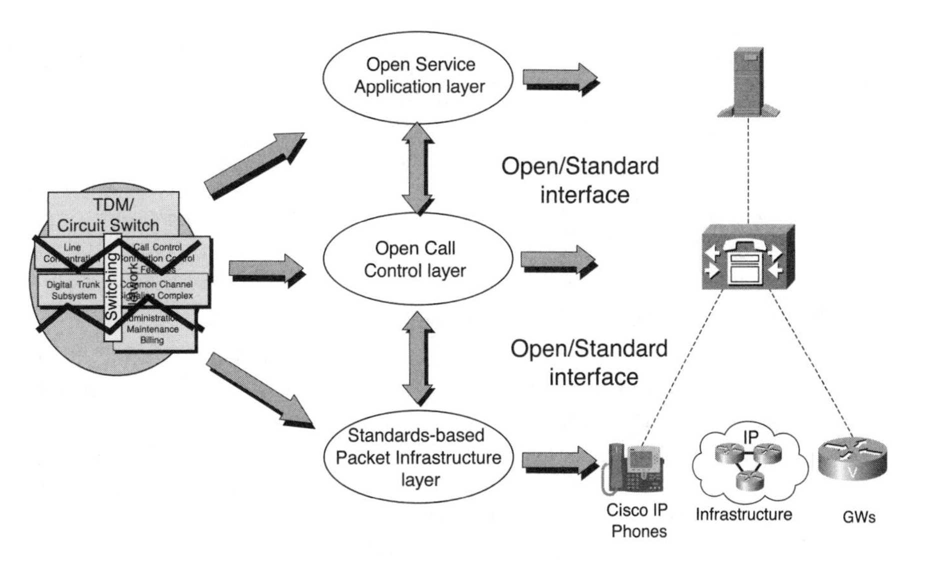

The standards-based infrastructure provides for client, gateway, and switching devices to operate in this converged network.

Cisco CallManager works in the Call Control layer, providing call processing and signaling for clients and gateways.

The Application layer is open, and innovative application services are rapidly being developed. Voice mail (VM), IVR, Audio Attendant (AA), and more applications are provided for in this layer.

CIPT Components

The CIPT network is a part of Cisco AVVID. The first "V" in Cisco AVVID designates the CIPT solution. Figure 1-5 shows a simple topology that highlights the CIPT components.

Figure 1-5 *CIPT Components*

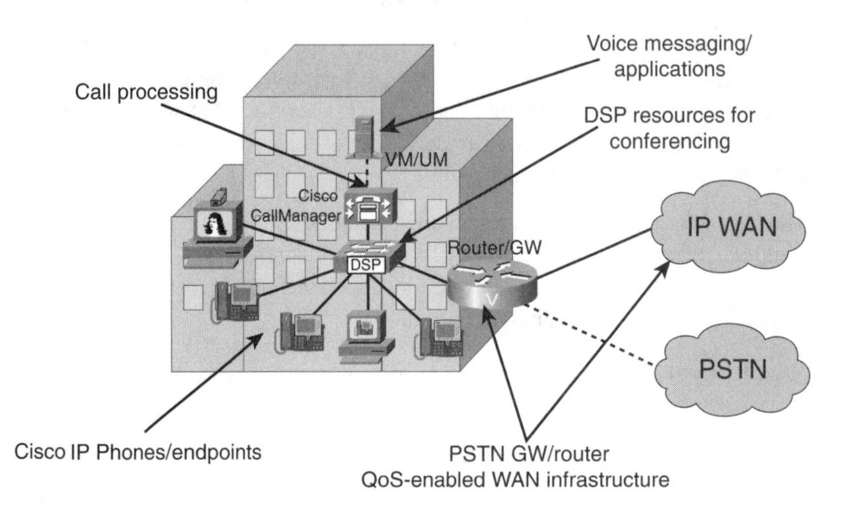

The following CIPT components are discussed in this section:

- **Cisco CallManager**—The software-based IP call processing engine of the CIPT network. Cisco CallManager extends enterprise telephony features and functions to packet telephony network devices such as Cisco IP Phones, media processing devices, VoIP gateways, and multimedia applications.
- **Cisco IP Phones**—The IP-based, full-featured telephones. Cisco IP Phones provide standard telephone functionality: call connection, hold, transfer, call park, conferencing, and more. Two of the available Cisco IP Phone models (7960 and 7940) enable you to deploy custom services that extend the phone's functionality beyond the traditional realm. With Cisco IP Phone services, you can extend the phone's purpose to more than a simple voice communication device.
- **Gateways**—The device that enables Cisco CallManager to communicate with non-IP telecommunications devices. Feature interoperability with versions of Cisco CallManager make these products perfect gateways for the PBX and PSTN for IP telephony, enabling features like call transfer, hold, and conferencing.
  - **Digital Gateways**—The Voice T1 and Services Module allows larger enterprises to connect the PSTN and legacy PBXs directly into the campus multiservice network.

- Analog Gateways (sub-item):
  - **Analog Gateways**—The Analog Gateway Module allows enterprises to connect legacy analog telephony equipment such as phones, speaker phones, and faxes.
- **Inline Power Modules**—The devices that enable inline power and simplify IP telephony deployment in the enterprise wiring closet and branch office. The Inline Power 10/100 Ethernet switching modules enable new Fast Ethernet features that are necessary for convergence in the wiring closet, such as phone discovery and auxiliary or voice VLANs.
- **DSP Resources**—Transcoding and conferencing are services for the multiservice network. Transcoding enables a full voice compression solution by offering transcoding services to endpoints not capable of supporting compressed voice or a different encoding type to the remote end. Cisco CallManager directs the media stream to one of several available DSP resources that bridge the media streams together while converting one codec to another dynamically.
- **Applications**—The various software that extends the functionality of the CIPT solution. Applications include VM, PA, IVR, IPCC, IP AA, and more.

NOTE Because there are many products that integrate with CIPT, the following URL lists all the products and part numbers to order from Cisco Systems, Inc.:

www.cisco.com/univercd/cc/td/doc/pcat/iptl__e1.htm#CHDDGAGB

Cisco CallManager

Cisco CallManager is the core call processing component of the CIPT solution. Cisco CallManager extends enterprise telephony features and functions to packet telephony network devices such as Cisco IP Phones, media processing devices, VoIP gateways, and multimedia applications. Additional data, voice, and video services, such as UM, multimedia conferencing, collaborative contact centers, and interactive multimedia response systems interact with the IP telephony solution through Cisco CallManager's open TAPI. Cisco CallManager is installed on the Cisco MCS and the following certified server platforms: Compaq ProLiant DL320 and DL380 servers and IBM xSeries 330 and 340 servers.

Cisco CallManager includes a suite of integrated voice applications that perform voice conferencing and manual attendant console functions. The salient benefit of all these voice applications is that special-purpose voice processing hardware is not required. Supplementary and enhanced services such as hold, transfer, forward, conference, multiple-line appearances, automatic route selection, speed dial, last-number redial, and other features are extended to Cisco IP Phones and gateways. Because Cisco CallManager is a software application, enhancing its capabilities in production environments is a matter of upgrading software on the server platform, thereby avoiding expensive hardware upgrade costs.

Further, Cisco CallManager and all phones, gateways, and applications can be distributed across an IP network, providing a distributed, virtual telephony network. The benefit of this architecture is improved system availability and scalability. Call admission control ensures that voice QoS is maintained across constricted WAN links and automatically diverts calls to alternate PSTN routes when WAN bandwidth is not available.

The MCS 7800 series (MCS-7800), Compaq Proliant DL320 and DL380, and IBM xSeries 330 and 340 are the supported hardware platforms for Cisco CallManager that provide for scalable hardware architecture.

MCS 7800 Series

The Cisco MCSs are powerful platforms for Cisco AVVID. The Cisco MCSs are an integral part of a complete, scalable architecture for a new generation of high-quality IP voice solutions that run on enterprise data networks. The MCS delivers the high performance and availability demanded by today's enterprise networks and represents a turnkey solution that is easy to deploy and highly cost effective.

NOTE No third-party software applications should run on an MCS series platform.

The sections that follow provide more detail about the MCS-7825-800 and MCS-7835-1000.

Cisco MCS 7825-800

The Cisco MCS 7825-800 (MCS-7825-800) is only one rack unit (1U) high and is the most space-efficient member of the MCS-7800 Series Server Family. You can configure the MCS-7825-800 to ship with Cisco CallManager release 3.1 or Cisco IP IVR, either of which can be loaded via a fast-running installation script to make the deployment of IP telephony simple and cost effective.

Cisco MCS 7835-1000

The Cisco MCS 7835-1000 (MCS-7835-1000) is only 3U high, the MCS-7835-1000 packs tremendous power in a low-profile chassis that minimizes rack space. The MCS-7835-1000 runs a variety of Cisco AVVID applications, such as Cisco CallManager and the Cisco IP IVR solution.

Compaq ProLiant DL320 and DL380 Servers

Software-only versions of CallManager and other CIPT applications are available for customers providing their own Cisco-approved server configuration of the Compaq ProLiant DL320 or DL380 server. Any deviation from the approved configuration will result in a non-supported system. The installation will not complete if the exact configuration is not followed. Customers can order the servers from their Compaq reseller or distributor or from the Compaq Web site at www.directplus.compaq.com (select **Specials and Promotions** and then enter Affiliate Pass code **4080**).

The sections that follow describe the approved configurations for the Compaq ProLiant DL320 and DL380.

Compaq DL320

The DL320 is a high-performance server platform designed for today's IP telephony applications. When clustered properly, this server can support 2000 users. The required processor is an Intel Pentium III 800MHz processor. No specific memory configuration is required as long as 512 MB of memory is achieved. Table 1-2 describes the Compaq DL320 parts.

Table 1-2 *Compaq DL320 Parts*

Qty	Compaq Part Number	Description
1	201501-001	ProLiant DL320-128 Model-1 ATA
1	212697-B21	20 GB ATA/100 7200 RPM Drive 1"
1	128278-B21	256 MB Registered 133MHz SDRAM DIMM
1	128277-B21	128 MB Registered 133MHz SDRAM DIMM
1	212691-B21	CD-ROM/Diskette Assembly
1	174574-B21	Third-Party Cabinet Rack Kit (Optional)

Compaq DL380

The DL380 is a robust server solution designed for CIPT software applications. When clustered properly, this server can support 10,000 users. The required processor is an Intel Pentium III 1GHz processor. No specific memory configuration is required as long as 1GB of memory is achieved. Table 1-3 describes the Compaq DL380 parts.

Table 1-3 *Compaq DL380 Parts*

Qty	Compaq Part Number	Description
1	193706-001	ProLiant DL380R01 P1000/133-128
2	142673-B22	18.2 GB Pluggable Wide Ultra SCSI3 Universal 10 KB Drive
1	128279-B21	512 MB Registered 133MHz SDRAM DIMM
1	128278-B21	256 MB Registered 133MHz SDRAM DIMM
1	128277-B21	128 MB Registered 133MHz SDRAM DIMM
1	143397-001	DL380 275 Watt HP RPS Module
1	295513-B22	12/24 GB DAT Drive (DDS-3)-Opal (Optional)

For more information about the Compaq servers go to

www.cisco.com/warp/public/779/largeent/avvid/products/cmpq_srvrs.html

IBM Servers

IBM provides a Cisco server solution for Cisco CallManager and other IP telephony applications as described in the sections that follow.

IBM xSeries 330 and 340 Servers

Software-only versions of CallManager and other IP telephony applications are available for customers providing their own Cisco-approved server configuration of the IBM xSeries 330 and 340 servers. Any deviation from the approved configuration will result in a non-supported system. The installation will not complete if the exact configuration is not followed. Customers can order the servers from their IBM reseller or distributor or directly from IBM.

Approved Configurations for the IBM xSeries 330 Servers

The xSeries 330 is a high-performance server platform designed for today's IP telephony applications. When clustered properly, this server can support 2000 users. The minimum required processor is an Intel Pentium III 800MHz processor. No specific memory configuration is required as long as 512 MB of memory is achieved. Table 1-4 shows the xSeries 330 parts.

Table 1-4 *IBM xSeries 330 Parts*

Qty	IBM Part Number	Description
1	865411Y	X330 800 256 256/Open 24X
1	37L7205	18.2 Ultra 160 SCI HS SL HD
1	33L3144	256MB ECC SDRAM RDIMM Memory
1	06P4792	xSeries Cable Chain Tech Kit
Other supported processors are listed below. Replace part number 865411Y.		
1	865431Y	X330 866 256 256/Open 24X
1	865441Y	X330 933 256 256/Open 24X
1	865451Y	X330 1000 256 256/Open 24X

Approved Configurations for the IBM xSeries 340 Servers

The xSeries 340 is a robust server solution designed for CIPT software applications. When clustered properly, this sever can support 10,000 users. The required processor is an Intel Pentium III 1GHz processor. No specific memory configuration is required as long as 1 GB of memory is achieved. Table 1-5 shows the xSeries 340 parts.

Table 1-5 *IBM xSeries 340 Parts*

Qty	IBM Part Number	Description
1	86566RY	X340 1000 256 128/Open 24X
2	37L7205	18.2 Ultra 160 SCSI HS SL HD
1	33L3123	128 MB ECC SDRAM RDIMM Memory
1	33L3125	256 MB ECC SDRAM RDIMM Memory
1	33L3127	512 MB ECC SDRAM RDIMM Memory
1	37L6091	ServerRAID-4L Ultra 160 SCSI
1	37L6880	270W Redundant Power Supply
1	00N7991	20/40 GB DDS/4 4mm Internal Tape Drive (Optional)

For more information about the IBM servers go to

www.cisco.com/warp/public/779/largeent/avvid/products/ibm_srvrs.html.

NOTE Cisco CallManager and other IP telephony applications ship on separate CD-ROMs and are designed for customers who want to provide their own server platform.

More and more server platforms are being tested and certified; check with your local Cisco representative for the most current list of supported server platforms.

Cisco IP Phones

Cisco IP Phones provide the "end" in end-to-end VoIP networks. The Cisco IP Phone series is a standards-based communication appliance. The Cisco IP Phone series can interoperate with IP telephony systems based on the Skinny Client Control Protocol (Skinny protocol), SIP, and (in the future) MGCP with system-initiated software updates. This multiple protocol capability is an industry first and provides investment protection and migration capability. Learn more about adding and configuring Cisco IP Phones in Chapter 6, "Cisco IP Telephony Devices," and 7, "Understanding and Using the Bulk Administration Tool (BAT)."

Cisco IP Phone 7960

The Cisco IP Phone 7960 shown in Figure 1-6 is a second-generation, full-featured Cisco IP Phone primarily designed for executives and managers. It provides six line or speed dial buttons and four interactive soft keys located beneath the display that guide a user through call features and functions such as hold, transfer, call park, conference, and more. The 7960 also features a large pixel-based LCD that provides the date and time, calling party name, calling party number, digits dialed, and feature and line status. A speaker is provided for hands-free communication along with a headset button and a mute button that controls the speaker, handset, or headset microphones.

Figure 1-6 *Cisco IP Phone 7960*

A cluster of buttons provide access to help (the **i** button), voice mail (the **messages** button), detailed configuration information, the ability to choose ringer sound, and more (the **settings** button), and access to standard and custom-configured directories (the **directories** button) and standard or customized phone services (the **services** button). You can learn more about creating customized directories or phone services in Chapter 5, "Cisco CallManager Administration Features and User Menus."

Cisco IP Phone 7940

The Cisco IP Phone 7940 shown in Figure 1-7 is a second-generation, full-featured Cisco IP Phone primarily designed for employees or individual contributors. It provides two line or speed dial buttons and four interactive soft keys located beneath the display that guide a user through call features and functions, such as hold, transfer, call park, conference, and more. Like the 7960, the 7940 features a large pixel-based LCD that provides features such as date and time, calling party name, calling party number, digits dialed, and feature and line status. A speaker is provided for hands-free communication along with a headset button and a mute button that controls speaker or handset or headset microphones.

Figure 1-7 *Cisco IP Phone 7940*

A cluster of buttons provide access to help (the **i** button), voice mail (the **messages** button), detailed configuration information, the ability to choose ringer sound, and more (the **settings** button), and access to standard and custom-configured directories (the **directories** button) and standard and customized phone services (the **services** button). You can learn more about creating customized directories or services in Chapter 5.

Cisco IP Conference Station 7935

The Cisco IP Conference Station 7935 shown in Figure 1-8 is a full-featured, IP-based, full-duplex, hands-free conference phone for use on desktops and offices and in small- to medium-sized conference rooms. This device easily attaches a Catalyst 10/100 Ethernet switch port with a simple RJ-45 connection. It features superior sound quality with a digitally tuned speaker and three microphones, allowing conference participants to move around while speaking or be situated around the Cisco 7935 device. In addition to the regular telephone keypad, the Cisco 7935 provides three soft keys and menu navigation keys that guide a user through call features and functions. The Cisco 7935 also features a pixel-based LCD providing the date and time, calling party name, calling party number, digits dialed, and feature and line status.

Figure 1-8 *Cisco IP Conference Station 7935*

Cisco IP Phone 7914

Cisco IP Phone 7914 Expansion Module shown in Figure 1-9 extends the functionality of the Cisco IP Phone 7960 by providing 14 additional line or speed dial buttons.

Figure 1-9 *Cisco IP Phone 7914*

The Cisco 7914 has an LCD to identify the function of the button and the line status. You can daisy chain up to two Cisco 7914 Expansion Modules to provide 28 additional line or speed dial buttons. The 7914 allows for an easier integration with Cisco WebAttendant. When the 7914 is added to a 7960, you will have total of 20 lines and speed dials.

NOTE The 7914 cannot utilize inline power through the phone and must be powered by a wall jack.

Cisco IP Phone 7910 and 7910+SW

The Cisco 7910 and 7910+SW shown in Figure 1-10 are basic telephones primarily for common-use areas that require only basic features, such as lobbies, break rooms, and hallways. The Cisco 7910+SW includes a Cisco three-port switch making it suitable for worker applications where basic phone functionality and a co-located Ethernet device such as a PC are desirable. The third port is not visible to the user because the port is inside the phone.

Figure 1-10 *Cisco IP Phone 7910 and 7910+SW*

This single-line phone also provides four dedicated feature buttons: line, hold, transfer, and settings, located prominently under the display. A cluster of six configurable feature buttons is located above the volume buttons. The factory default configuration for the six buttons is messages (*msgs*), conference (*conf*), forward, speed dial (*speed 1, speed 2*), and redial. Using a custom button template you create, the buttons can be configured to perform other functions, such as call park and call pick up, as well as additional speed dials and other traditional telephone features.

The 7910 provides a pixel-based, 2x24 character LCD, showing the date and time, calling party name, calling party number, and digits dialed, as well as call state indicators, a settings menu, and other information. No speakerphone capability is provided, but there is an on-hook dialing feature and call monitor mode. The phone also has a mute button for the handset and headset microphones. In addition, the phone has a rocker switch for controlling volume for ringer, handset, and call monitor. The phone supports a headset when you disconnect the handset and plug the headset into the jack formerly used by the handset.

Cisco IP SoftPhone

Cisco IP SoftPhone shown in Figure 1-11 is a Windows-based application for the PC. Used as a stand-alone endpoint or in conjunction with the Cisco IP Phone, it provides the following features:

- **Mobility**—With Cisco IP SoftPhone running on a laptop, you can take your extension with you and receive calls wherever you are connected to the corporate network. Even dial-up connections while on the road can be used to check voice mail and place calls while online.
- **Directory Integration**—Integration with LDAP3 directories enables you to quickly place or transfer calls by looking up people by name or e-mail address. Corporate and public directories as well as a personal address book are supported.
- **User Interface**—Cisco IP SoftPhone has an intuitive user interface and context sensitive controls replace the keystroke combinations of legacy phones. Extensive use of drag-and-drop enables you to quickly place calls using VCards, directory entries, and text fields from most Windows programs.
- **Virtual Conference Room**—With Cisco IP SoftPhone setting up conference calls is quick and intuitive. Participants can be invited by dragging and dropping directory entries onto the SoftPhone's user interface to create a virtual conference room. Once a voice conference is established you can share applications running on your desktop with all participants by selecting them from a list or dragging associated documents onto the virtual conference room.

Figure 1-11 *Cisco IP SoftPhone*

NOTE Technically Cisco IP SoftPhone is a CIPT application; however, most users think of Cisco IP SoftPhone as a phone.

Gateways

There are over 20 Cisco voice gateway candidates to choose from. Gateways range from specialized, entry-level, stand-alone voice gateways to the high-end, feature-rich integrated router and Catalyst gateways.

You should choose a gateway by combining common or core requirements with site- and implementation-specific features. The three common or core requirements for a CIPT gateway are

- DTMF relay capabilities
- Support for supplementary services
- The capability to handle clustered CallManagers

Any gateway selected for a large campus deployment should have the capability to support these features. Additionally, every CIPT implementation will have its own site-specific or implementation-specific feature requirements.

Each of the gateways uses one of the supported protocols. The first protocol is the Skinny Gateway Control Protocol (SGCP or simply Skinny gateway). Skinny gateways are a series of digital gateways that include the DT-24+ and the DE-30+ (these are legacy gateways and, as such, are not shown in Table 1-6).

The next gateway protocol is traditional H.323. Cisco IOS integrated router gateways use H.323 to communicate with Cisco CallManager.

The final gateway protocol used in Cisco gateways is MGCP. Cisco CallManager uses MGCP to control the VG200 analog gateway, the 2600- and 3600-series gateways, Catalyst 4224, Catalyst 4000 Access Gateway Module, DT-24+, DE30+, WS-X6608x1, and WS-X6624-FXS Catalyst Voice Module.

Each of the protocols used by these gateways follows a slightly different methodology to provide support for the core gateway features. Adding and configuring gateways is covered in Chapter 6, "Cisco IP Telephony Devices." Table 1-6 lists the gateways supported in a CIPT network.

Table 1-6 *Gateways for a CIPT Network*

Gateway	MGCP	H.323	Skinny Gateway Protocol
VG200	Yes, for FXS/FSO	Yes,with Cisco IOS Software 12.1(5)-XM1, the VG-200 uses H.323 to support a wider range of digital and analog interfaces	No
DT-24+	Yes with Cisco CallManager release 3.1	No	Yes
827	No	Yes for FXS	No
1750	No	Yes	No
3810 V3	Cisco IOS Software 12.1(3)T and Cisco CallManager release 3.0(5)	Yes	No
2600	Cisco IOS Software 12.1(3)T and Cisco CallManager release 3.0(5) Analog Interfaces only no E&M T1 CAS 12.1(5)XM & 12.2.1T Q.931 PRI Backhaul 12.2.2T**	Yes	No
3600	Cisco IOS Software 12.1(3)T and Cisco CallManager release 3.0(5) Analog Interfaces only no E&M T1 CAS 12.1(5)XM & 12.2.1T Q.931 PRI Backhaul 12.2.2T**	Yes	No
7200	Cisco IOS Software 12.2.(1)T***	Yes	No

Table 1-6 *Gateways for a CIPT Network (Continued)*

Gateway	MGCP	H.323	Skinny Gateway Protocol
7500	Undecided	Yes (Cisco IOS Software 12.1.5)	No
5300*	Yes (Cisco IOS Software 12.1(1)T	Yes	No
Catalyst 4000 WS-X4604-GWY Gateway Module	Cisco CallManager release 3.1	Yes, for PSTN interfaces	Yes, for conferencing and MTP/ transcoding services
Catalyst 6000 WS-X6608-x1 Gateway Module & FXS Module WS-X6624	In Cisco CallManager release 3.1: T1/E1 module supporting PRI and CAS FXS module	No	Yes, for FXS module and T1/E1 prior to Cisco CallManager release 3.1
Catalyst 4224	Projected for Cisco CallManager release 3.1	Yes	No

* While the 5300 supports MGCP, it is as a trunk gateway module using SS7 signaling, which is not supported in CallManager

** Cisco IOS Software Release 12.2.2T PRI Backhaul support for 26xx/36xx uses RUDP and is not compatible with Cisco CallManager. PRI Backhaul with Cisco CallManager release 3.1 as the Call Agent is scheduled for Cisco IOS Software Release 12.2.4T and uses TCP as the transport.

*** Not supported in Cisco CallManager

Also note prior to any deployment consideration, it would be prudent to check the IOS Release Notes to confirm feature or interface support.

For more information about choosing a gateway, read "Choosing a Voice-over-IP Gateway—A 3.0 Gateway Solution Guide" at

www.cisco.com/warp/public/cc/pd/ga/prodlit/gatwy_wp.htm

Inline Power Switches

The family of inline power switches extends the CIPT networking capabilities of the Catalyst backbone to the enterprise wiring closet and branch office. Many of the new Fast Ethernet modules support a feature called inline power, which is 48-volt DC power provided over standard Category 5 or higher UTP cable up to 100 meters. The inline power Fast Ethernet module enables the modular wiring closet infrastructure to provide centralized power for CIPT networks.

Today there are two different implementations of inline powered Ethernet ports in Cisco Catalyst switching products. Each of these mechanisms allows Cisco Power Sourcing Equipment (PSE) to discover a phone and supply it operating power. Chapter 9, "LAN Infrastructure for Cisco IP Telephony," discusses more about Cisco IP Phone discovery.

The inline power modules prepare the network infrastructure for IP-based converged business applications that provide seamless communication and collaboration between branch and corporate sites. Not only do some of these switches provide inline power, some support single port multiple VLANs (voice and data subnets).

The Inline Power 10/100BaseT Ethernet switching module supports up to 48 ports per module (RJ-45 interfaces). Along with phone discovery, these modules support auto-sensing/auto-negotiation to determine the speed and duplex mode of the attached device. The Catalyst 4003 uses the Catalyst Inline Power Patch Panel to provide inline power. The Catalyst 4006, with support for up to 240 multiservice ports, directly provides inline power. To support the new demand for phone power on the Catalyst 4006, Cisco has developed a new auxiliary DC power shelf that supplies the Catalyst 4006 with the 48 volts needed for inline power.

Catalyst 6000

Cisco is using the Catalyst 6000 family to lead its customers to campus convergence. The first product features to be introduced are Fast Ethernet enhancements delivered by the new 48-port Inline Power 10/100BaseT Ethernet Switching Module. Instead of using wall power, terminal devices such as Cisco IP Phones use power provided from the Catalyst 6000 switch. This capability gives the network administrator centralized power control, which translates into greater network availability. By deploying the Catalyst 6000 family of switches with UPS systems in secured wiring closets, network administrators can ensure that building power outages will not affect network telephony connections.

Catalyst 4224

The Catalyst 4224 can provide 48-volt DC power over standard Category 5 or higher UTP cable. Instead of requiring wall power at every desktop, terminal devices, such as Cisco IP Phones, use power supplied by the Catalyst 4224. With the Phone Discovery feature, the Catalyst 4224 automatically detects the presence of a Cisco IP Phone and supplies inline power. By deploying Catalyst 4224 switches with UPS systems, network administrators can ensure that power outages do not affect a branch office's telephony connections.

Catalyst 3524

The Catalyst 3524-PWR XL switch has 24 10/100 switched ports with integrated inline power and two Gigabit Interface Converter (GBIC)-based Gigabit Ethernet ports. Integrated inline power provides DC current to devices that can accept power over traditional UTP cabling (for example, the Cisco 79*xx* family of Cisco IP Phones or Cisco Aironet wireless access points). The dual GBIC-based Gigabit Ethernet implementation provides tremendous deployment flexibility, enabling you to implement one type of stacking and uplink configuration today while preserving the option to migrate that configuration in the future.

Patch Panel

The Catalyst Inline Power Patch Panel enables inline power for Cisco multiservice-enabled Catalyst switches that do not have inline power blades available within their product line. This capability gives the network administrator centralized power control, which translates into greater network availability. By deploying Catalyst gear with UPS systems in secured wiring closets, network administrators can ensure that building power outages will not affect network telephony connections.

DSPs

DSPs in a CIPT solution provide a means for transcoding between different codecs (for example, G.711, G.723, and G.729), converting TDM to packet, and conferencing resources.

The following products have DSP resources that support transcoding, conferencing, and/or converting TDM to packet:

- Catalyst 4000 Access Gateway Module
- Catalyst 6000 T1/E1 Gateway Modules
- Catalyst 4224 Access Gateway Switch (TDM to packet only and conferencing and transcoding in the future)
- VG200 (TDM to packet only and conferencing and transcoding in the future)

NOTE At the time of the writing of this book, the Catalyst 4224 and the VG200 DSPs support only TDM to packet and are targeted to support conferencing and transcoding in the future. Check with your local Cisco representative or the following link for the current status of these products:

www.cisco.com/univercd/cc/td/doc/product/voice/c_callmg/3_1/sys_ad/adm_sys/ccmsys/a05dsp.htm

---

Applications

Applications extend the CIPT solution beyond call processing and call routing. Applications bring voice and UM, IVR, AA, contact center, productivity services, phone services,

Cisco WebAttendant, PA, and more. Because the CIPT solution is based on open standards, applications are being developed and integrated quickly.

Check the following URL for the most up-to-date list of applications:

www.cisco.com/warp/public/779/largeent/avvid/products/iptel_apps.html

NOTE Cisco IP SoftPhone is an application but is discussed in the Cisco IP Phones section in this chapter.

Call Processing Design Concept

Figure 1-12 illustrates a call processing design concept, which varies depending on your company's direction. In the figure, a Cisco CallManager cluster is located at the headquarters or the regional center. A design goal of IP telephony is to have primary connectivity to the regional center, branch office, and telecommuter through the IP WAN and in the future to the rest of the world. The PSTN is for backup use if the IP WAN should go down or bandwidth is unavailable.

Branch office call processing is performed at headquarters and phone calls between the branch office and headquarters are placed over the IP WAN. If the IP WAN goes down, the calls can use the PSTN to connect using the voice-enabled access routers.

Figure 1-12 *Call Processing Design Concept*

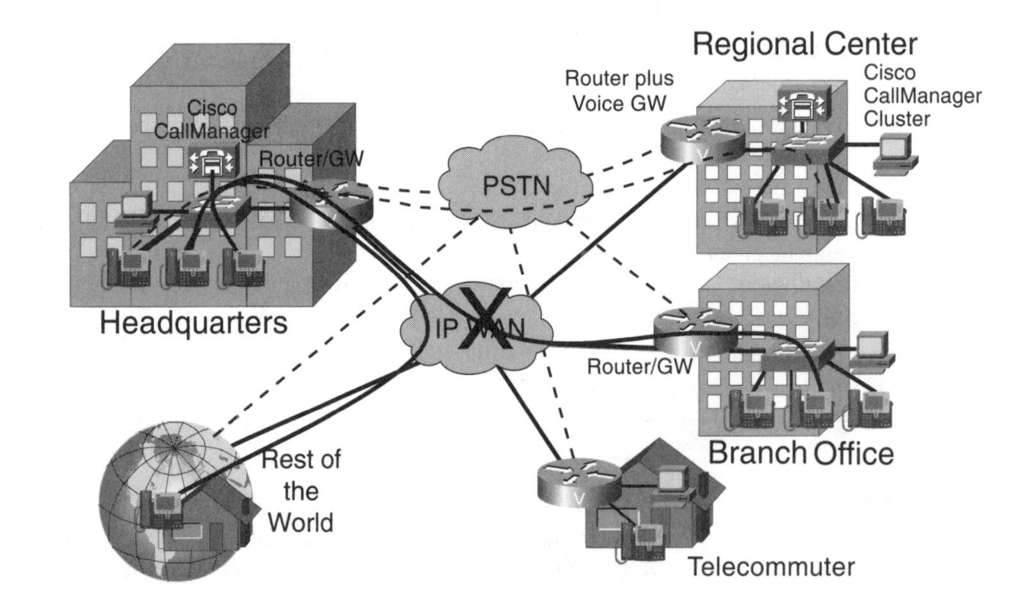

Summary

Because the CIPT solution provides a converged network of voice over data, it is important for implementers of the CIPT solution to have a foundation of IP and the OSI model.

Cisco AVVID is an architecture based on convergence and open standards. CIPT is an integral part of Cisco AVVID. The primary focus of this book is the installation, configuration, support, and maintenance of the CIPT components. Cisco CallManager is the core component of the CIPT solution. The primary task of the Cisco CallManager is call processing. The 7960, 7940, 7910, 7910+SW, and 7935 are the five Cisco IP Phone 79*xx* series models.

The Cisco IP Phone 7914 Expansion Module enables users to add up to 28 lines and speed dials to their 7960 model phones. The Cisco 7914 enables an easier integration with Cisco WebAttendant.

When deploying a CIPT solution, the infrastructure power options, IP addressing, DSP resources, and gateways need to be part of the design phase before implementation and configuration. Some of the inline power options include the Catalyst 6000 48-port Ethernet module, the Catalyst 4000 48-port Ethernet module, the Catalyst 3524, the Catalyst 4224, and the Inline Power Patch panel. Gateways enable the CIPT solution to connect to other sites via the WAN or to the PSTN.

Basic Configuration

Upon completion of this chapter, you will be able to complete the following tasks:

- Given a list of menu items from Cisco CallManager Administration, list and provide a brief description of the menu items.
- Given a Cisco CallManager Administration, access the online help included with the system.
- Given a list of requirements and guidelines of different deployment models, match those requirements and guidelines with the correct deployment model.
- Given a newly installed Cisco CallManager cluster, configure the system parameters to prepare the cluster for adding phones, users, and gateways.

CHAPTER 2

Navigation and System Setup

This chapter is designed to familiarize you with Cisco CallManager Administration (CCMAdmin). Cisco CallManager is a software-based call processing engine that can be configured via a Web interface. The Web interface layer interacts with the underlying database layer in Cisco CallManager. If you are going to be an administrator of a Cisco CallManager cluster, you need to be familiar with CCMAdmin and how the changes made in CCMAdmin affect the database, the cluster, and the users. This chapter covers the following topics:

- Abbreviations
- Navigation
- Intra-Cluster Communication
- System Setup

Pre-Test

Do you already know this information? The pre-test is designed to help you gauge your knowledge about this chapter. Of the 10 questions, if you answer one to three questions correctly, we recommend that you read this chapter. If you answer four to seven questions correctly, we recommend that you skim through this chapter, reading those sections that you need to know more about. If you answer 8 to 10 questions correctly, you probably understand this information well enough to skip this chapter. You can find the answers to the pre-test for this chapter in Appendix B, "Answers to Chapter Pre-Test and Post-Test Questions."

1. Why is it recommended to change the server name to an IP address in the CCMAdmin user interface?

   ____

   ____

   ____

2. Assuming DHCP and your network are configured correctly and besides un-checking the box **auto-registration Disabled on this Cisco CallManager**, what must you provide so that auto-registration works properly?

   ____

   ____

   ____

3. Based on configuring one Cisco CallManager group in a Cisco CallManager cluster that has two or three servers, how many users can this cluster support?

   ____

   ____

   ____

4. If you expected some growth in your organization, what would be a better way to support 2500 users? How many servers? What would your Cisco CallManager Groups be like? (If it is easier, you may draw your answer.)

   ____

   ____

   ____

5. When would you assign a Date/Time Group other than CMLocal to a device?

   ____

   ____

   ____

6. What is the main reason for using G.711 between devices as much as possible?

   ____

   ____

   ____

7 A device pool is made of characteristics. List the three required characteristics of a device pool.

____

____

____

8 After configuring a device pool and assigning it to a phone, what must you do to the phone so that those changes take effect?

____

____

____

9 You have two clusters, Cluster One and Cluster Two. Can you set the enterprise parameters in Cluster One and also apply those changes to Cluster Two?

____

____

____

10 Which deployment model uses Locations?

____

____

____

Abbreviations

Table 2-1 lists the abbreviations used in this chapter. In the telecom and networking industry, there are many abbreviations. Abbreviation definitions may depend on the industry or technology being discussed. For more information about terms and abbreviations used in this chapter refer to the *IP Telephony Network Glossary* at the following URL:

www.cisco.com/univercd/cc/td/doc/product/voice/evbugl4.htm

Table 2-1 *Abbreviations Used in the Chapter*

Abbreviation	Complete Term
AST	Admin Serviceability Tool; in future releases, this name will be changed to Real-Time Monitoring
BAT	Bulk Administration Tool
CAC	call admission control
CCM	Cisco CallManager
CDR	call detail record
CM	CallManager
CMLocal	CallManager local time
DFW	airport code for Dallas/Fort Worth
DLL	dynamically linked library
DSP	digital signal processor
GW	gateway
MGCP	Media Gateway Control Protocol
MOH	music on hold
PC	personal computer
PSTN	public switch telephone network
RTP	Real-time Transport Protocol
SQL	Structured Query Language
TCP	Transmission Control Protocol
TCP KA	Transmission Control Protocol KeepAlive
TFTP	Trivial File Transfer Protocol
UM	Unified Messaging
VM	voice mail
WAN	wide-area network

Navigation

This section focuses on navigation within Cisco CallManager Administration. As with any Web page, the ease of moving between items and shortcuts to sections you most frequently use can cut down on clicks and time.

Administering Cisco CallManager can be like sailing around the world or navigating through your local neighborhood. Landmarks, signs, and general north, south, east, and west can help you get from point A to point B. Once you become familiar with local landmarks and signs, you can easily get from your house to the grocery store. This section provides you with landmarks (main menus) and signs (menu items) to get, for example, from the Cisco CallManager Administration main page to the Route Pattern Configuration page.

Another helpful tool when driving around your neighborhood may be how the roads, pathways, and freeways are organized. For instance, in the U.S., the freeway system is large and follows a general rule. The rule is that all odd numbered freeways travel north/south and all even numbered freeways travel east/west. A helpful tool in CCMAdmin is that when there are only a few items, they are listed in the left column. If there are a lot of items, a search engine and search parameters are offered. This is the difference between high-count items and low-count items.

When there are high-count items in CCMAdmin, a search engine is provided on the page. You will be able to identify the different search criteria for each page you want to use.

Cisco CallManager Administration Menus

CCMAdmin is organized so that you can start from the left and proceed to the right in setting up a Cisco CallManager system and cluster. Figures 2-1 through 2-8 illustrate the menus and menu items of CCMAdmin.

The sections that follow describe the main menus and the corresponding menu items in CCMAdmin.

System Menu

The System menu enables system-wide or cluster-wide configuration. The default configuration for most items under the System menu is very simple but will get you operational. If there are certain items you want to customize, you should do so prior to configuring Cisco CallManager features. Figure 2-1 shows System menu items in CCMAdmin.

Figure 2-1 *Cisco CallManager Administration System Menu*

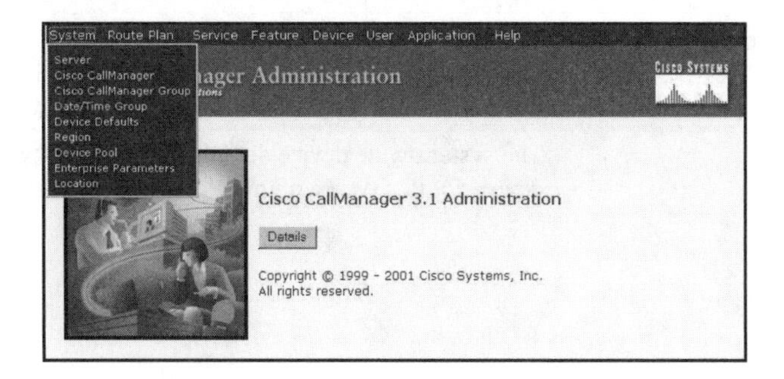

Table 2-2 lists some brief descriptions of System menu items.

Table 2-2 *Cisco CallManger Administration System Menu Items*

Menu Item	Functionality
Server	Use server configuration to specify the unique IP address or Domain Name System (DNS) name of the server on which Cisco CallManager is installed. Names or addresses must be unique throughout the network. Using the IP address of the server is recommended. When a new subscriber is added to the cluster, the server is automatically added to the list of servers. Because of this you should never have to specify a server in this screen, but you can use this screen to change the server name to an IP address. Do not specify a fully qualified domain name; DNS/IP address should always just be either an IP address or a hostname.
Cisco CallManager	Use Cisco CallManager configuration to specify the TCP port numbers and other properties for each Cisco CallManager installed in the same cluster. Cisco recommends against changing TCP port numbers without a compelling reason. An entry should be specified for every Cisco CallManager server that provides call processing (as opposed to a stand-alone database Publisher or TFTP server). However, you should never have to manually add an entry to this menu item. The name of the Cisco CallManager server listed here does not have to be a DNS name or IP address and can be left as the default.
Cisco CallManager Group	A Cisco CallManager group is a prioritized list of up to three Cisco CallManagers. The first Cisco CallManager in the list serves as the primary Cisco CallManager for that group, and the other members of the group serve as secondary (backup) Cisco CallManagers.
Date/Time Group	Use date/time groups to define time zones for the various devices connected to Cisco CallManager. Each device exists as a member of only one device pool, and each device pool has only one assigned date/time group.
Device Defaults	Use device defaults to set the system-wide default values for load information, device pool, and phone template for each type of device that registers with a Cisco CallManager. The system-wide device defaults for a device type apply to all devices of that type within a Cisco CallManager cluster.

Table 2-2 *Cisco CallManger Administration System Menu Items (Continued)*

Menu Item	Functionality
Region	Use regions to specify the voice compression/decompression (codec) used for calls within and between regions. The voice codec determines the maximum amount of bandwidth used per call and specifies the technology used to compress and decompress voice signals. The choice of voice codec determines the compression type and amount of bandwidth used per call. The default voice codec for all calls through Cisco CallManager is G.711. If you do not plan to use any other voice codec, you do not need to use regions.
Device Pool	Use device pools to define sets of common characteristics for devices. Device pools includes the following values: • Cisco CallManager group • Date/time group • Region • Media resource group list • User MOH audio source • Network MOH audio source • Calling search space for auto-registration Not all values are required to create a device pool.
Enterprise Parameters	Enterprise parameters provide default settings that apply to all devices and services in the same cluster. When you install a new Cisco CallManager, it uses the enterprise parameters to set the initial values of its device defaults.
Location	Use locations to implement CAC in a centralized call processing system. CAC enables you to regulate the amount of bandwidth used by calls over links between the locations.

Route Plan Menu

The Route Plan menu enables you to configure Cisco CallManager route plans using route patterns, route filters, route lists, and route groups. Figure 2-2 shows Route Plan menu items in Cisco CallManager Administration.

Figure 2-2 *Cisco CallManager Administration Route Plan Menu*

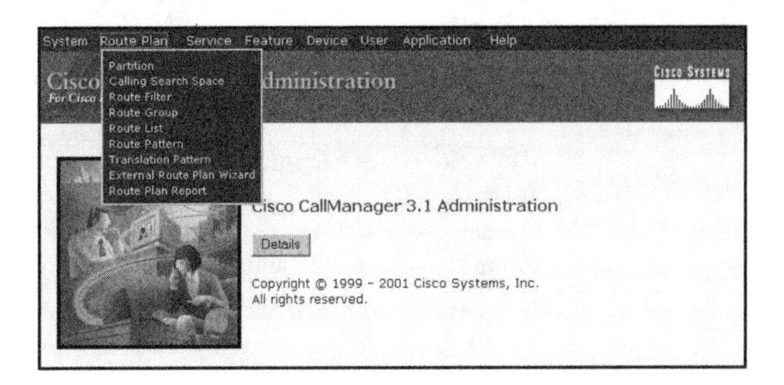

Table 2-3 lists some brief descriptions of Route Plan menu items.

Table 2-3 *Cisco CallManger Administration Route Plan Menu Items*

Menu Items	Functionality
Partition	A partition is a list of route patterns and directory numbers. Partitions facilitate call routing by dividing the route plan into logical subsets based on organization, location, and call type.
Calling Search Space	A calling search space (CSS) comprises an ordered list of route partitions. CSSs determine the partitions where calling devices search when attempting to complete a call.
Route Filter	Route filters, along with route patterns, use dialed-digit strings to determine how a call is handled. You can only use route filters with North American Numbering Plan (NANP) route patterns—that is, route patterns that use an "at" symbol (@) wildcard.
Route Group	A route group enables you to designate the order in which gateways are selected. It enables you to prioritize a list of gateways and ports for outgoing trunk selection.
Route List	A route list associates a set of route groups with a route pattern and determines the order in which those route groups are accessed. The order controls the progress of the search for available trunk devices for outgoing calls.

Table 2-3 *Cisco CallManger Administration Route Plan Menu Items (Continued)*

Menu Items	Functionality
Route Pattern	A route pattern comprises a string of digits (an address) and a set of associated digit manipulations that can be assigned to a route list or a gateway. Route patterns provide flexibility in network design. They work in conjunction with route filters and route lists to direct calls to specific devices and to include, exclude, or modify specific digit patterns.
Translation Pattern	Cisco CallManager uses translation patterns to manipulate dialed digits before routing a call. In some cases, the system uses a translation pattern to modify the dialed number. You configure translation patterns almost exactly like route patterns. They use the same calling and called party transformations and the same wildcard notation. But unlike route patterns, translation patterns do not correspond to a physical or logical destination. Instead, a translation pattern relies on the calling and called party transformations to perform its function. Use to configure Private Line Automatic Ring-down (PLAR).
External Route Plan Wizard	The External Route Plan Wizard enables you to quickly configure external routing to the PSTN, to private branch exchanges (PBXs), or to other Cisco CallManager systems.
Route Plan Report	The route plan report lists all call park numbers, call pickup numbers, conference numbers, route patterns, and translation patterns in the system. This information proves useful when you are developing or expanding the route plan, and for troubleshooting purposes. The route plan report enables you to view either a partial or full list and to go directly to the associated configuration pages by selecting a route pattern, partition, route group, route list, call park number, call pickup number, conference number, or gateway.

Service Menu

The Service menu enables you to configure media resources, Cisco Messaging Interface, Cisco WebAttendant, and service parameters. Figure 2-3 shows Service menu items in CCMAdmin.

Figure 2-3 *Cisco CallManager Administration Service Menu*

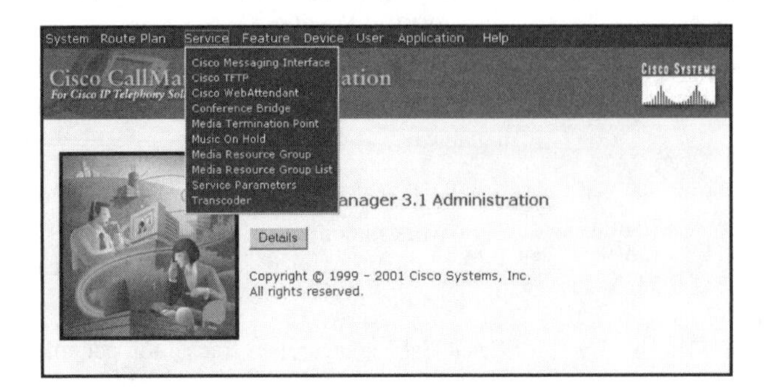

Table 2-4 lists some brief descriptions of Service menu items.

Table 2-4 *Cisco CallManger Administration Service Menu Items*

Menu Items	Functionality
Cisco Messaging Interface	The Cisco Messaging Interface (CMI) service enables you to interface Cisco CallManager with an SMDI-compliant voice mail system.
Cisco TFTP	Cisco TFTP is a Windows NT service that builds and serves files consistent with the TFTP. TFTP is a simplified version of the File Transfer Protocol (FTP). Cisco TFTP serves embedded component executable files, ringer files, and configuration files.
Cisco WebAttendant	Cisco WebAttendant is a client/server application that enables a user, administrative assistant, or receptionist to manage incoming and outgoing phone calls. Cisco WebAttendant provides quick directory access to look up phone numbers, the ability to monitor line states, and direct calls.
Conference Bridge	Conference bridge for Cisco CallManager (Release 3.0 and later) is a hardware solution and software application designed to enable both Ad Hoc and Meet-Me voice conferencing. Each conference bridge, whether hardware or software, is capable of hosting several simultaneous, multi-party conferences.

Table 2-4 *Cisco CallManger Administration Service Menu Items (Continued)*

Menu Items	Functionality
Media Termination Point	A media termination point (MTP) is invoked on behalf of H.323 endpoints involved in a call in order to enable supplementary call services. In some cases, H.323 gateways may require that calls use an MTP to enable supplementary call services such as hold or transfer, but normally Cisco gateways do not. Cisco gateways running Cisco IOS Software Release 12.0(5)T or later do not need MTP for supplementary services. Only software MTPs can be configured in this menu item. Hardware MTPs are configured in **Service > Transcoder**.
Music On Hold	The integrated MOH feature provides the ability to play music from a streaming source for OnNet and OffNet calls.
Media Resource Group	A media resource group (MRG), is a logical grouping of conference bridges, MOH servers, and transcoders. These groups can then be assigned to devices to distribute the resources by geographic location or to restrict the use of these resources to a subset of the user community. You can also form MRGs to control the usage of servers or the type of service (Unicast or multicast) desired.
Media Resource Group List	A media resource group list (MRGL) provides a prioritized grouping of MRGs. An application selects the required media resource, such as an MOH server, from among the available media resources according to the priority order defined in an MRGL.
Service Parameters	Service parameters for Cisco CallManager (Release 3.0 and later) enable you to configure different services on selected servers. You can insert and delete services on a selected server, as well as modify the service parameters for those services. Service parameters are global parameters for each Cisco CallManager in the cluster.
Transcoder	A transcoder is a device that takes the output stream of one codec and transcodes (converts) it from one compression type to another compression type. Transcoders transcode between G.711, G.723, and G.729a codecs. In addition, a transcoder provides MTP capabilities, and may be used to enable supplementary services for H.323 endpoints when required.

Feature Menu

The Feature menu enables you to configure call park, call pickup, and Cisco IP Phone services Meet-Me Number/Pattern (not described below and covered in the "Conference Bridge (ConfBr)" section in Chapter 4.). Figure 2-4 shows Feature menu items in CCMAdmin.

Figure 2-4 *Cisco CallManager Administration Feature Menu*

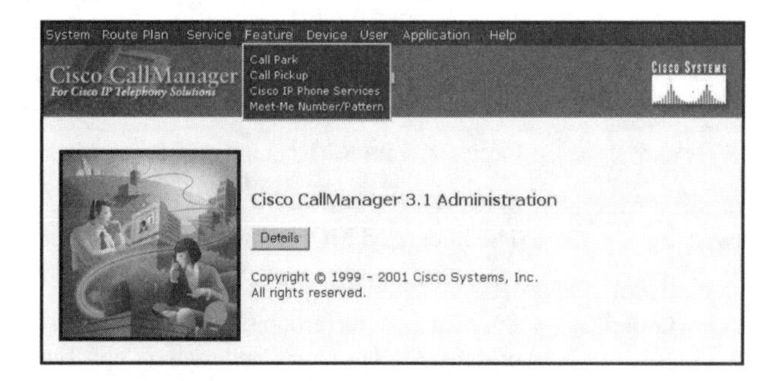

Table 2-5 lists some brief descriptions of Feature menu items.

Table 2-5 *Cisco CallManger Administration Feature Menu Items*

Menu Items	Functionality
Call Park	Call park enables you to place a call on hold at a specific extension. The parked call can be retrieved by any phone in the Cisco CallManager cluster by dialing the extension at which the call has been parked.
Call Pickup	Call pickup enables you to answer a call that comes in on a directory number other than your own. When you hear an incoming call ringing on another phone, you can redirect the call to your phone by using the call pickup feature.
Cisco IP Phone Services	Cisco IP Phone Services provides an area where you can define and maintain the list of Cisco IP Phone services to which users can subscribe. Cisco IP Phone services are XML applications that enable interactive content with text and graphics to be displayed on Cisco IP Phone models 7960 or 7940.
Meet-Me Number/Pattern	This item is described in the "Conference Bridge (ConfBr)" section in Chapter 4.

Device Menu

The Device menu enables you to add and configure devices such as CTI route points, voice mail ports, gateways, gatekeeper, phones, and phone button templates. Figure 2-5 shows Device menu items in CCMAdmin.

Figure 2-5 *Cisco CallManager Administration Device Menu*

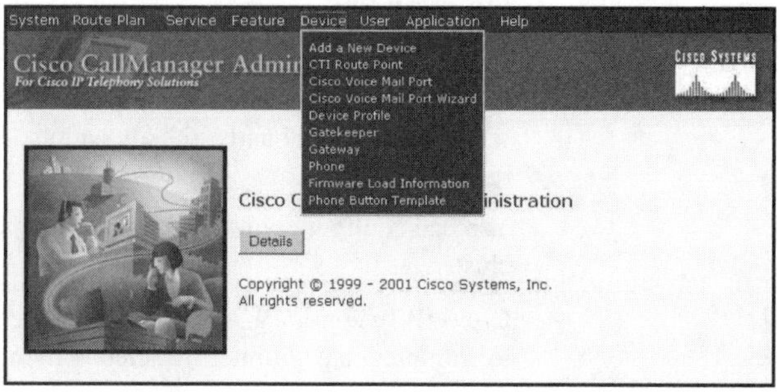

Table 2-6 lists some brief descriptions of Device menu items.

Table 2-6 *Cisco CallManger Administration Device Menu Items*

Menu Items	Functionality
Add a New Device	Use to add a new device (CTI route point, Cisco voice mail port, gatekeeper, gateway, phone) to CCMAdmin.
CTI Route Point	A computer telephony integration (CTI) route point is a virtual device that can receive multiple simultaneous calls for application-controlled redirection. For first-party call control, you must add a CTI port for each active voice line. Applications that use CTI route points and CTI ports include Cisco IP SoftPhone, Cisco IP AutoAttendant, and Cisco IP Interactive Voice Response System.
Cisco Voice Mail Port	The Cisco Unity software, available as part of Cisco IP Telephony Solutions, provides voice messaging capability for users when they are unavailable to answer calls. Use this menu item to add and delete ports associated with a Cisco voice mail server when you do not want to use the Cisco Voice Mail Port Wizard.
Cisco Voice Mail Port Wizard	The wizard enables you to add and delete ports associated with a Cisco voice mail server.
Device Profile	A device profile comprises the set of attributes (services and/or features) associated with a particular device for use with the extension mobility feature. A user device profile contains device information to be used when a user logs in to a device.

continues

Table 2-6 *Cisco CallManger Administration Device Menu Items (Continued)*

Menu Items	Functionality
Gatekeeper	A gatekeeper device (also known as a Cisco Multimedia Conference Manager or MCM) supports the H.225 RAS message set used for CAC, bandwidth allocation, and dial pattern resolution. You can configure only one gatekeeper per Cisco CallManager cluster.
Gateway	Cisco IP Telephony gateways enable Cisco CallManager devices to communicate with non-IP telecommunications devices. Cisco CallManager supports several types of H.323 and MGCP-based gateways. Use this menu item to add, update, delete, or search for gateways.
Phone	Use this menu item to add or update configuration information for Cisco IP Phones and CTI ports or delete or search for Cisco IP Phones and CTI ports.
Firmware Load Information	The Firmware Load Information page in CCMAdmin enables you to quickly locate devices that are not using the default firmware load for their device type.
Phone Button Template	Cisco CallManager includes several default phone button templates. When adding phones, you can assign one of these templates to the phones or create a new template.

User Menu

The User menu enables you to add, configure, and search for users in the cluster. Figure 2-6 shows User menu items in CCMAdmin.

Figure 2-6 *Cisco CallManager Administration User Menu*

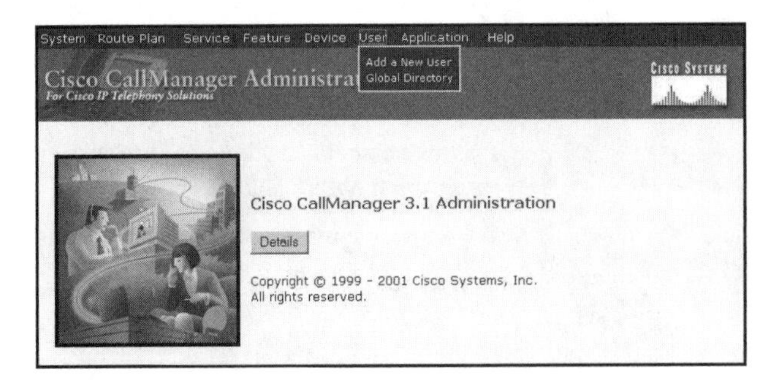

Table 2-7 lists some brief descriptions of User menu items.

Table 2-7 *Cisco CallManger Administration User Menu Items*

Menu Items	Functionality
Add a New User	In this screen you can define the properties of a new user, including name, password, PIN, job details, and whether the user requires CTI application use. Information that you enter here is also accessed by Directory Services, Cisco WebAttendant, and the Cisco IP Phone User Options Web page. If you want user information to be available to these components, you must complete the information in the User area for all users and their directory numbers and also for resources such as conference rooms or other areas with phones (this is useful for Cisco WebAttendant).
Global Directory	You can use the Global Directory for Cisco CallManager (Release 3.0 and later) to search for a user in a Cisco CallManager directory. Cisco CallManager uses the Lightweight Directory Access Protocol (LDAP) to interface with a directory that contains user information. This is an embedded directory supplied with Cisco CallManager. Its primary purpose is to maintain the associations of devices with users.

Application Menu

The Application menu enables you to install and configure plug-ins. It also links to the Cisco CallManager Serviceability (CCMServiceability) page that can be used for monitoring and troubleshooting. If you have downloaded and installed them, links are also provided for the Administrative Reporting Tool (ART) and BAT. In releases subsequent to Cisco CallManager Release 3.1(1), ART changed its name to CDR Analysis & Reporting and is available as part of CCMServiceability, and BAT is available as a plug-in. Figure 2-7 shows Application menu items in CCMAdmin.

Figure 2-7 *Cisco CallManager Administration Application Menu*

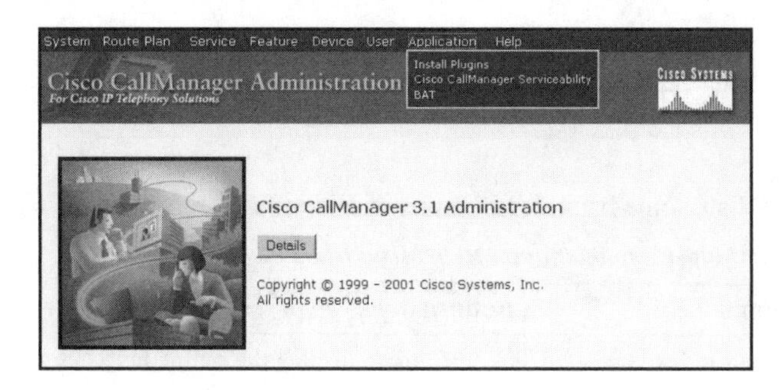

Table 2-8 lists some brief descriptions of Application menu items.

Table 2-8 *Cisco CallManger Administration Application Menu Items*

Menu Items	Functionality
Install Plug-ins	Application plug-ins extend the functionality of Cisco CallManager. For example, the Cisco WebAttendant plug-in enables a receptionist to rapidly answer and transfer calls within an organization; the JTAPI plug-in enables a computer to host applications that access Cisco CallManager via the Java telephony application programming interface (JTAPI). Plug-ins are modular by nature so that deployments that do not need or want them do not utilize valuable resources for functionality they will not be using.
Cisco CallManager Serviceability	CCMServiceability helps you monitor and troubleshoot system problems. CCMServiceability provides its own Web interface. This menu item links directly to the CCMServiceability page.

Help Menu

The Help menu enables you to access help on a topic or for a specific page. It also enables you to check the component versions and make sure they are in-sync. Figure 2-8 shows help menu items in CCMAdmin.

Figure 2-8 *Cisco CallManager Administration Help Menu*

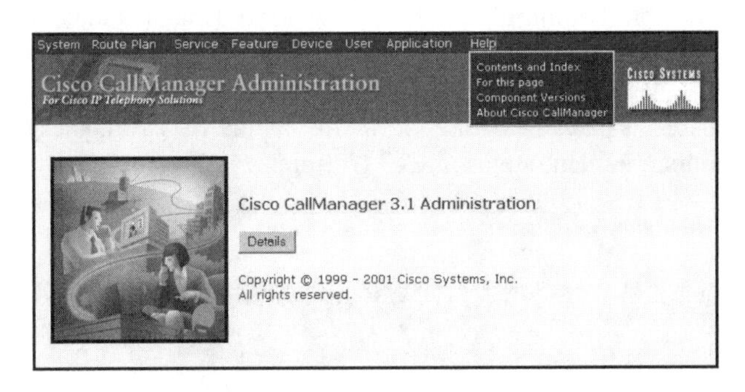

Table 2-9 lists some brief descriptions of Help menu items.

Table 2-9 *Cisco CallManger Administration Help Menu Items*

Menu Items	Functionality
Contents and Index	Click this link to display the online help.
For this Page	Click this link to display context-sensitive help for the page currently displayed.

Table 2-9 *Cisco CallManger Administration Help Menu Items (Continued)*

Menu Items	Functionality
Component Versions	The Component Versions page displays system component version information for any Cisco CallManager server, lists servers in the cluster with out-of-sync system components, and displays latest installed component version information across all Cisco CallManager servers in the cluster.
About Cisco CallManager	Click this link to display the main CCMAdmin page.

Cisco CallManager Serviceability Menus

Access Serviceability from the CCMAdmin window by choosing **Applications > Cisco CallManager Serviceability**. CCMServiceability is automatically available with Cisco CallManager.

Use CCMServiceability to monitor and troubleshoot system problems. This Web-based tool provides the following services:

- **Alarm**—Saves Cisco CallManager services alarms and events for troubleshooting and provides alarm message definitions.
- **Trace**—Saves Cisco CallManager services trace information to various log files for troubleshooting. Administrators can configure, collect, and analyze trace information.
- **AST (also known as Real-Time Monitoring)**—Monitors real-time behavior of the components in a Cisco CallManager cluster and enables you to set alarms for notification when specific events occur.
- **Control Center**—Enables you to view the status of Cisco CallManager services. Use Control Center to start and stop services.

The sections that follow describe the menus and menu items in CCMServiceability.

Alarm Menu

The Alarm menu enables you to configure alarms and events and view alarm message definitions. Figure 2-9 shows Alarm menu items in CCMServiceability.

Figure 2-9 *Cisco CallManager Serviceability Alarm Menu*

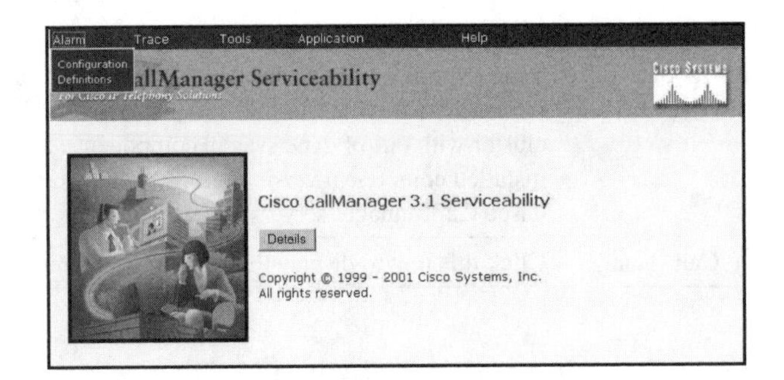

Table 2-10 lists some brief descriptions of Alarm menu items.

Table 2-10 *Cisco CallManger Serviceability Alarm Menu Items*

Menu Items	Functionality
Configuration	You configure alarm parameters to be used for tracing. You can configure alarms for Cisco CallManager servers that are in a cluster and services for each server, such as Cisco CallManager, Cisco TFTP, and Cisco CTIManager.
Definitions	Cisco CallManager stores alarm definitions and recommended actions in an SQL server database. You can search the database for definitions of all the alarms. The alarm definitions include the alarm name, description, recommended action, severity, parameters, and monitors.

Trace Menu

The Trace menu enables you to configure trace parameters, collect trace parameters, and analyze trace data for troubleshooting problems. Figure 2-10 shows Trace menu items in CCMServiceability.

Figure 2-10 *Cisco CallManager Serviceability Trace Menu*

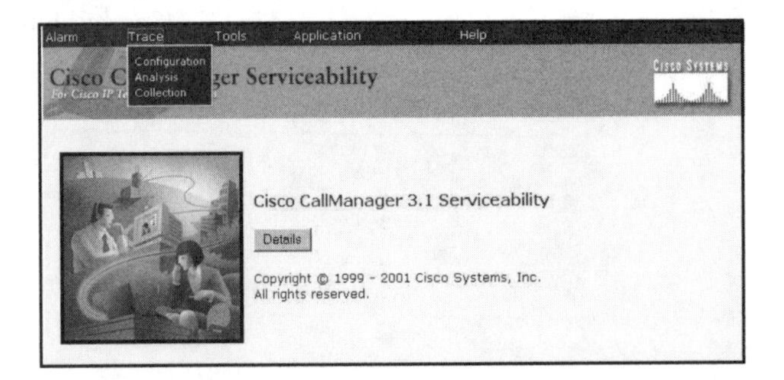

Table 2-11 lists some brief descriptions of Trace menu items.

Table 2-11 *Cisco CallManger Serviceability Trace Menu Items*

Menu Items	Functionality
Configuration	Use trace configuration to specify the parameters you want to trace for troubleshooting Cisco CallManager problems. You specify the following trace parameters: the Cisco CallManager server (within the cluster), the Cisco CallManager service on the server, the debug level, the specific trace fields, and the location to which trace files should be saved. You can also specify device name-based tracing and enable XML-formatted trace results.
Analysis	Cisco CallManager system administrators and Cisco engineers use Trace Analysis to debug system problems. After the trace is configured and collected, you request a list of SDI or SDL log files. From the list, you can choose a specific log file and request information from that log file such as host address, IP address, trace type, and device name. SDI traces are the same as CCM traces. In most but not all cases, the abbreviation SDI has been replaced by CCM (Cisco CallManager).
Collection	After trace parameters are configured, you can select trace information to collect for analysis. You can base the collection of information on SDL or SDI trace, type of Cisco CallManager service, and time and date of trace.

Tools Menu

The Tools menu enables you to configure and monitor real-time behavior of the components in a Cisco CallManager cluster. It also enables you to stop and start services on servers in the cluster. Figure 2-11 shows the Tools menu items in Cisco CallManager Serviceability.

Figure 2-11 *Cisco CallManager Serviceability Tools Menu*

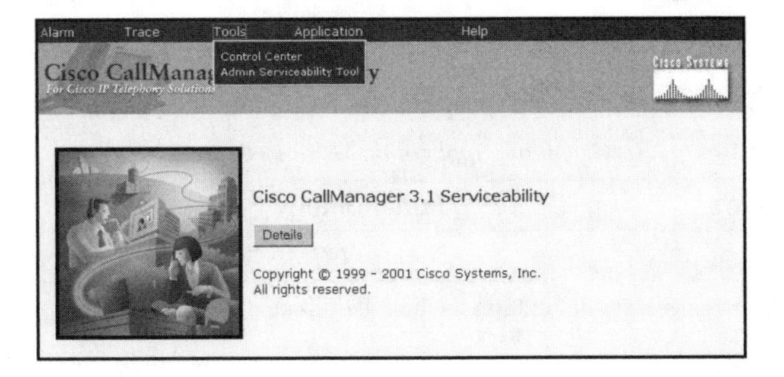

Table 2-12 lists some brief descriptions of Tools menu items.

Table 2-12 *Cisco CallManger Serviceability Tools Menu Items*

Menu Items	Functionality
Control Center	Use the Control Center to view status and to start and stop Cisco CallManager services for a particular server or all servers in the cluster.
CDR Analysis & Reporting	Use this Web-based tool, which is the ART renamed to CAR, for system and device reporting and database purge functionality.
AST or Real-Time Monitoring Tool	Use this Web-based tool to monitor real-time behavior of the components in a Cisco CallManager cluster. This tool monitors device status, system performance, and device discovery. It also enables you to connect directly to devices using HTTP for troubleshooting system problems and to set up event notification via e-mail, popup message, or pager.

Application Menu

The Application menu enables you to install and configure plug-ins. It also links to CCMAdmin. Figure 2-12 shows Application menu items in CCMServiceability.

Figure 2-12 *Cisco CallManager Serviceability Application Menu*

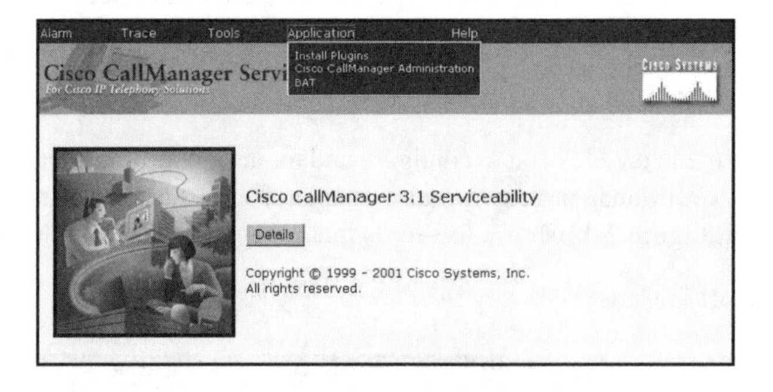

Table 2-13 lists some brief descriptions of Application menu items.

Table 2-13 *Cisco CallManger Serviceability Application Menu Items*

Menu Items	Functionality
Install Plug-ins	Same as CCMAdmin.
Cisco CallManager Administration	Click the link to go to CCMAdmin.
ART	Click the link to go the Administrative Reporting Tool, if installed.
BAT	Click the link to go to BAT, if installed.

Help Menu

The Help menu enables you to access help in general or for a specific page. It also enables you to check the component versions and make sure they are in-sync. Figure 2-13 shows Help menu items in CCMServiceability.

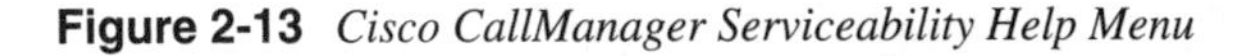

Figure 2-13 *Cisco CallManager Serviceability Help Menu*

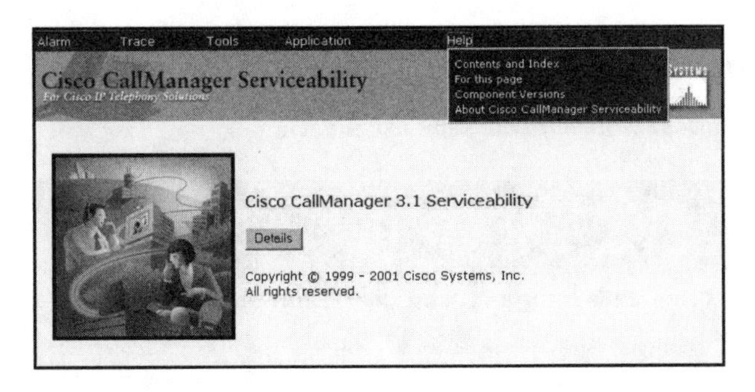

Table 2-14 lists some brief descriptions of Help menu items.

Table 2-14 *Cisco CallManger Serviceability Help Menu Items*

Menu Items	Functionality
Contents and Index	Same as CCMAdmin.
For this Page	Same as CCMAdmin.
Component Versions	Same as CCMAdmin.
About Cisco CallManager Serviceability	Same as CCMAdmin, but it takes you to the main page for CCMServiceability.

Low- to Medium-Count Items Versus High-Count Items

In Cisco CallManager, the number of items in the database for a particular device type, such as phones or gateways, can be very high. This means that any list of all items would be very long, making it difficult to find a particular device. To make the response time of CCMAdmin more efficient, some pages automatically list all the items related to that page and other pages provide a search engine that enables you to search for devices that have specific characteristics. The list of servers (under the System menu) is usually a low-count item, meaning the number of servers displayed is usually small. There may be two to eight servers in a cluster, which will all be listed in the left column of the page. On the other hand, the list of phones (under the Device menu) is usually a high-count item, meaning the number of phones displayed in a list is quite large. You could have up to 10,000 phones in a single cluster. If all the phones appeared when you clicked **Device > Phone**, you could be waiting a long time for the list of phones to appear, and then the list would be so long as to be unmanageable. Because phones are considered a high-count item, search parameters are provided to ease the output to the Phone page.

Low- to Medium-Count Items

Figure 2-14 shows how low- to medium-count items appear in a left frame on the CCMAdmin pages. Select an item from the left column to edit, update, or delete that item.

Figure 2-14 *Locating Low- to Medium-Count Items to Configure*

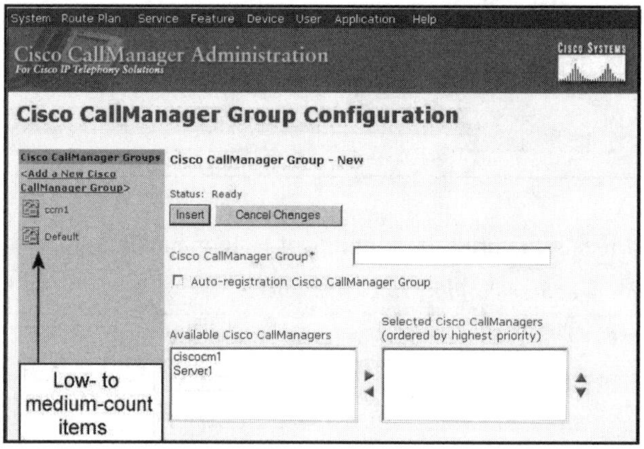

High-Count Items

Figure 2-15 shows how high-count items include a search engine that can help you find the device you are looking for without listing every device. Use the search options to create a query for devices. Phones and gateways are considered high-count items. When search results are returned, click on the link to an item to edit it, or click on its icon to perform actions (copy, delete, reset) from the list. When you select **Find** without establishing any search criteria, all registered phones appear.

Figure 2-15 *Locating High-Count Items to Configure*

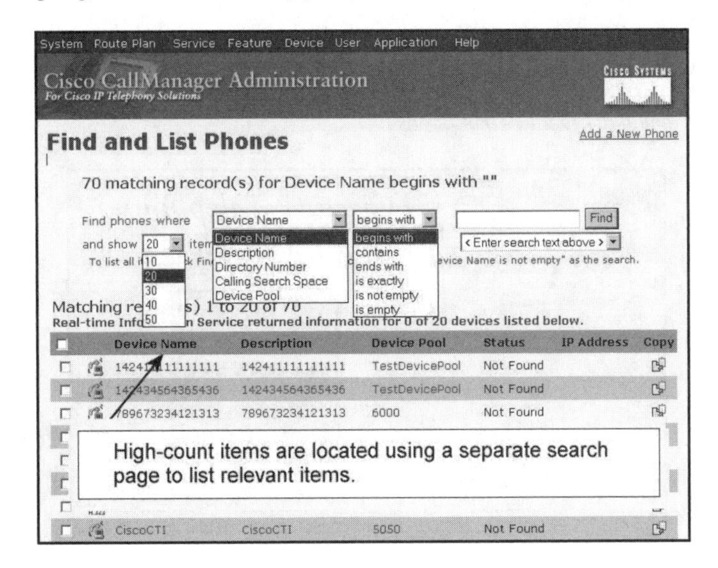

Online Help

The online documentation is extremely useful and can help you understand concepts and learn how to configure and monitor the components of a Cisco IP Telephony solution. Figure 2-16 shows the Help menu in CCMAdmin and the component versions page.

Figure 2-16 *Online Help*

Cisco CallManager Component Versions

Component Versions for DADA:

Component	Version	Installation ID
ace.dll	5.1.12.0	3.1(1.030)
aced.dll	5.1.12.0	3.1(1.030)
astisapi.dll	3.1.0.29	3.1(1.030)
aupair.exe	3.1.0.66	3.1(1.030)
aupairps.dll	3.0.1.84	3.1(1.030)
avvidcustomerdirectoryconfigurationplugin.exe	2.11.15.0	3.1(1.030)
bootp.exe	2.0.2.2	3.1(1.030)
ccm.exe	3.1.0.30	3.1(1.030)
ccmperfmon.dll	3.1.0.30	3.1(1.030)
ccntest.exe	3.1.0.30	3.1(1.030)
cdpintf.dll	2.1.0.0	3.1(1.030)
Cisco CallManager Administration	3.1(0.5)	CCM3.1(1.030)
Cisco CallManager Serviceability	3.1(0.2)	CCM3.1(1.030)
Cisco CallManager Trace Filter Extension	3.1(0.1)	CCM3.1(1.030)

You can learn more about each of the menu items in Table 2-9 in the section "Help Menu" earlier in this chapter.

Contents and Index Menu Item

By selecting Contents and Index from the Help menu, a new browser opens, displaying the window in Figure 2-17.

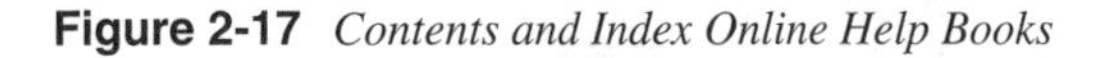

Figure 2-17 *Contents and Index Online Help Books*

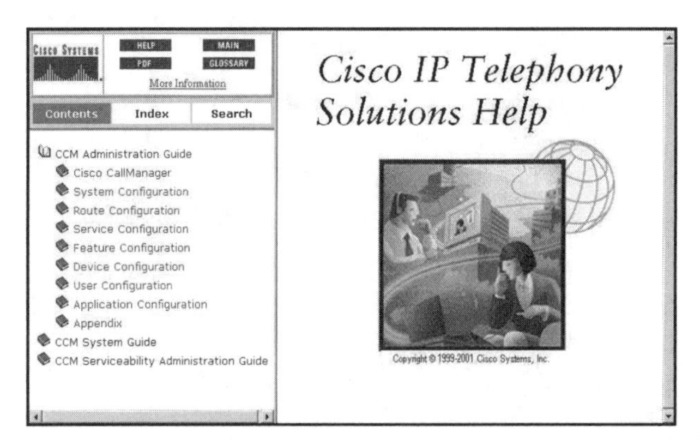

The Cisco IP Telephony Solutions Help includes some or all of the following books:

- **Cisco CallManager Administration Guide**—Follows the menu items of CCMAdmin and provides step-by-step instruction to configure each item. Tables and charts help to define and assist with configuring items in CCMAdmin.
- **Cisco CallManager System Guide**—Provides a broad look at the entire Cisco IP Telephony Solution. This section provides background information about a Cisco CallManager cluster and other related components.
- **Cisco CallManager Serviceability Administration Guide**—Describes how to use CCMServiceability, a tool used to help monitor and troubleshoot a Cisco IP Telephony system.
- **Administrative Reporting Tool User Guide**—Describes how to configure reports on device utilization, system utilization, system overview, and more, as well as how to use database purge, configure notifications, and much more.
- **BAT User Guide**—Describes how to add, update, or delete large numbers of devices, users, and combinations of users and devices by creating BAT templates and using them in conjunction with comma-separated values (CSV) files.

Intra-Cluster Communication

This section discusses Cisco CallManager cluster concepts. Figure 2-18 illustrates the two types of communication that occur within a cluster.

Figure 2-18 *Intra-Cluster Communication*

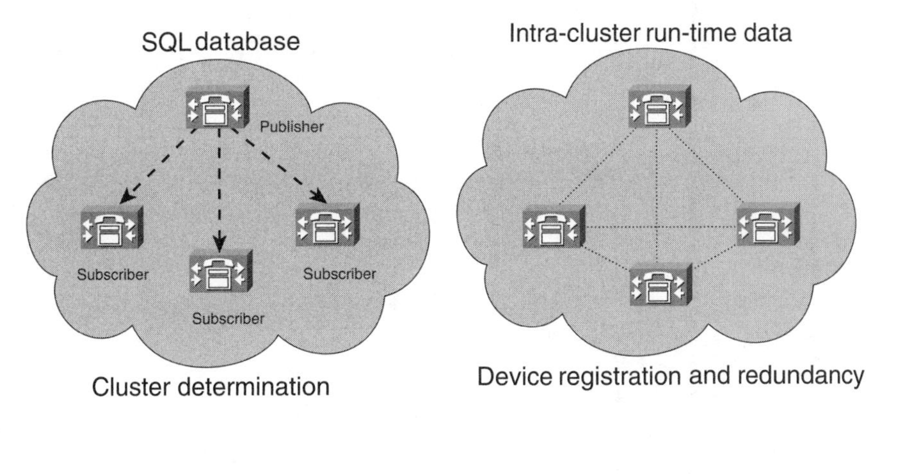

Following is a high level view of the two types of intra-cluster communication.

The first is the database component packets that contain device configuration information. The database relationship (Publisher and Subscribers) is what determines a cluster. All changes are made in the Publisher's database and those changes are replicated to the Subscriber databases. The database used is Microsoft SQL Server 7.0; configuration is stored and modified on the Publisher server's database and replicated to all members of the cluster. This ensures that the configuration is consistent across the members of the cluster as well as facilitating spatial redundancy of the database.

The second intra-cluster communication is the propagation and replication of run-time data, such as Cisco IP Phone registration and the registration of gateways and DSP resources. This information is shared across all members of a cluster and assures optimum routing of calls between members of the cluster and associated gateways.

You should install and configure the Publisher database server before building any Subscriber servers. In general, Cisco recommends you create a Publisher server that does not do any call processing, meaning Cisco CallManager is not installed on the server running the Publisher database. For smaller deployments of 2500 phones or less, however, a Publisher server may also be the backup Cisco CallManager for devices. In larger deployments, you should have a stand-alone Publisher server and a stand-alone Cisco TFTP server. Table 2-15 provides a summary of the recommended Cisco CallManager components to install based on the size of the deployment.

Table 2-15 *Recommended CallManager Components Based on System Size*

Deployment Size and Server	Components						
	Cisco CallManager Components	*Cisco CallManager*	*Cisco TFTP Server*	**Cisco CallManager Web Components**	**Cisco Optional Components**	*Cisco Messaging Interface*	*Cisco IP Voice Media Streamer*
Small 2500 or fewer							
Publisher	W	Y	X	W			
Subscriber	W	X			Z	Z	Z
Medium 2501 to 5000							
Publisher	W	X	X	W			
Subscriber	W	X			Z	Z	Z
Subscriber	W	X			Z	Z	Z
Subscriber	W	Y			Z	Z	Z
Large 5001 to 10,000							
Publisher				W			
TFTP				W			
Subscriber	W	X			Z	Z	Z
Subscriber	W	X	W		Z	Z	Z
Subscriber	W	Y			Z	Z	Z
Subscriber	W	X			Z	Z	Z
Subscriber	W	X			Z	Z	Z
Subscriber	W	Y			Z	Z	Z

W – Must install

X – Part of the Cisco CallManager component installation

Y – Serves as a backup Cisco CallManager server

Z – Optional install

If you have a small deployment of 2500 or fewer and are using the Publisher to be a backup Cisco CallManager, install all the Cisco CallManager components, Cisco CallManager Web components, and any optional components you think you may need.

If you have a medium deployment of 2500 to 5000 and are using the Publisher as a Cisco TFTP server, install Cisco TFTP and Cisco CallManager Web components.

If you have a large deployment of 5000 to 10,000 and are going to have a stand-alone Cisco TFTP server, install Cisco CallManager Web components. You also have to build a TFTP server, and at this dialog box during installation, you need only select Cisco TFTP.

Because database changes can only occur if the Publisher server is up, you do not need to install Cisco CallManager Web components or Cisco TFTP on the Subscriber servers.

Later in this chapter we will discuss cluster recommendations and supported deployment models.

Data Source

The database used is Microsoft SQL Server 7.0 plus Service Pack 2.0.

The Publisher server reads its local database for information. The Publisher's database is one-way replication or uni-directional replication.

The database users should read from the same source that is written to. The Subscriber servers look to the Publisher server's database for their information.

Figure 2-19 shows which database that servers within a cluster access.

Figure 2-19 *Data Source*

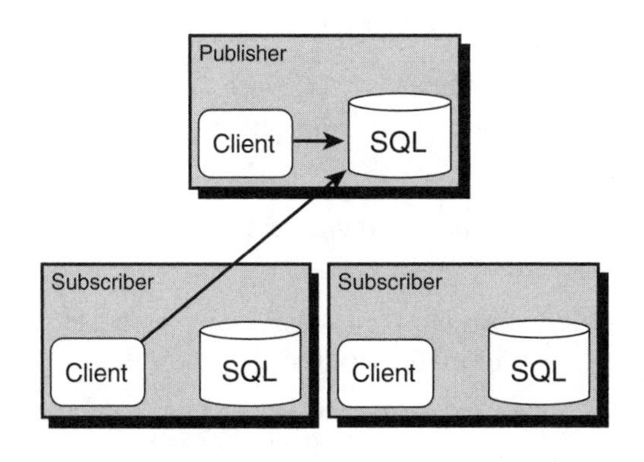

Database information is written to the Publisher database only. If the Cisco CallManager Web components are installed on the Subscriber and you browse into the Subscriber, then all the entries you make using the CCMAdmin page of the Subscriber gets written to the database on the Publisher server. If the Publisher is down, any entry made in the CCMAdmin page of the Subscriber server will not be entered and will not be queued in a log for transmission at a later time. The updates are lost.

When you are building a Publisher server, you may choose to have Cisco CallManager installed or not. When the Publisher server does not have Cisco CallManager installed on it, it can be referred to as the glass house. Sometimes (usually in smaller clusters) the Publisher has Cisco CallManager installed and is used as a backup Cisco CallManager. Figure 2-20 shows the direction of replication within a cluster.

Figure 2-20 *Data Source Replication*

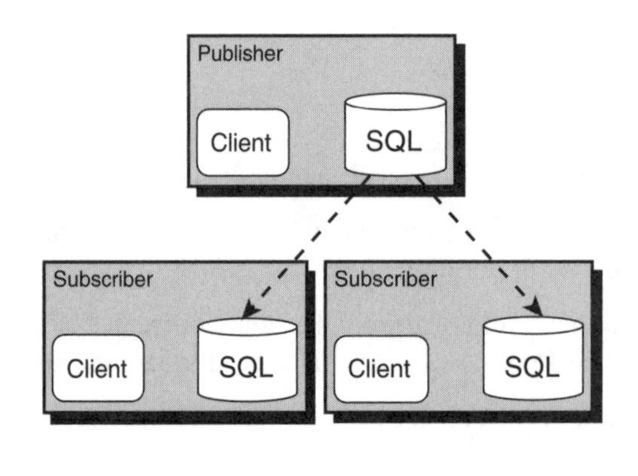

The replication of the database within a Cisco CallManager cluster is one-way (or unidirectional) and works like this:

- Changes made to the Publisher's database are replicated out via transactional process. Changes happen immediately unless the link is down.
- If the link is down, SQL keeps a transaction log and replicates the data when possible.
- The Subscriber database is read-only.

Figure 2-21 shows the order a Subscriber accesses a database.

Figure 2-21 *Database Reads*

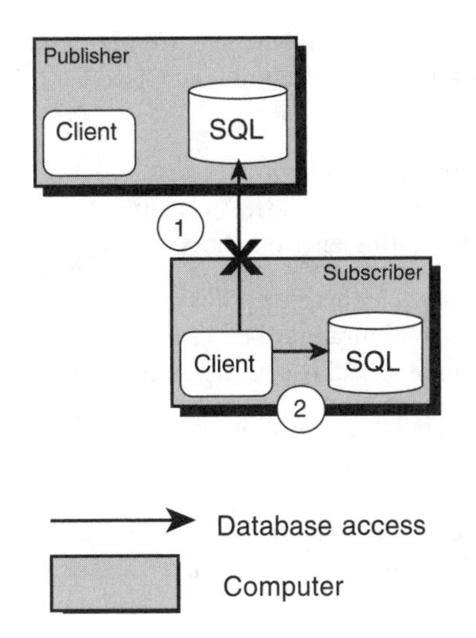

The relationship between Cisco CallManager Publisher and Subscriber databases within a cluster works in the following way:

- The Publisher database reads its own local database for information.
- The Subscriber (or attached Cisco CallManagers in the same cluster) servers look to the Publisher database for their information.
- In the event of a failure, Subscribers read their local copy of the database.

In a small deployment, there may be confusion as to whether you are on a Publisher or a Subscriber. To ensure you know which server you are working on (either a Publisher or Subscriber), go to SQL Server Enterprise Manager and expand all the way down to the database. Select the SQL Server Group <*Server_Name*> Databases CCM030X (or whichever database number is the highest; these database numbers increment each time a Cisco CallManager upgrade occurs on the same server). If this is the Publisher, you see a folder labeled Publications; if this is a Subscriber, you see a folder labeled Pull Subscriptions.

Usually there is no confusion when deploying large systems because the Publisher server is a stand-alone server only used for database changes and updates.

A Subscriber server is dependent on the Publisher for database information within a cluster. If a catastrophic failure occurs to that Subscriber server, you rebuild the server and then remove all references to the failed Subscriber on the Publisher server. When you re-install the Subscriber, all the database replication configuration is completed as part of the reinstall.

System Setup

This section focuses on the first menu in CCMAdmin, the System menu. The System menu is where you will begin to configure your Cisco CallManager system after all the servers in the cluster are installed.

Assuming that DHCP is configured correctly, setting up auto-registration and configuring a few system items can be the fastest way to get two phones to communicate. This is often done in laboratory testing environments.

In larger deployments, either green field systems or existing systems, this is only a small stepping-stone of configurations before getting phones to communicate. We are going to work in CCMAdmin as if this is a large deployment.

The following is a quick description of tasks covered in this section:

- Server
  - — Use server configuration to specify the IP address of the server. DNS host name can also be used, but Cisco recommends using DHCP and IP addresses.
- Cisco CallManager
  - — Clean up the name of the Cisco CallManager (make the name more descriptive).
  - — Set up directory numbers for auto-registration.
- Cisco CallManager Group
  - — Create Cisco CallManager groups for device redundancy.
- Date/Time Group
  - — Use date/time groups to define time zones for the various devices connected to Cisco CallManager.
- Device Defaults
  - — Use device defaults to set the system-wide default values of each type of device that registers with a Cisco CallManager.
  - — The system-wide device defaults for a device type apply to all devices of that type within a Cisco CallManager cluster.
- Region
  - — Use regions to specify the voice codec used for calls within a region and between regions.
  - — The voice codec determines the maximum amount of bandwidth used per call.
- Device Pool
  - — Use device pools to define sets of common values for devices.

- — You can specify the following device settings for a device pool: region, date/time group, Cisco CallManager group, MRGL, user hold MOH audio source, network hold MOH audio source, CSS for auto-registration, and auto-answer enable for Cisco IP Phones that support this feature.
- Enterprise Parameters
  - — Enterprise parameters provide default settings that apply to all devices and services in the same cluster.
  - — When you install a new Cisco CallManager, it uses the enterprise parameters to set the initial values of its device defaults.
- Location
  - — Use locations to implement CAC in a centralized call processing system.
  - — CAC enables you to regulate voice quality by limiting the amount of bandwidth available for calls over links between the locations.

Server Menu Item

This section discusses the Server menu item on the System menu in CCMAdmin. Figure 2-22 shows the Server Configuration page in CCMAdmin.

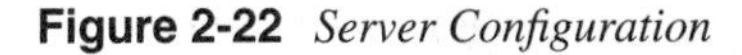

Figure 2-22 *Server Configuration*

During the installation of the operating system and backup recovery CD, you will assign a name to the server. Names of servers in the cluster can be very helpful for monitoring and troubleshooting. Once a server name has been selected during installation, it cannot be changed.

A server name should include the following three characteristics:

- **Location**—The physical location of the server, usually a town or city or if the city you are in has an airport, you could use the airport code.
- **Cluster Number**—In one location you may have multiple clusters. For example, at Cisco Systems in San Jose, there are multiple clusters each available to serve 10,000 users. Cluster numbers can be as simple as 1, 2, and 3 or more descriptive, such as office codes of phone numbers such as 853, 525, 526, and 527.
- **Server Function**—In the cluster, a server can perform different functions based on the deployment model and number of users. The function the server will perform in the cluster can be noted by using letters, such as A, B, and C. More about cluster recommendations and deployment models later in this chapter.

Planning and design are important before building a cluster. Refer to the following link on CCO for more information about design:

www.cisco.com/univercd/cc/td/doc/product/voice/

During the installation of Cisco CallManager, the default server name is the name of the server provided during installation of the first CD. Cisco recommends you change the name of the server to the IP address of the server.

DNS is commonly used to provide name resolution to an IP address of a device on the network. DNS is helpful because now we can just remember names and not IP addresses of Web sites. If we go into a DOS prompt and ping for www.cisco.com, we see a reply of an IP address (192.133.219.5). DNS was used to correlate the name Cisco to an IP address.

As for the CIPT solution, when devices need to communicate and DNS is used, the devices encounter another transaction that must be completed before being able to communicate with Cisco CallManager. Be advised: this is also a potential point of failure. When using DNS resolution for the Cisco CallManager server name, a failure of your DNS server results in the inability for phones to contact Cisco CallManager upon bootup.

Cisco CallManager Menu Item

This section discusses configuration of the menu item Cisco CallManager of the System menu in CCMAdmin. Figure 2-23 shows the Server Configuration page in CCMAdmin.

The default name of Cisco CallManager at this menu item level is CM_<*server name*>. Cisco recommends deleting the prefix CM_ and leaving the server name for the Cisco CallManager name. Removing the CM_ reduces the number of characters in the name, which is helpful when displaying the server names in a list. For the description, enter the function that this Cisco CallManager is performing in the cluster.

Figure 2-23 *Server Configuration*

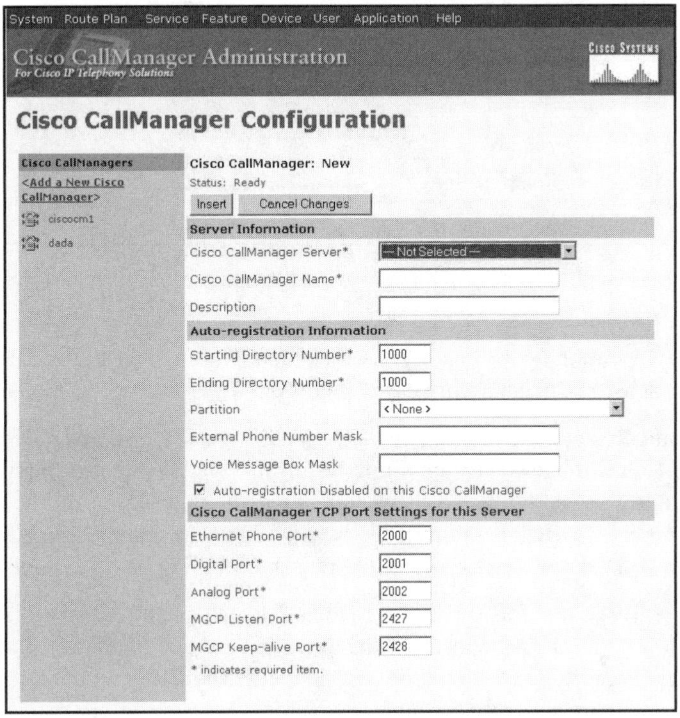

After making those changes and clicking Update, the following appears at the top of the Cisco CallManager Configuration page in CCMAdmin:

> *<server name>* (description*<function>*) on *<IP address>*
>
> DFWCLSTR1A (Publisher) on 172.28.16.3

On the Cisco CallManager Configuration page, you can enable and configure auto-registration. Auto-registration is disabled by default.

You can also view the TCP port settings for the server selected. Cisco CallManager uses the TCP port settings to communicate with the associated devices shown in Table 2-16. Cisco recommends you do not change default port settings unless you have a clear understanding of the ramifications of doing so.

Table 2-16 *TCP Port Settings*

Port Name	TCP Port
Ethernet phone port	2000
Digital port	2001
Analog port	2002
MGCP listen port	2427
MGCP KeepAlive port	2428

Auto-Registration

Auto-registration is a fast and easy way to add phones to the network. Auto-registration is used extensively in laboratory environments.

To configure auto-registration, uncheck the box **Auto-registration Disabled on this Cisco CallManager**. Then provide a starting and ending directory number and click **Insert** or **Update**. The partition and external phone number mask can also be configured for auto-registration. If DHCP is setup correctly, you can now begin plugging Cisco IP Phones into the network, and within two minutes you can be calling between phones.

Auto-registration is mostly used for green field and small deployments. If you are setting up a new office and have received the direct inward dial (DID) range from the local telephone company, you could use the first and last number of the DID range for the auto-registration directory number range. Auto-registration directory numbers are assigned to devices consecutively based on when each phone registers on the network. When all the auto-registration numbers are assigned, Cisco CallManager goes back to the first available directory number in the configured range. If you are using auto-registration and must configure a large number of new phones, consider using the Tool for Auto-Registered Phone Support (TAPS). TAPS, an optional component of BAT, enables you to update dummy MAC addresses by simply dialing into a TAPS directory number and following a few simple prompts. Use BAT to create a phone template and a CSV file, check the **Use Dummy MAC Address** option (which saves you the labor of typing in each and every MAC address), plug the phones into the network, and then dial into TAPS on each new phone to update the dummy MAC address in the Cisco CallManager database with the actual MAC address. Dialing into TAPS is so simple that you can pass that task onto the new phone owners.

NOTE To use TAPS, you must have purchased a Cisco Customer Response Applications server and configured a TAPS directory number.

Another use for auto-registration is to prevent rogue or unauthorized Cisco IP Phones from being plugged into your network to make calls. To manage this you can set up a partition called auto-registration and the directory number range outside of the DID range provided by the local telephone company.

Auto-registration can also be configured in the Cisco CallManager group. It is good to enable this at a group level so that the rogue phones are registered to a backup Cisco CallManager and will not affect the call processing of the primary Cisco CallManagers. If auto-registration is enabled at the Cisco CallManager group level, you can also apply a CSS at the device pool level.

Being able to apply a partition and CSS to an unauthorized phone can be very powerful. The rogue phone can be turned into a Private Line Automatic Ring-down (PLAR) phone. PLAR is a feature where a phone goes off-hook and is automatically routed to a predetermined directory number. When the rogue phone goes off-hook, the caller is automatically routed to a specific person who can collect information from the caller to determine whether to activate the phone. Partitions and CSSs are what make PLAR possible; this is further described in Chapter 3, "Cisco CallManager Administration Route Plan Menu."

Configuring a Server and Cisco CallManager

Use server configuration to specify the address of the server where Cisco CallManager is installed. If your network uses DNS services, you can specify the DNS name of the server. If your network does not use DNS services, you must specify the IP address of the server. It is recommended to change the server name to an IP address.

Use Cisco CallManager configuration to specify the TCP ports and other properties for each Cisco CallManager installed in the same cluster. This is a good page to identify the server name to an IP address and a function of the server in the cluster. Cisco CallManager configuration is also used to set up for auto-registration.

The following are quick steps to configure a server and Cisco CallManager:

Step 1 Open CCMAdmin and click **System > Server**.

Step 2 Change the server name to the server's IP address.

Step 3 Configure Cisco CallManager. Go to **System > Cisco CallManager** in CCMAdmin.

Step 4 Edit the Cisco CallManager name and enter the function of the server in the description box.

Step 5 Enable a Cisco CallManager for auto-registration. Enter starting and ending directory numbers and enable auto-registration.

Cluster Recommendations

This section discusses the recommendations to follow while building a cluster based on how many users you are going to support.

Review

Figure 2-24 shows the two types of communication that go on within a cluster.

Figure 2-24 *Intra-Cluster Communication*

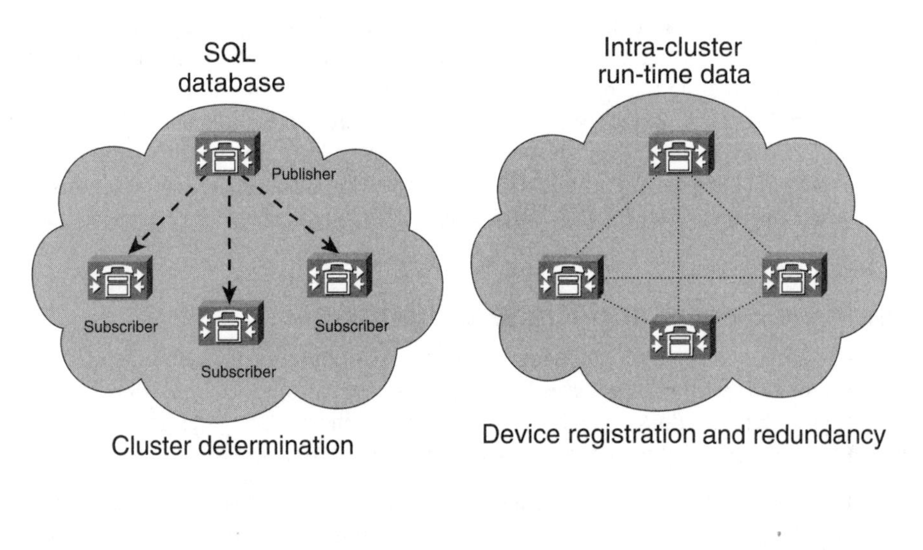

- - - - - One-way database replication
.... Two-way real-time communication

Remember that within a cluster there are two types of communication taking place. One deals with database communication and the other deals with device registration and redundancy. After discussing cluster recommendations, you will be able to configure Cisco CallManager groups, which determine device registration and redundancy.

Up to 2500 Users

This section discusses the recommendations to support up to 2500 users. Figure 2-25 shows cluster recommendations to support up to 2500 users.

Figure 2-25 *Cluster Recommendations: Up to 2500 Users*

A Cluster of Two Cisco CallManagers

- Publisher is also the secondary
- Single active Cisco CallManager

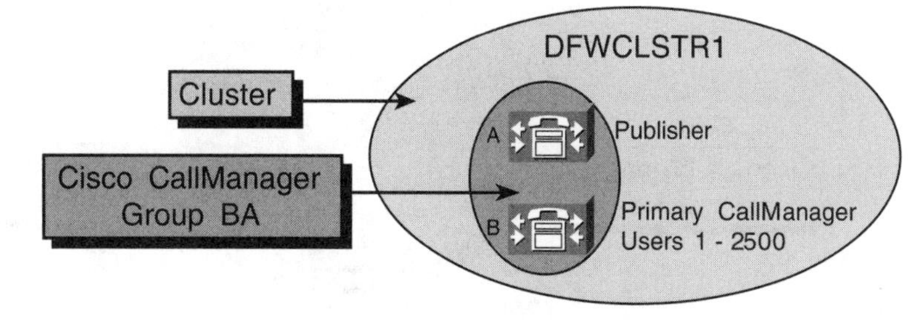

To support up to 2500 users you should have at least two servers. As you can see from Figure 2-25, one server will be the Publisher and the secondary or backup Cisco CallManager. The second server will be a Subscriber server and the primary Cisco CallManager to handle all the call processing.

If you anticipate growth beyond 2500 users, Cisco recommends using three servers to support 2500 users. One server will be the Publisher and performs no call processing. The second server will be the primary Cisco CallManager for users 1–1250 and the backup for users 1251–2500. The third server will be the primary Cisco CallManager for users 1251–2500 and the backup for users 1–1250. This enables for growth to up to 5000 by adding only one more server.

2501 to 5000 Users

To support 2501 to 5000 users, the cluster should be built with four servers. Figure 2-26 shows cluster recommendations to support up to 5000 users.

As the figure above shows, you will have a stand-alone Publisher (A), also known as the "glass house" because it does not provide call processing; two primary Cisco CallManagers (B and C); and a dedicated backup Cisco CallManager (D).

Cisco recommends that a single server have no more than 2500 users. The question here is, "What if the two primary Cisco CallManagers go down and my backup now has 5000 users to support and provide call processing for?" This is a possible situation. For the best server performance and providing call processing it is best to keep the number at or below 2500, so 5000 users on one server is too much.

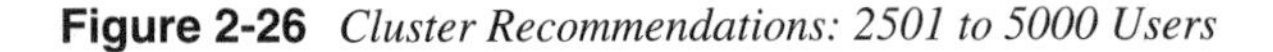

Figure 2-26 *Cluster Recommendations: 2501 to 5000 Users*

- Stand-alone Publisher
- Two primary Cisco CallManagers (B and C)
- One backup Cisco CallManager (D)

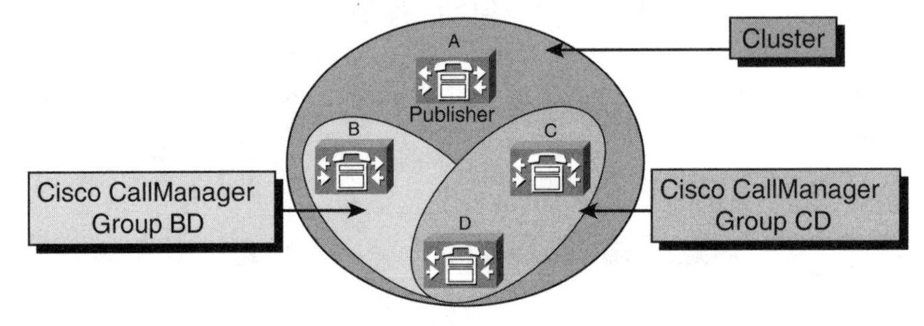

When there are over 2500 users on one server, the time required to get dial tone and call processing is extended dramatically. It may take two to three seconds after you go off-hook for you to hear a dial tone. If this happens, focus on getting one primary server up as quickly as possible and then the next server. Use the information from alarms and logs to troubleshoot the cause of the problem.

5001 to 10,000 Users

This section discusses the recommendations to support from 5001 to 10,000 users. Figure 2-27 shows cluster recommendations for up to 10,000 users.

To support 10,000 users, add four more servers and duplicate the primary backup scenario for 5001 users. The device redundancy scheme is a little different and now we can have a secondary and tertiary. In the cluster to support 10,000 users there are eight servers. One server is the Publisher (glass house) that performs database processing but does not provide call processing. One server is the Cisco TFTP server that serves device configuration files and does not provide call processing. Four servers are primary Cisco CallManagers, and two servers are backup Cisco CallManagers.

The redundancy scheme uses both backup servers as secondary and tertiary for the primary Cisco CallManagers. As shown in the Figure 2-27, B is the primary, D is the secondary, and H is the tertiary for a device that is assigned to the BDH Cisco CallManager group.

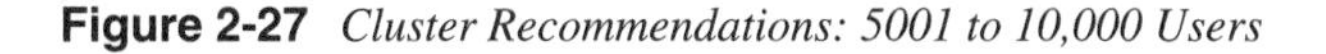

Figure 2-27 *Cluster Recommendations: 5001 to 10,000 Users*

- Stand-alone Publisher and Cisco TFTP
- Four primary Cisco CallManagers (B, C, E, and G)
- Two backup Cisco CallManagers (D and H)

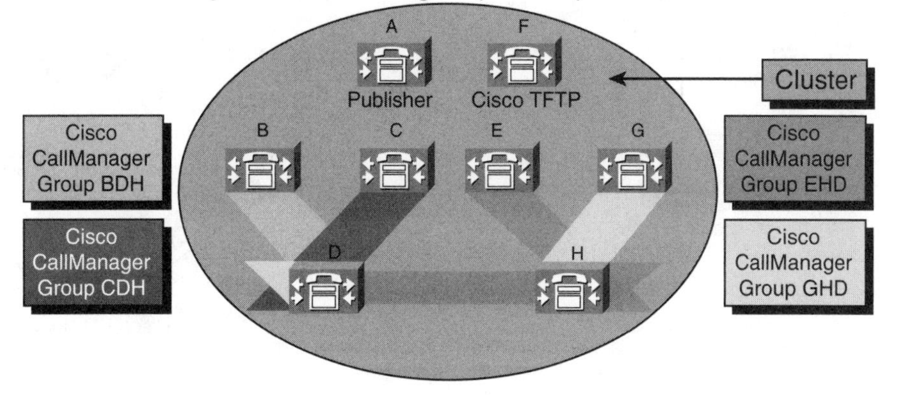

Cisco CallManager Groups

This section discusses how to configure Cisco CallManager groups.

To ensure that only a single CallManager is active at a time, all devices should be assigned to a single CallManager redundancy group. For example, in a small deployment, you could assign all phones to the same device pool to be sure that all phones move to the backup Cisco CallManager when a failure on the primary Cisco CallManager occurs, and then all phones re-home together to the primary Cisco CallManager when service is restored. The devices are assigned a Cisco CallManager group when they are assigned to a device pool. This CallManager redundancy group consists of a prioritized list of up to three Cisco CallManagers. Cisco CallManager groups provide two important features for your system:

- **Distributed Call Processing**—Enables you to distribute the control of devices across a cluster of Cisco CallManagers.
- **Redundancy**—Enables you to designate a primary, a secondary, and a tertiary Cisco CallManager in each group.

Figure 2-28 shows two simple calls between two Cisco IP Phones registered to two different Cisco CallManagers in the same cluster. Cisco CallManager groups enable devices to be registered to different Cisco CallManagers in the cluster and, depending on which server the device is registered to, that Cisco CallManagers will handle the call processing.

Figure 2-28 *Distributed Call Processing*

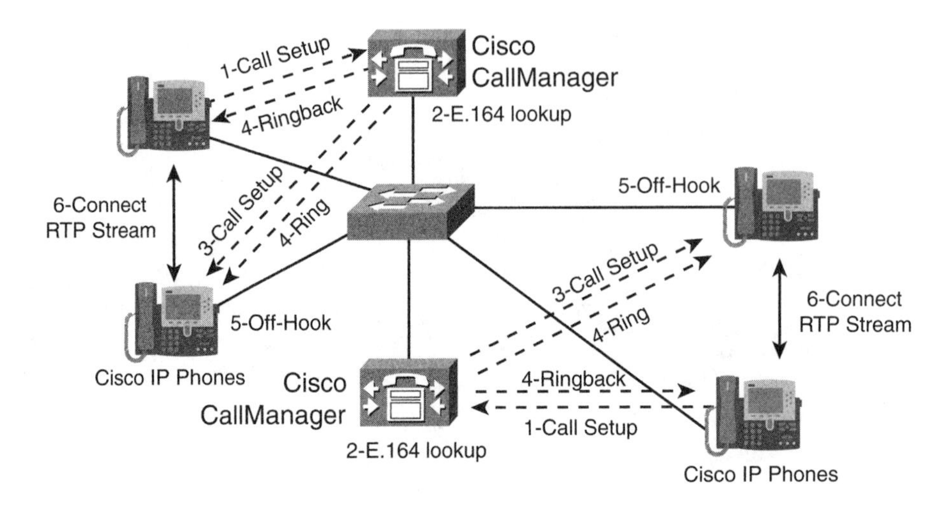

Cisco CallManager groups enable Cisco CallManager nodes in the cluster to share call processing while accessing the same database.

As Figure 2-28 illustrated, two calls are made simultaneously from two Cisco IP Phones registered to two Cisco CallManager nodes. Each call is processed at the same time because of the distributed call processing provided by Cisco CallManager groups in a cluster.

After the Cisco CallManager groups are configured, devices have redundant Cisco CallManager nodes in case of failure or loss of connection. A device can only be registered with one Cisco CallManager at a time and can have up to three Cisco CallManagers for redundancy. The example in Figure 2-29 uses a phone for the device that registers to Cisco CallManager.

NOTE In a centralized call processing model, the default router's IP address is in the list of Cisco CallManagers for fail-over just in case the WAN link is down. This is the Survivable Remote Site Telephony (SRST), which is covered later in this book.

The device uses a specific TCP port to connect to its registered Cisco CallManager as shown in the figure above. The device sends a TCP connect register (active) message to its primary Cisco CallManager and a TCP connect (standby) to its secondary Cisco CallManager. While the primary Cisco CallManager is active, the device sends a TCP KeepAlive to its primary Cisco CallManager every 30 seconds.

Figure 2-29 *Redundancy, Part 1*

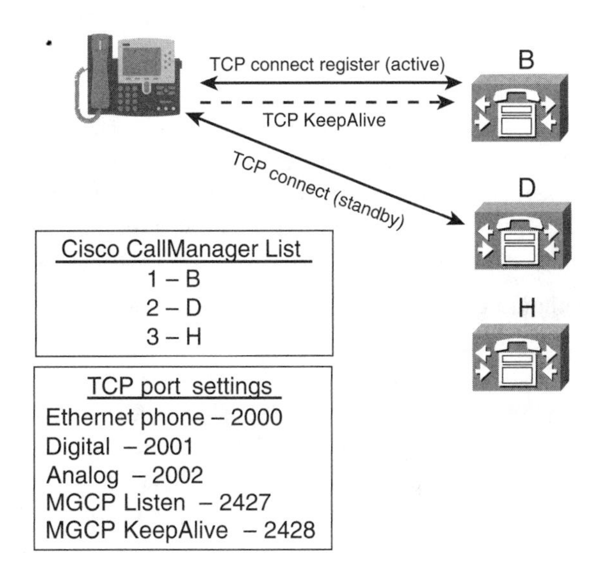

If the link between the primary Cisco CallManager goes down as shown in Figure 2-30, the device sends a TCP connect register message to its secondary and a TCP connect (standby) to its tertiary.

Figure 2-30 *Redundancy, part 2*

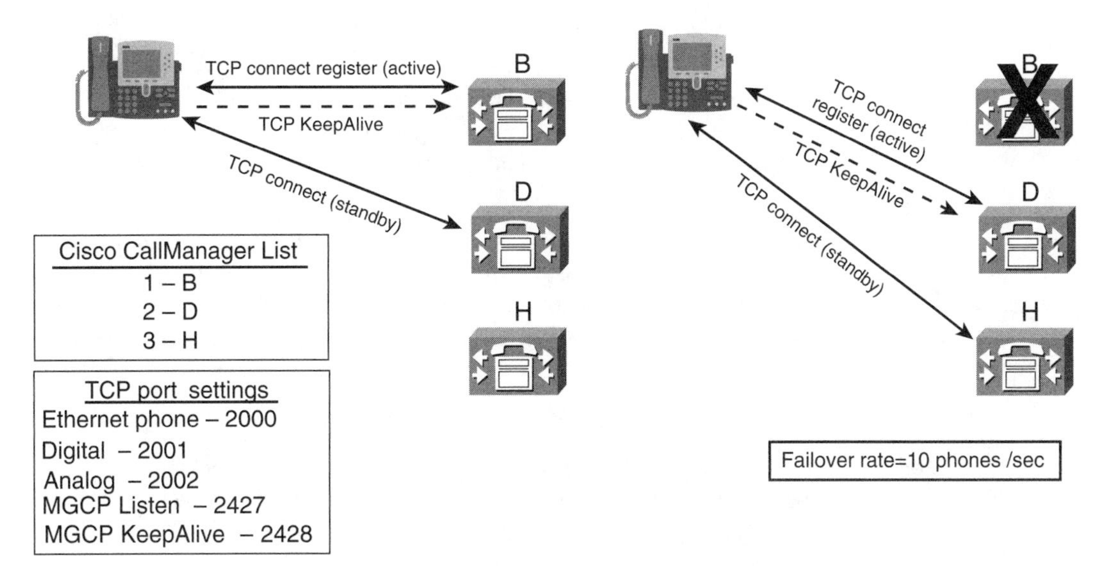

While registered to the secondary, the device sends TCP KeepAlive messages to the secondary until it can re-establish a connection with its primary or until the primary comes back on-line.

The fail-over rate is 10 phones per second. This enables 2500 phones to fail-over in a little over four minutes. This is a configurable parameter. The parameter can be found in the Cisco CallManager service parameters; the parameter name is MaxStationsInitPerSecond, and the default setting is 10. This means that 10 phones per second can register with a Cisco CallManager, and other phones are queued for registration later. This stress-tested service parameter should not be changed unless advised to do so by Cisco TAC.

When configuring Cisco CallManager groups, be aware of some issues:

- When devices are added in CCMadmin, they are assigned to Cisco CallManager groups manually. Being assigned to a Cisco CallManager group is different from a device registering to a Cisco CallManager. When a device pool is assigned to a device, it creates a configuration file in the TFTP server. When the device is connected to the network it gets its configuration file and in that file it gets a Cisco CallManager group. The first Cisco CallManager listed in the group is where the device tries to register to first.
- There is no easy way to identify which group devices are members of and no easy way to ensure Cisco IP Phones are evenly distributed (no **reshuffle** command).
- Proper design, planning, and using BAT can be helpful to evenly distribute devices. The descriptive naming of Cisco CallManager groups and device pools can help to identify which Cisco CallManager a device is registered to. The AST/Real-Time Monitoring Tool (the name is determined by your Cisco CallManager release) can also help in quickly identifying how devices are assigned to a Cisco CallManager group.

Cisco CallManager groups provide redundancy for devices registered in the cluster. Depending on the cluster size, you may have a primary, secondary, and/or a tertiary list of Cisco CallManagers in your Cisco CallManager groups.

In **System > Cisco CallManager** you enabled a Cisco CallManager for auto-registration and provided a directory number range. In **System > Cisco CallManager Group** you can enable a Cisco CallManager group for auto-registration. Unless you are doing a green field or small deployment, you want to enable auto-registration to a backup Cisco CallManager so you do not affect call processing when rogue (un-administered) phones register in the cluster.

The following are quick steps to configure a Cisco CallManager group:

Step 1 Open the Cisco CallManager Group Configuration page by clicking **System > Cisco CallManager Group** in CCMAdmin. Figure 2-31 shows the Cisco CallManager Group Configuration page.

Figure 2-31 *CallManager Group Configuration*

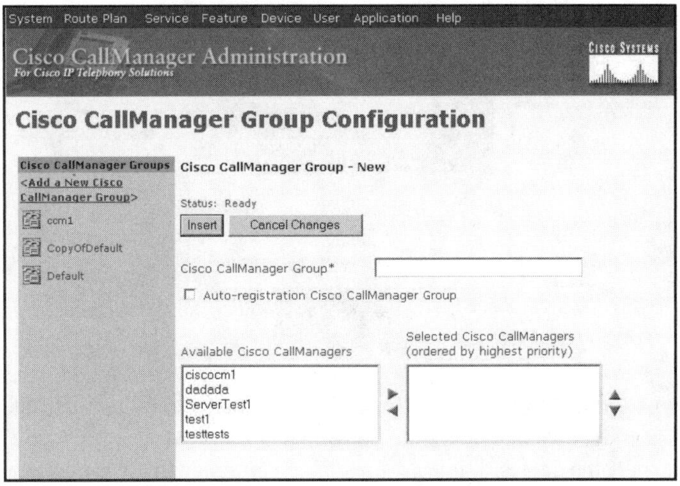

Step 2 Enter a descriptive name for the **Cisco CallManager Group** and check or leave un-checked the box for **Auto-registration Cisco CallManager Group**.

Step 3 Configure the list of Cisco CallManagers in the Cisco CallManager Group. Highlight a Cisco CallManager from the list of **Available Cisco CallManagers** and move it to the **Selected Cisco CallManagers** box.

Step 4 Highlight another Cisco CallManager from the list of **Available Cisco CallManagers** and move it to the **Selected Cisco CallManagers** box.

Step 5 Order the list of Cisco CallManagers using the highest as priority. Use the arrows to the right of the **Selected Cisco CallManagers** box to move Cisco CallManagers up or down in the list.

Step 6 Click **Insert** to add the group.

Date/Time Group Menu Item

You configure date/time groups using the Date/Time Group menu item in the System menu of CCMAdmin. Figure 2-32 shows how the Date/Time Group is displayed on a Cisco IP Phone 7960 or 7940.

Figure 2-32 *Date/Time Group*

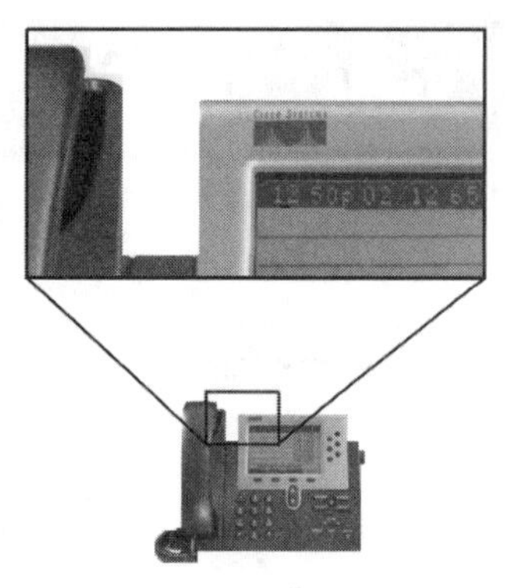

In CCMAdmin, click **System > Date/Time Group** to configure date and time groups. Each device is assigned a date and time group when it registers. The Cisco IP Phone 7960 and 7940 display the date and time group they are assigned in the top left corner of the LCD. In the Date/Time Group configuration you can configure formats and different time zones.

To configure a date/time group, do the following:

Step 1 Open the Date/Time Group Configuration page in CCMAdmin by clicking **System > Date/Time Group**. Figure 2-33 shows the Date/Time Group Configuration page.

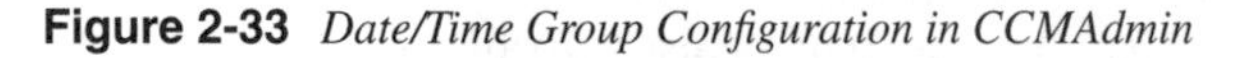

Figure 2-33 *Date/Time Group Configuration in CCMAdmin*

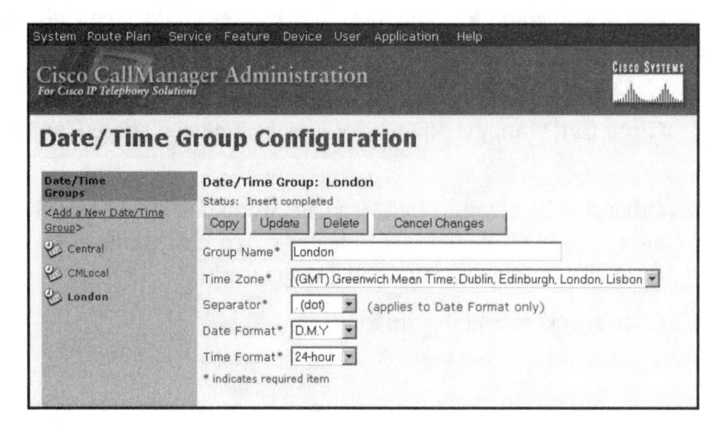

Step 2 Enter a time zone name for the **Date/Time Group** and select the time zone from the **Time Zone**.

Step 3 Configure the **Separator** for date format, and time format for this date/time group.

Step 4 Select **Insert**.

Cisco CallManager uses the CMLocal date and time for its CDR.

Do not change the CMLocal name. CMLocal gets its date and time information from the operating system of the server. CMLocal resets to the operating system date and time whenever you restart Cisco CallManager or upgrade the Cisco CallManager software to a new release.

Cisco recommends that if you are going to deploy devices in multiple time zones, create a date/time group for each time zone you are going to register devices in. For example, if your cluster is in the Pacific time zone and you know you are going to have devices in the Central time zone registering to this cluster, create a date/time group called Central and assign the Central time zone to this group.

Device Defaults

Use device defaults to set the system-wide default characteristics of each type of device that registers with a Cisco CallManager. The system-wide device defaults for a device type apply to all devices of that type within a Cisco CallManager cluster. The following list gives default settings for devices:

- Device load
- Device pool
- Phone button template

When a device auto-registers with a Cisco CallManager, it acquires the system-wide device default settings for its device type. After a device registers, you can update its configuration individually to change the device settings.

Installing Cisco CallManager automatically sets device defaults. You cannot create new device defaults or delete existing ones, but you can change the default settings. Cisco CallManager upgrades automatically change the device defaults if new device load files are included. Cisco recommends that you change device defaults only when recommended to do so by a Cisco TAC support engineer.

Region Menu Item

You configure regions using the Region menu item in the System menu of CCMAdmin.

Figure 2-34 shows a very simple region configuration example for deployment with a central site and two remote branches. In the example, an administrator configures a region for each site. The G.711 codec equals the maximum bandwidth codec used for calls within each site, and the G.729a codec equals the maximum bandwidth codec used for calls between sites across the WAN link.

Figure 2-34 *Region Configuration Example*

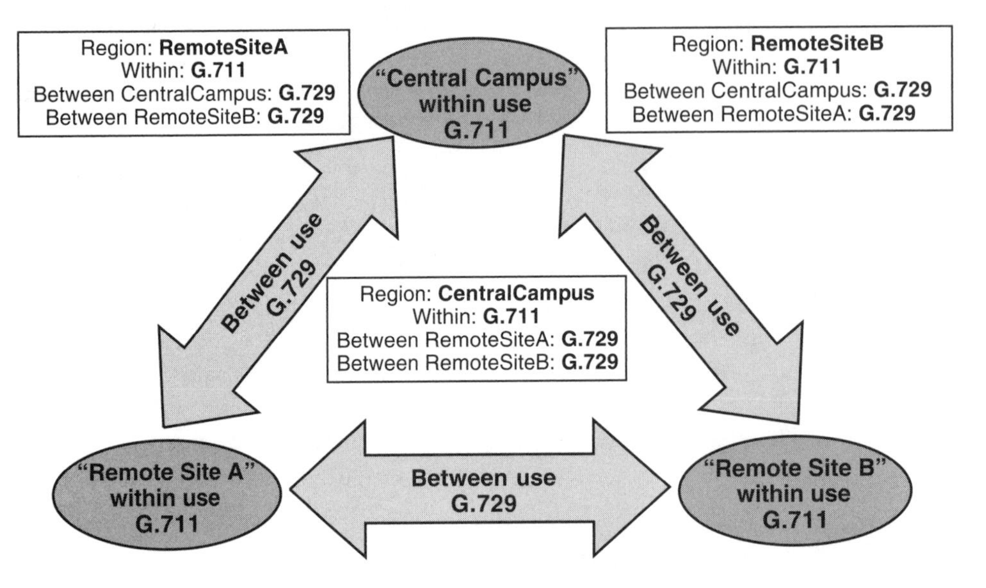

After region configuration, the administrator assigns devices to the following sites:

- The Central Campus site to device pools that specify CentralCampus as the region setting.
- Remote Site A to device pools that specify RemoteSiteA as the region setting.
- Remote Site B to device pools that specify RemoteSiteB for the region setting.

Use regions to specify the voice codec used for calls within a region and between existing regions. The codec determines the maximum amount of bandwidth used per call. By default, there is one region named Default that uses G.711 for all calls. If you do not plan to use any other codec, you can use the Default region for all devices.

When you create a region, you specify the codec that can be used for calls within that region and between the other regions.

The codec type specifies the technology used to compress and decompress voice signals. The choice of codec determines the compression type and amount of bandwidth used per call. See the Table 2-17 for specific information about bandwidth usage for available codecs.

Chapters 4, "Cisco CallManager Administration Service Menu" and 11, "Media Resources," have more information about what codecs are and what they do.

Regions prove useful for Cisco CallManager multiple site deployments where need to limit the bandwidth for calls sent across a WAN link, but where you want to use a higher bandwidth for internal calls.

Supported Voice Codecs and Bandwidth Usage

Cisco CallManager supports the following voice codecs for use with regions:

- **G.711**—Default codec for all calls through Cisco CallManager. G.711 is a high bit-rate codec with 64-kbps compression supported by Cisco IP Phone 79*xx* Family models.
- **G.729a**—Low bit-rate codec with 8-kbps compression supported by Cisco IP Phone 79*xx* Family models. Typically, you would use low bit-rate codecs for calls across a WAN link because they use less bandwidth. For example, a multiple-site WAN with centralized call processing can set up a G.711 and a G.729a region per site to permit placing intra-branch calls as G.711 and placing inter-branch calls as G.729a.
- **G.723**—Low bit-rate codec with 6-kbps compression for older Cisco IP Phone models including the 12 SP+ and 30 VIP.
- **GSM**—The Global System for Mobile Communications (GSM) codec enables the MNET system for GSM wireless handsets to operate with Cisco CallManager. Assign GSM devices to a device pool that specifies GSM as the codec for calls within the GSM region and between other regions. Depending on device capabilities, this includes GSM EFR (enhanced full rate) and GSM FR (full rate). Cisco IP Phones support GSM-FR but not GSM-EFR.
- **Wideband**—Currently only supported within a Cisco CallManager system or cluster for calls from Cisco IP Phone to Cisco IP Phone, the Wideband audio codec, uncompressed with a 16-bit, 16-kHz sampling rate, works with phones with handsets, acoustics, speakers, and microphones that can support high-quality audio bandwidth, such as Cisco IP Phone 79*xx* model phones.

  Regions that specify Wideband as the codec type must have a large amount of network bandwidth available because Wideband uses four times the amount of bandwidth that G.711 uses.

The total bandwidth used per call depends on the codec type as well as factors such as data packet size and overhead (packet header size), as indicated in Table 2-17.

Table 2-17 *Bandwidth Used per Call by Each Codec Type*

Voice Codec	Bandwidth Used for Data Packets Only (Fixed Regardless of Packet Size)	Bandwidth Used per Call (Including IP Headers) with 30-ms Data Packets	Bandwidth Used per Call (Including IP Headers) with 20-ms Data Packets
G.711	64 kbps	80 kbps	88 kbps
G.723	6 kbps	22 kbps	Not applicable
G.729a	8 kbps	24 kbps	32 kbps
Wideband[1]	256 kbps	272 kbps	280 kbps
GSM	12.8 kbps	28.8 kbps	36.8 kbps
GSM-EFR[2] GSM-FR[3]	13.2 kbps	29.2 kbps	37.2 kbps

1 Uncompressed. Cisco CallManager supports Wideband audio from Cisco IP Phone to Cisco IP Phone for Cisco IP Phone 79*xx* Family model phones only.

2 Enhanced full-rate GSM

3 Full-rate GSM (Cisco IP Phones support full-rate GSM, but not enhanced full-rate GSM)

Configuring a Region

To configure a region, do the following:

Step 1 Open the Region Configuration page in CCMAdmin by clicking **System > Region**. Figure 2-35 shows the Region Configuration page in CCMAdmin.

Figure 2-35 *Region Configuration in CCMAdmin*

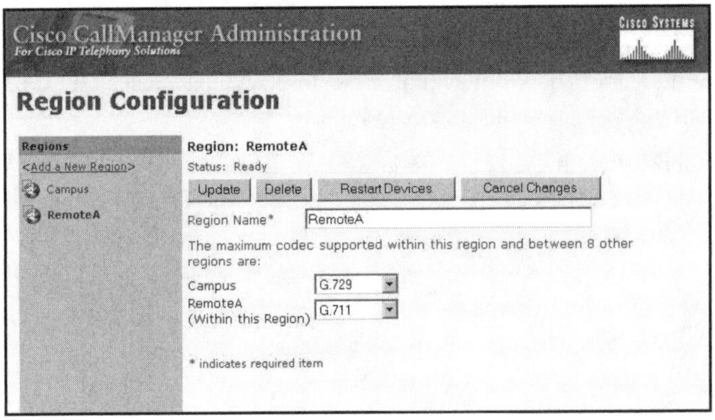

Step 2 Create different region names. For example, enter **Campus** for the region name and click **Insert**. Enter **RemoteA** for a new region name and click **Insert**.

Step 3 Configure the codec to be used within and between the regions. Select the Campus region and for **Within this Region** select G.711 and between RemoteA select G.729.

Step 4 Click **Update**.

Step 5 Verify the changes. Select the RemoteA region from the left column menu and verify that within the RemoteA region it is using G.711 and between the Campus it is using G.729a.

Device Pool Menu Item

Device pools provide a convenient way to define a set of common characteristics that can be assigned to devices. A good practice in naming a device pool is to provide information about the characteristics of the device. After configuring the following characteristics they can be specified in a device pool:

- **Cisco CallManager group**—Specifies a prioritized list of up to three Cisco CallManagers to facilitate redundancy. The first Cisco CallManager in the list serves as the primary Cisco CallManager for that group, and the other members of the group serve as secondary (backup) Cisco CallManagers. See the "Cisco CallManager Groups" section earlier in this chapter for more details.
- **Date/time group**—Specifies the date and time zone for a device. See the "Date/Time Group" section earlier in this chapter for more details.
- **Region**—Specifies the voice codecs used within and between regions. Regions are required only if you use multiple codecs within an enterprise. See the "Region" section earlier in this chapter for more details.
- **MRGL (optional)**—Specifies a prioritized list of MRGs. An application selects the required media resource (for example, an MOH server, transcoder, or conference bridge) from the available MRGs according to the priority order defined in the MRGL.
- **User Hold MOH Audio Source (optional)**— Specifies the audio source to use for MOH when a user places a call on hold.
- **Network Hold MOH Audio Source (optional)**—Specifies the audio source to use for MOH when the network initiates a hold action.
- **Calling search space for auto-registration (optional)**—Specifies the partitions that an auto-registered device can reach when placing a call.

- **Auto-Answer Feature Enable (optional)**—Globally enables or disables the auto-answer feature for all phones in the device pool that can support this feature. The auto-answer feature automatically delivers calls to agents who are available and ready to take calls. Agents hear a notification that the call has arrived (for example, a zip tone or a beep tone), but they do not have to press a button to answer the call. Auto-answer must be used in conjunction with a headset.

NOTE The preceding items must be configured first if you want to assign them as characteristics of a device pool.

After adding a new device pool to the database, you can use it during device configure for devices such as Cisco IP Phones, gateways, conference bridges, transcoders, MTPs, voice mail ports, CTI route points, and so on.

If you have a new system and want to assign all devices of a given type (for example, all Cisco IP Phone 7960s) to a device pool, use the Device Defaults page in CCMAdmin in combination with auto-registration. New devices auto-registering added after you have made this designation in the Device Defaults page will use the specific device pool (any existing devices will not be affected).

The sections that follow describe the different deployment models and provide tables that facilitate estimating the number of device pools you will be configuring based on your deployment model.

Device Pools in a Single-Site Cluster with No WAN Voice Interconnectivity

With this deployment model, device pools are configured only based on CallManager redundancy groups. The use of regions in this scenario is not required as all calls are G.711. Table 2-18 shows how to calculate approximately how many device pools are necessary for a single-site cluster with no WAN voice interconnectivity.

Table 2-18 *Single-Site Cluster with No WAN Voice Interconnectivity—Device Pools Needed Based on Cluster Size*

Users and Servers per Cluster	Cisco CallManager Groups	Number of Device Pools = Number of Cisco CallManager Groups
2500 users 2-server cluster (A, B)	BA	1

continues

Table 2-18 *Single-Site Cluster with No WAN Voice Interconnectivity—Device Pools Needed Based on Cluster Size (Continued)*

Users and Servers per Cluster	Cisco CallManager Groups	Number of Device Pools = Number of Cisco CallManager Groups
2500 users 3-server cluster (A, B, C)	BC CB	2
5000 users 4-server cluster (A, B, C, D)	BD CD	2
10,000 users 8-server cluster (A, B, C, D, E, F, G, H)	BDH CDH EHD GHD	4

Device Pools in a Multiple-Site WAN with Centralized Call Processing

In this type of deployment, only a single Cisco CallManager redundancy group exists. However, G.711 and G.729a region are required per location to permit intra-branch calls to be placed as G.711 and inter-branch calls to be placed as G.729a, for example. Another reason to use regions is for QoS, so that certain phones can be assured the highest quality voice. For example, using regions, phones assigned to executives can be assured G.711, while phones assigned to employees use G.729a. Table 2-19 shows how to calculate approximately how many device pools are necessary for a cluster (based on cluster size) in a multiple-site WAN centralized call processing solution.

Table 2-19 *Multiple-Site WAN with Centralized Call Processing—Device Pools per Cluster Size*

Users and Servers per Cluster	Regions	Locations X = Number of Locations	Number of Device Pools = Number of Regions * Number of Locations
2500 users 2-server cluster (A, B)	Default (G.711) G.729a	X	2*X
2500 users 3-server cluster (A, B, C)	Default (G.711) G.729a	X	2*X

continues

Table 2-19 *Multiple-Site WAN with Centralized Call Processing—Device Pools per Cluster Size (Continued)*

Users and Servers per Cluster	Regions	Locations X = Number of Locations	Number of Device Pools = Number of Regions * Number of Locations
5000 users 4-server cluster (A, B, C, D)	Default (G.711) G.729a	X	2*X
10,000 users 8-server cluster (A, B, C, D, E, F, G, H)	Default (G.711) G.729a	X	2*X

Device Pools in a Multiple-Site WAN with Distributed Call Processing

In this deployment model, device pools are configured as in the previous model, but with the additional complexity of regions for codec selection. Each cluster could potentially have a G.711 and G.729a region per Cisco CallManager group. Table 2-20 shows how to calculate approximately how many device pools are necessary for a cluster (based on cluster size) in a multiple site WAN distributed call processing solution.

Table 2-20 *Multiple-Site WAN with Distributed Call Processing—Device Pools per Cluster Size*

Users and Servers per Cluster	Cisco CallManager Groups	Regions	Number of Device Pools = Number of Regions * Number of Cisco CallManager Groups
2500 users 2-server cluster (A, B)	BA	Default (G.711) G.729a	2
2500 users 3-server cluster (A, B, C)	BC CB	Default (G.711) G.729a	4
5000 users 4-server cluster (A, B, C, D)	BD CD	Default (G.711) G.729a	4

continues

Table 2-20 *Multiple-Site WAN with Distributed Call Processing—Device Pools per Cluster Size (Continued)*

Users and Servers per Cluster	Cisco CallManager Groups	Regions	Number of Device Pools = Number of Regions * Number of Cisco CallManager Groups
10,000 users	BDH	Default (G.711)	8
8-server cluster	CDH	G.729a	
(A, B, C, D, E, F, G, H)	EHD		
	GHD		

Configuring a Device Pool

The following are quick steps to configure a device pool:

Step 1 Open the Cisco CallManager Group Configuration page by clicking **System > Device Pool** in CCMAdmin. Figure 2-36 shows the Device Pool Configuration page in CCMAdmin.

Figure 2-36 *Device Pool Configuration*

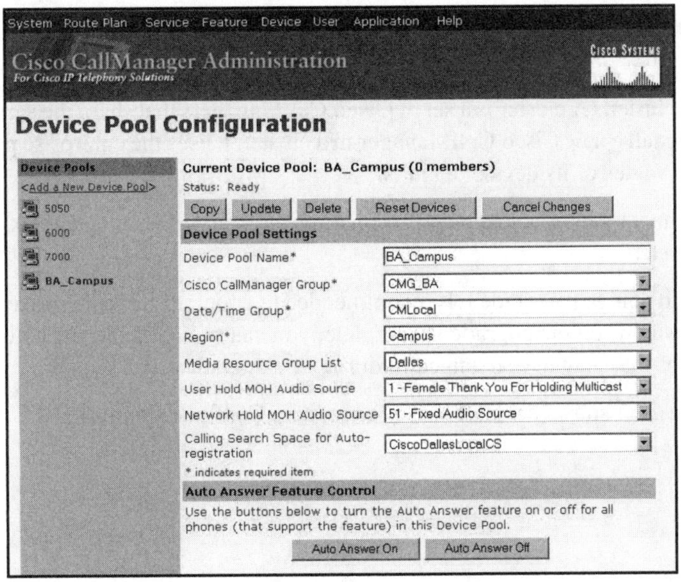

Step 2 Enter a descriptive name for the Device Pool. Use abbreviations of the characteristics in a device pool as part of the device pool name. For example, the device pool name is BA_Campus and the characteristics are the following:

— **Cisco CallManager Group**—CMG_BA

— **Date/Time Group**—CMLocal (because it is in the Campus Region)

— **Region**—Campus

Step 3 Configure the characteristics of a device pool. At each characteristic (Cisco CallManager Group, Date/Time Group, Region, MRGL, User Hold MOH Audio Source, Network Hold MOH Audio Source, and Calling Search Space for Auto-registration), use the drop-down menu to assign characteristics to the device pool.

Step 4 Click **Insert**.

Step 5 Learn how to create another device pool by copying and updating the properties of an existing device pool. Click **Copy** and change the device pool name, assign characteristics using the drop-down menu, and then click **Insert**.

Enterprise Parameters Menu Item

Enterprise parameters provide default settings that apply to all devices and services in the same cluster. (A cluster is a set of Cisco CallManagers that share the same database.) When you install a new Cisco CallManager in a cluster, it uses the enterprise parameters to set the initial values of its device defaults.

You cannot add or delete enterprise parameters, but you can update existing enterprise parameters.

Prior to doing an upgrade it is recommended to document the enterprise parameter settings. That way when the upgrade is complete, you can re-configure any enterprise parameter settings that may have changed in during the upgrade.

Figure 2-37 shows the Enterprise Parameters Configuration page.

Figure 2-37 *Enterprise Parameters Configuration*

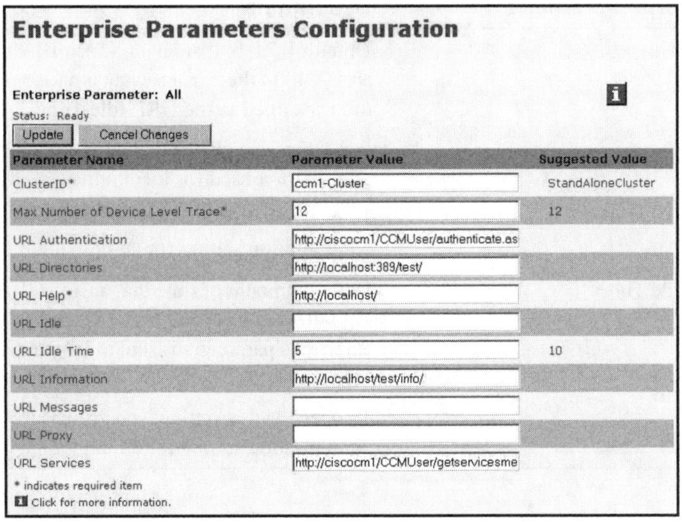

Table 2-21 defines the settings on the Enterprise Parameters Configuration page in CCMAdmin.

Table 2-21 *Enterprise Parameter Settings*

Enterprise Parameter	Description
Cluster-ID	Descriptive name for the Cisco CallManager cluster that identifies the cluster in CDRs, generated reports, traces, and so on.
Max Number of Device Level Trace	Maximum number of device levels for tracing a call. The default is 12.
URL Directories	The URL used for the **directories** button on the Cisco IP Phone models 7960 and 7940.
URL Information	The URL used for the information (**i**) button on the Cisco IP Phone models 7960 and 7940.
URL Messages	The URL used for the **messages** button on the Cisco IP Phone models 7960 and 7940.
URL Services	The URL used for the **services** button on the Cisco IP Phone models 7960 and 7940.

continues

Table 2-21 *Enterprise Parameter Settings (Continued)*

Enterprise Parameter	Description
URL Idle	Default URL to display on Cisco IP Phone models 7960 and 7940 in the cluster when a phone is not used for the time specified in the URL Idle Time enterprise parameter. For example, you can display a logo on the LCD when the phone is not used for five minutes. To override this setting, specify a value in the Idle Timer configuration setting for an individual phone.
URL Idle Time	Default amount of time that a Cisco IP Phone model 7960 or 7940 is idle before the URL specified by the URL Idle enterprise parameter is sent to the phone. The default is 10 seconds. To override this setting, specify a value in the Idle Timer configuration setting for an individual phone.

Step 1 Open the Enterprise Parameters Configuration page in CCMAdmin by clicking **System > Enterprise Parameters**.

Step 2 Change the server name in URL to an IP address. Because we are not using DNS, each default URL parameter needs to be changed from the server name (http://XXXCLSTR1A/CCMUser/…) to the IP address (http://172.16.x0.3/CCMUser/…). Edit the following URL enterprise parameters:

— URL Directories

— URL Information

— URL Services

Step 3 Select **Update**.

Deployment Models

This section discusses different supported deployment models. The last menu item in the System menu in CCMAdmin, Locations will make more sense if you understand the different deployment models.

The overall goals of an IP telephony network are as follows:

- End-to-end IP telephony
- IP WAN as the primary voice path and the PSTN as the secondary voice path between sites

- Lower total cost of ownership with greater flexibility
- Flexibility through new applications

There are four general deployment models that apply to the majority of implementations:

- Single-site deployment
- Isolated deployment
- Multiple-site deployments that use distributed call processing
- Multiple-site deployments that use centralized call processing

For more information about deployment models read the *Cisco IP Telephony Network Design Guide* at the following URL:

www.cisco.com/univercd/cc/td/doc/product/voice/ip_tele/network/index.htm

Single-Site Deployments

This section discusses a single-site deployment model. Figure 2-38 shows a single site deployment.

Figure 2-38 *Single-Site Deployment*

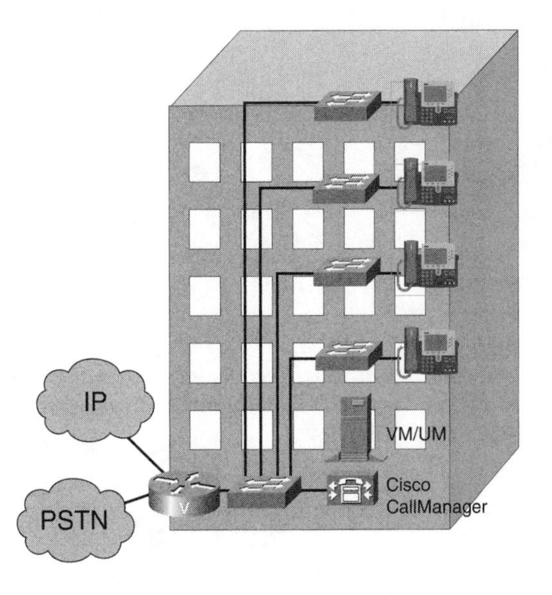

The single-site model has the following design characteristics:

- Single Cisco CallManager or Cisco CallManager cluster
- Maximum of 10,000 users per cluster

- Maximum of eight servers in a Cisco CallManager cluster (four servers for primary call processing, two for backup call processing, one database Publisher, and one TFTP server)
- Maximum of 2500 users registered with a Cisco CallManager at any time
- PSTN only for all external calls
- DSP resources for conferencing
- Voice mail and UM components
- G.711 codec for all Cisco IP Phone calls (80 kbps of IP bandwidth per call, uncompressed)
- Cisco LAN switches with a minimum of two queues to guarantee voice quality

Isolated Deployments

This section discusses the isolated deployment model. Figure 2-39 shows isolated deployment.

Figure 2-39 *Isolated Deployment*

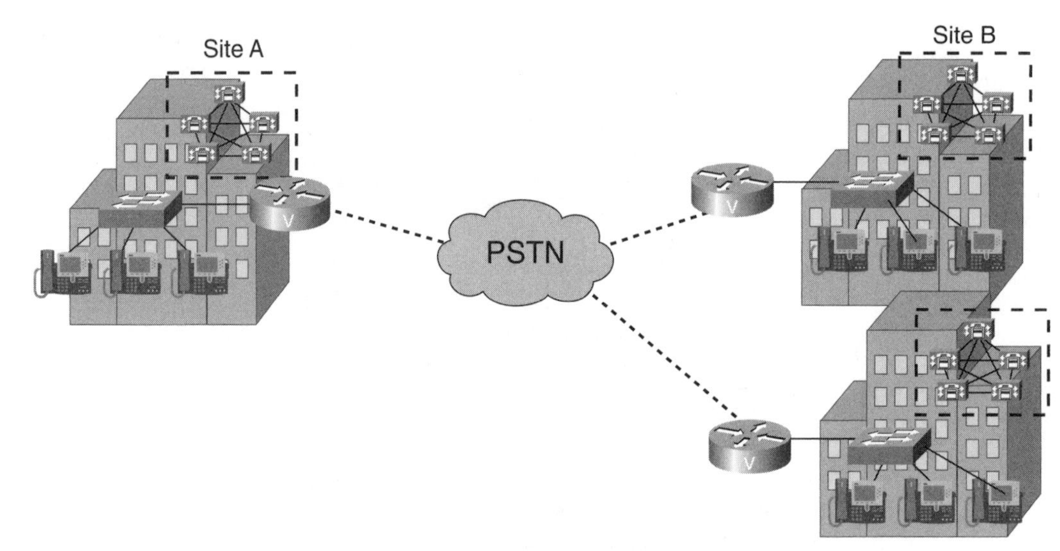

In this model, each site has its own Cisco CallManager or Cisco CallManager cluster to handle call processing for that site.

The model for independent multiple sites has the following design characteristics:

- Cisco CallManager server or Cisco CallManager cluster at each site to provide scalable call control
- Maximum of 10,000 Cisco IP Phones per cluster
- No limit to number of clusters
- Use of PSTN for networking multiple sites and for all external calls
- DSP resources for conferencing at each site
- Remote sites can use any type of gateway
- Voice message or UM components at each site
- Voice compression not required
- Uniform dial plan

Multiple-Site Deployments with Distributed Call Processing

This section discusses a multiple-site deployment with distributed call processing. Figure 2-40 shows a multiple-site WAN deployment with distributed call processing.

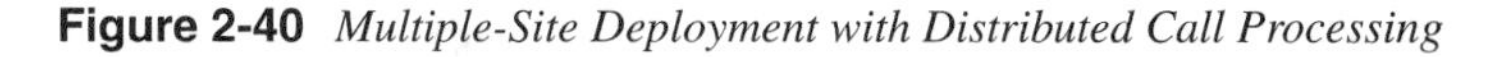

Figure 2-40 *Multiple-Site Deployment with Distributed Call Processing*

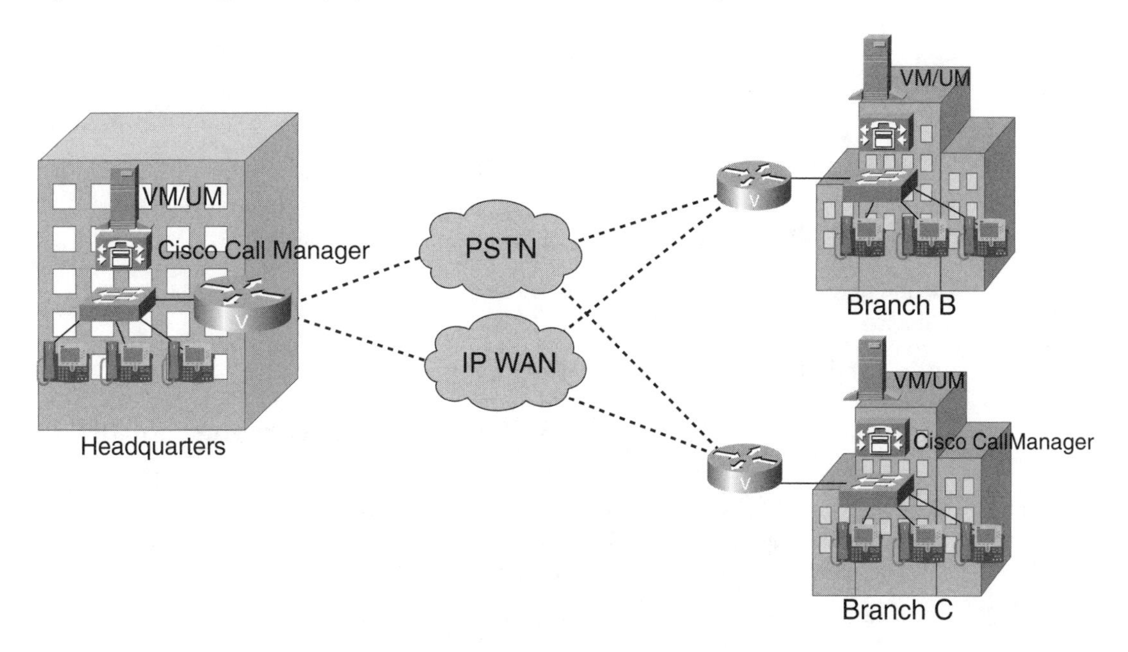

The multiple-site IP WAN with distributed call processing has the following design characteristics:

- Cisco CallManager or Cisco CallManager cluster at each location (10,000 users maximum per site).
- Cisco CallManager clusters are confined to a single campus and may not span the WAN.
- IP WAN as the primary voice path between sites, with the PSTN as the secondary voice path.
- Transparent use of the PSTN if the IP WAN is unavailable.
- Cisco IOS gatekeeper for E.164 address resolution.
- Cisco IOS gatekeeper for admission control to the IP WAN.
- Maximum of 100 sites interconnected across the IP WAN using hub-and-spoke topologies.
- DSP resources for conferencing and WAN transcoding at each site.
- Voice mail or UM components at each site or at a centralized site.
- Minimum bandwidth requirement for voice and data traffic is 56 kbps. For voice, interactive video, and data, the minimum requirement is 768 kbps. In each case, the bandwidth allocated to voice, video, and data should not exceed 75 percent of the total capacity.
- Remote sites can use any type of gateway.

Multiple-Site Deployments with Centralized Call Processing

This section discusses a multiple-site deployment with centralized call processing. Figure 2-41 shows multiple-site WAN deployment with centralized call processing.

The multiple-site IP WAN with centralized call processing has the following design characteristics:

- Central site supports only one active Cisco CallManager. A cluster can contain a secondary and tertiary Cisco CallManager as long as all Cisco IP Phones served by the cluster are registered to the same Cisco CallManager at any given time. This is called a centralized call processing cluster.
- Each centralized call processing cluster supports a maximum of 2500 users (no limit on number of remote sites). Multiple centralized call processing clusters of 2500 users at a central site can be interconnected using H.323.
- The CAC mechanism is based on bandwidth by location.
- Compressed voice calls across the IP WAN are supported.

- Manual use of the PSTN is available if the IP WAN is fully subscribed for voice traffic. (PSTN access code must be dialed after a busy signal.)
- Remote Survivability IOS is required for Cisco IP Phone service across the WAN in case the IP WAN goes down.
- Voice mail, UM, and DSP resource components are available at the central site only.
- Minimum bandwidth requirement for voice and data traffic is 56 kbps. For voice, interactive video, and data, the minimum requirement is 768 kbps. In each case, the bandwidth allocated to voice, video, and data should not exceed 75 percent of the total capacity.
- Remote sites can use any type of gateway.
- If using voice mail, each site must have unique internal dial plan number ranges. You cannot overlap internal dial plans among remote sites if voice mail is required. (For example, no two sites can share 1XXX.)

Figure 2-41 *Multiple Site Deployment with Centralized Call Processing*

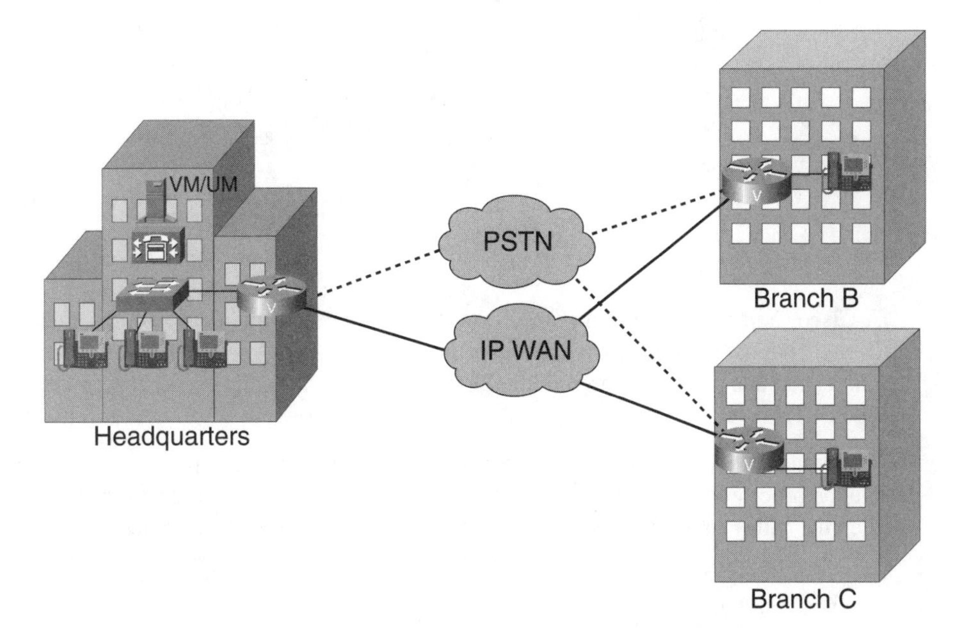

Figure 2-42 shows Cisco IP Telephony design goals.

Figure 2-42 *Cisco IP Telephony Design Goals*

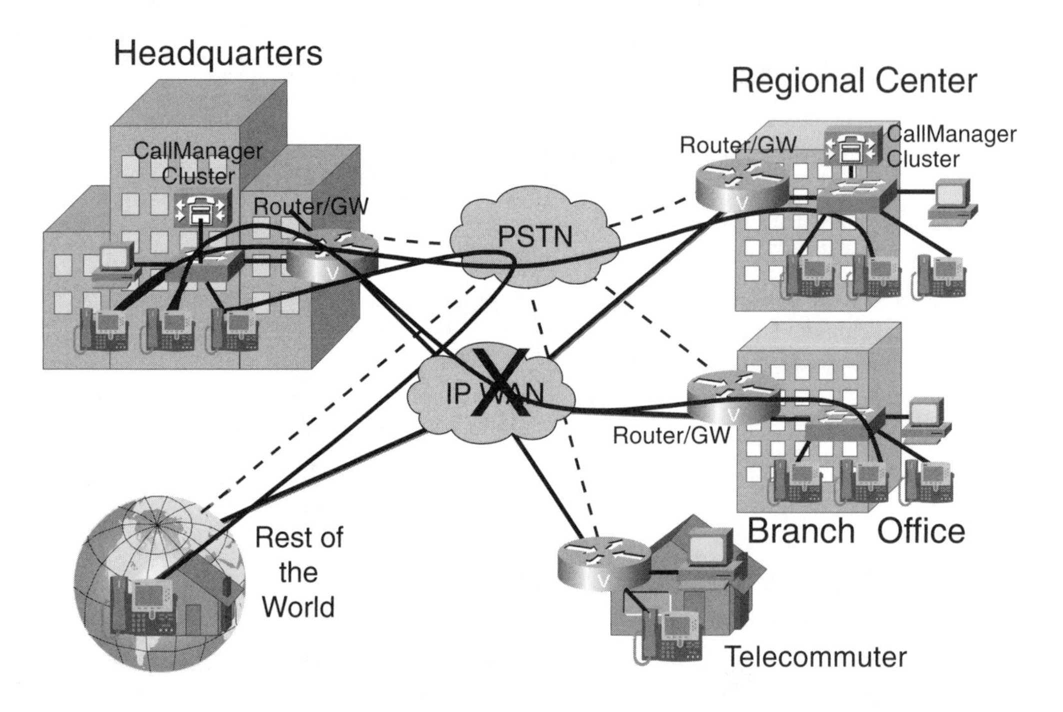

A Cisco CallManager cluster is located at the headquarters and the regional center. The design goal of IP telephony is to have primary connectivity to the regional center, branch office, and telecommuter through the IP WAN and, in the future, to the rest of the world. The PSTN is for back up use if the IP WAN should go down or if bandwidth is unavailable.

The branch office call processing is performed at headquarters and phone calls between the branch office and headquarters will be placed over the IP WAN. If the IP WAN goes down, then the calls can use the PSTN to connect using the voice-enabled routers.

With the abundance of IP to the home, now the rest of the world can access the IP WAN to call headquarters.

Location Menu Item

This section discusses the Location menu item on the System menu in CCMAdmin. Figure 2-43 shows the Location configuration page.

Figure 2-43 *Location Configuration*

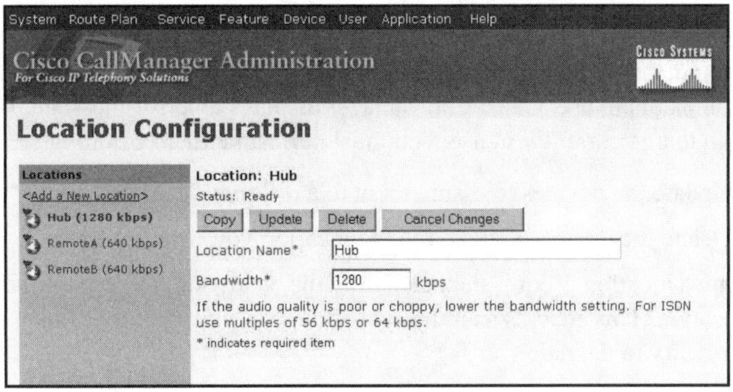

The Locations feature provides CAC for centralized call processing systems. A centralized system uses a single primary Cisco CallManager to control all the locations.

Because the available bandwidth is considered to be unlimited for calls between devices on the same LAN, CAC does not apply in that case. However, calls between locations travel over a WAN link that has limited available bandwidth. If the number of calls placed over a WAN link is high enough to deplete the amount of available bandwidth on the link, the audio quality of all the calls over that WAN link is degraded due to over-subscription. To avoid oversubscribing the WAN link, you can use the Locations feature to decrease the number of calls that are completed to a particular location.

Cisco CallManager uses a hub-and-spoke topology for locations. The main location, or hub, is the location of the primary Cisco CallManager controlling the network. Using CCMAdmin, you can define the spoke locations and assign devices to those locations. You can also specify how much bandwidth to allocate for calls between each spoke location and the hub.

For example, assume that the locations are configured as shown in Table 2-22

Table 2-22 *Location and Bandwidth Settings*

Location	Bandwidth to Hub (kbps)
Hub	Unlimited
Austin	100
Dallas	200

For calculation purposes, assume that calls using G.711 compression consume 80 kbps of bandwidth and calls using G.723 or G.729a compression consume 24 kbps. Cisco CallManager continues to admit new calls to a link as long as sufficient bandwidth is still available. Thus, the link to the Austin location could support one G.711 call at 80 kbps or four G.723 or G.729a calls at 24 kbps each. If any additional calls try to exceed the bandwidth limit, the system rejects them, the calling party receives a reorder tone, and a text message appears on the phone.

You cannot delete a location that has any devices assigned to it. If you try to delete a location that is in use, Cisco CallManager displays an error message. Before deleting a location that is currently in use, you must perform either or both of the following tasks:

- Update the devices to assign them to a different location.
- Delete the devices assigned to the location you want to delete.

Deleting a location is equivalent to setting the bandwidth to zero for the links connected to that location. This enables an unlimited number of calls on those links, and it can cause the voice quality to degrade.

Step 1 Open the Location Configuration page by clicking **System > Location**.

Step 2 Create a location. Enter a **Location Name** (Hub) and assign **1280 kbps** for **Bandwidth**. Select **Insert**.

Step 3 Create another location. Select **Copy** and change the **Location Name** to RemoteA and assign **640 kbps** for **Bandwidth**, and then click **Update**.

Step 4 Create another location. Select **Copy** and change the **Location Name** to RemoteB and assign **640 kbps** for **Bandwidth**, and then click **Update**.

Summary

In this chapter, you learned about the menu items available under the System menu.

- Server
- Cisco CallManager
- Cisco CallManager Group
- Date/Time Group
- Device Defaults
- Region
- Device Pool
- Enterprise Parameters
- Location

You also learned about the various deployment models and how the locations feature is used in those models.

- Single site
- Isolated
- Multiple site using distributed call processing
- Multiple site using centralized call processing

Post-Test

Did you get all that? Test yourself by taking this post-test to assess or reinforce your knowledge on this topic. The results of the quiz also help you determine whether you need to read this chapter again. You can find the answers to the post-test for this chapter in Appendix B, "Answers to Chapter Pre-Test and Post-Test Questions."

1 Why is it recommended to change the server name to an IP address in the CCMAdmin user interface?

____

____

____

2 Assuming DHCP and your network are configured correctly and besides un-checking the box **auto-registration Disabled on this Cisco CallManager**, what must you provide so that auto-registration works properly?

____

____

____

3 Based on configuring one Cisco CallManager group in a Cisco CallManager cluster that has two or three servers, how many users can this cluster support?

____

____

____

4 If you expected some growth in your organization, what would be a better way to support 2500 users? How many servers? What would your Cisco CallManager Groups be like? (If it is easier, you may draw your answer.)

____

____

____

5 When would you assign a Date/Time Group other than CMLocal to a device?

____

____

____

6 What is the main reason for using G.711 between devices as much as possible?

____

____

____

7 A device pool is made of characteristics. List the three required characteristics of a device pool.

____

____

____

8 After configuring a device pool and assigning it to a phone, what must you do to the phone so that those changes take effect?

____

____

____

9 You have two clusters, Cluster One and Cluster Two. Can you set the enterprise parameters in Cluster One and also apply those changes to Cluster Two?

____

____

____

10 Which deployment model uses Locations?

____

____

____

Upon completion of this chapter, you will be able to complete the following tasks:

- Given the components of a route plan, identify and define each component and how they relate in a route plan.
- Given the components of a route plan, construct a flow chart of a simple route plan.
- Given a list of common route pattern wildcards, match the wildcards with the digits or definition of digits that the wildcard represents.
- Given a route group, route list, route filter, and route pattern, configure the route plan using the Cisco CallManager Administration.
- Given the terms partition and calling search space (CSS), define these two terms and how they relate to route patterns and devices.
- Given some partitions and CSSs, configure the partitions and CSSs to permit or restrict devices access to route patterns.

CHAPTER 3

Cisco CallManager Administration Route Plan Menu

The Route Plan menu in Cisco CallManager Administration allows you to configure route plans. Route plans provide access to external devices. The route plan within Cisco CallManager can be broken up into the following pieces: route patterns, route filters, route lists, and route groups. Route lists and route groups permit you to introduce routing flexibility into your Cisco IP Telephony (CIPT) network design.

The Route Plan menu in Cisco CallManager Administration is also used to configure partitions and calling search spaces (CSSs) to restrict or permit devices to contact other devices in a CIPT network. This chapter covers the following topics:

- Abbreviations
- Understanding Route Plans
- Route Patterns
- Translation Patterns
- External Route Plan Wizard
- Route Plan Reports
- Partitions and CSSs

Route Plan: Visual Objective

Figure 3-1 depicts the goal of a well-designed and configured route plan. A well-designed and configured route plan enables the enterprise to avoid long distance charges by automatically routing a user's calls across the IP WAN when the destination that the user dials is accessible via the IP WAN. Cisco CallManager sends the call over the IP WAN (first choice) or over the PSTN (second choice).

Figure 3-1 *Route Plan Design*

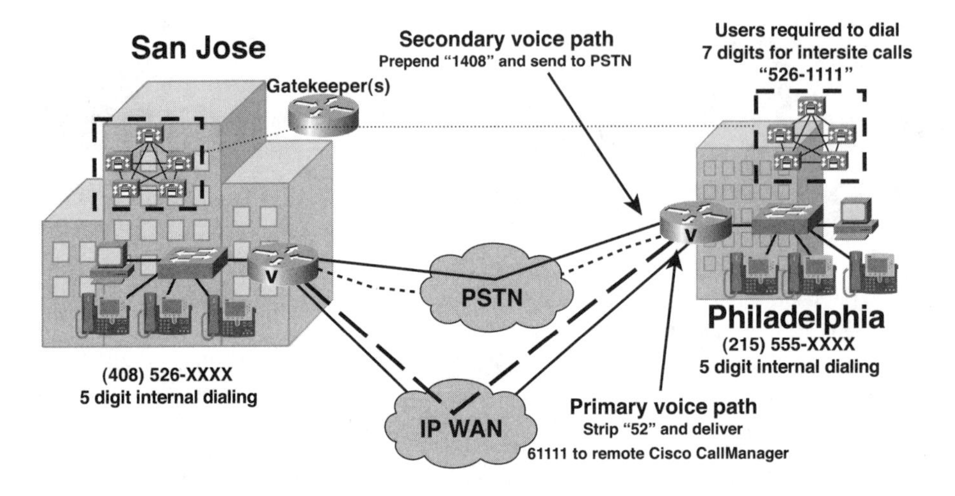

Dial plan design is usually done far in advance of configuring a route plan. This chapter provides you with some background knowledge about route plans so that when you are configuring route plans you will be able to understand the flow of a call when the user dials.

CAUTION Dial plan design and planning are critical to a CIPT roll-out. A well thought out dial plan needs to be designed before configuring anything to ensure that it meets all the requirements of the customer. In most cases, it is considerably more difficult to go back and change a poor dial plan than it is to just make a new one.

Pre-Test

Use this section to verify what you already know. Use the following scale to decide whether to read, skim, or skip this chapter. If you answer one to six questions correctly, we recommend that you read the chapter. If you answer 7 to 12 questions correctly, we recommend you skim the headers and read the sections related to the questions you missed. If you answer 13 to 16 questions correctly, you can skim the headers or skip this chapter completely. You can find the answers to the pre-test for this chapter in Appendix B, "Answers to Chapter Pre-Test and Post-Test Questions."

1 Write or draw the process used for configuring a route plan.

____

____

____

2 Match the most commonly used route pattern wildcards to its definition.

Route Pattern Wildcards						
X	!	#	[x-y]	.	@	[^x-y]

a. One or more digits (0 – 9)

b. Generic range notation

c. Terminates access code

d. Exclusion range notation

e. North American Numbering Plan

f. Terminates inter-digit timeout

g. Single digit (0 – 9)

3 What are the two things a route pattern can be assigned to?

____

____

____

4 Write or draw the flow of a call when a user dials digits.

____

____

____

5 If you are using an "@" in a route pattern, what is automatically recognized as an end-of-dialing character for international calls?

____

____

____

6 What removes a portion of the dialed digit string before passing the number on to the adjacent system?

____

____

____

7 Which one of the transformation settings changes the dialed number?

____

____

____

8 If dialed digits match a translation pattern, what does Cisco CallManager do with those digits?

____

____

____

9 What are the prerequisites for using the External Route Plan Wizard?

____

____

____

10 List five of the route plan components generated by the External Route Plan Wizard.

____

____

____

11 Match the gateway-type naming convention used by the External Route Plan Wizard to the gateway type.

Gateway types naming convention

AA	DA	HT	MS	MT

a. Digital Access

b. MGCP Trunk

c. MGCP Station

d. Analog Access

e. H.323 Trunk

12 The Route Plan Report can be saved as what type of file?

13 What does Cisco CallManager use to apply class of service to devices?

14 What are directory numbers and route patterns assigned to?

15 What does a CSS provide for the device it is assigned to?

16 What are the three specific problems that partitions and CSSs are designed to address?

Abbreviations

This section defines the abbreviations used in this chapter. For more information about terms and acronyms used in this chapter refer to the *IP Telephony Network Glossary* at the following URL:

www.cisco.com/univercd/cc/td/doc/product/voice/evbugl4.htm

Table 3-1 provides the abbreviations with the complete terms used frequently in this chapter.

Table 3-1 *Abbreviations Used in This Chapter*

Abbreviation	Complete Term
AA	analog access
CLID	Calling Line Identification
CSS	Calling Search Space
DA	digital access
DDI	Discard Digit Instruction
DN	directory number
GW	gateway
HT	H.323 Trunk
IP	Internet Protocol
IPV4	Internet Protocol Version 4 Addressing
IP WAN	Internet Protocol wide-area network
LD	long distance
MT	MGCP Trunk
MGCP	Media Gateway Control Protocol
ms	milliseconds
NANP	North American Numbering Plan
PSTN	Public Switched Telephone Network
PT	partition
RG	route group
RL	route list
SJ	San Jose

Understanding Route Plan

The Route Plan menu enables Cisco CallManager route plan configuration using route patterns, route filters, route lists, route groups, partitions, and CSSs. Figure 3-2 shows and defines the main components of a route plan: route pattern, route list, and route group.

Cisco CallManager uses route patterns to route or block calls. A directory number is a specific type of route pattern applied to a Cisco IP Phone. Route patterns are assigned to route lists or gateways.

Route filters are used in conjunction with a specific type of route pattern used for national dialing (the @ wildcard) and permit routing national calls by type (emergency, local, LD, and international). More about route filters in the section "Route Pattern Wildcards and Special Characters."

Route groups are a prioritized list of gateway ports. A route group can contain one or more gateways. Note that a gateway cannot be put into a route group if it has been directly assigned to a route pattern. It is highly recommended that gateways be assigned to route groups and route lists rather than being assigned directly to a route pattern. Doing this offers the greatest flexibility for making future changes.

Route lists determine the order of preference for route group usage. If a route list is configured, at least one route group must be configured. One or more route lists can point to one or more route groups. Route lists can also perform digit manipulation on a per-route group basis.

A route pattern represents a string of digits that can be routed to a particular device or group of devices. Not all route patterns are explicitly configured on the Route Pattern Configuration page. To Cisco CallManager, all routable numbers that you configure are route patterns. Directory numbers, Meet-Me conference numbers, and call park numbers are really just specialized route patterns. The Route Pattern Configuration page is used to configure route patterns that will point to gateways such as Cisco Access Analog Trunk gateways, Cisco Access Digital Trunk gateways, Cisco MGCP gateways, and H.323-compliant gateways. Cisco gateways can route ranges of numbers with complex restrictions and manipulate directory numbers before Cisco CallManager passes them on to an adjacent system. The adjacent system can be a Central Office (CO), a private branch exchange (PBX), or a device on another Cisco CallManager system.

Figure 3-2 *Understanding Route Plans*

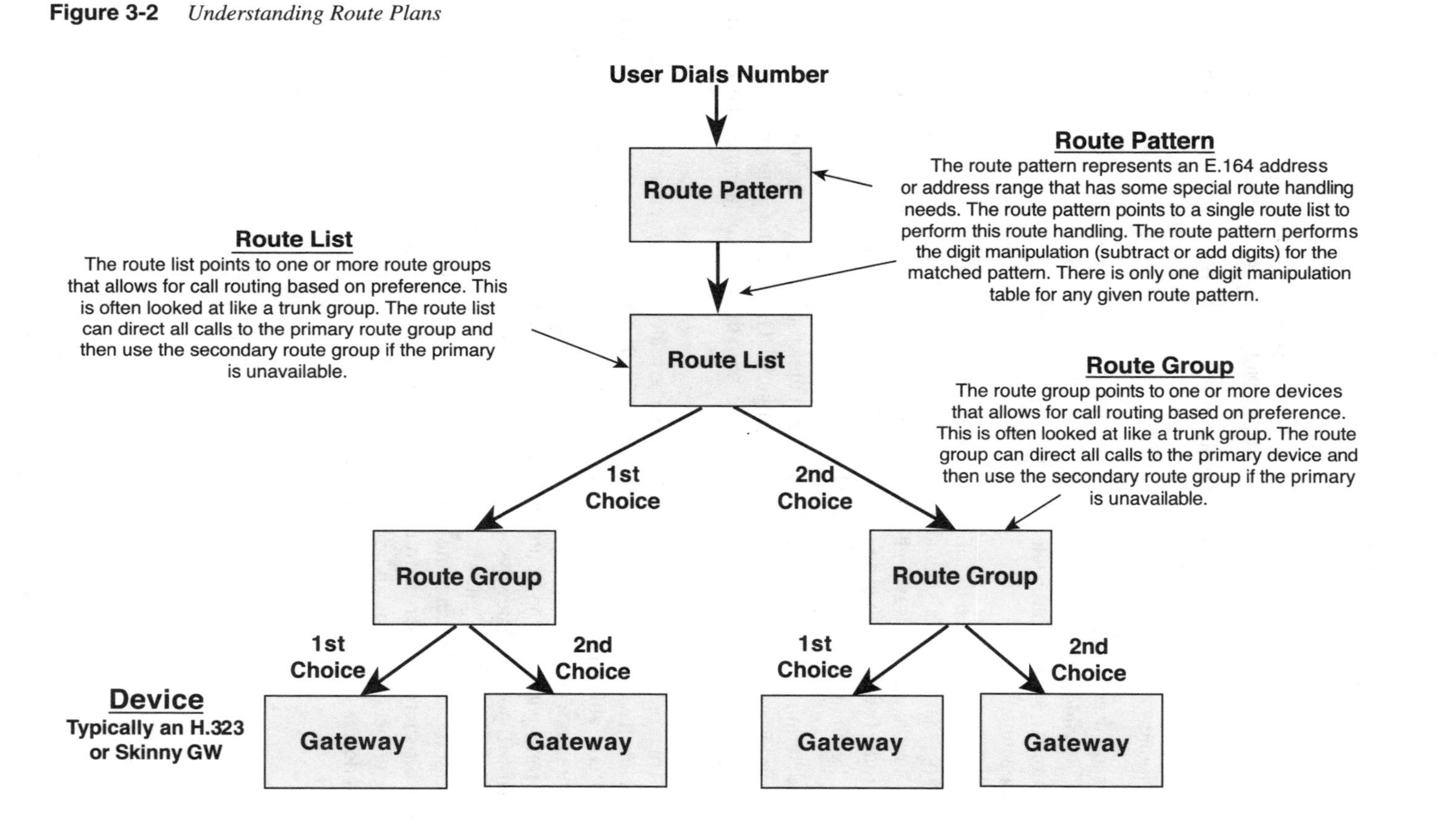
User Dials Number
Route Pattern
Route Pattern
The route pattern represents an E.164 address or address range that has some special route handling needs. The route pattern points to a single route list to perform this route handling. The route pattern performs the digit manipulation (subtract or add digits) for the matched pattern. There is only one digit manipulation table for any given route pattern.
Route List
The route list points to one or more route groups that allows for call routing based on preference. This is often looked at like a trunk group. The route list can direct all calls to the primary route group and then use the secondary route group if the primary is unavailable.
Route List
1st Choice
2nd Choice
Route Group
The route group points to one or more devices that allows for call routing based on preference. This is often looked at like a trunk group. The route group can direct all calls to the primary device and then use the secondary route group if the primary is unavailable.
Route Group
Route Group
1st Choice
2nd Choice
1st Choice
2nd Choice
Device
Typically an H.323 or Skinny GW
Gateway
Gateway
Gateway
Gateway

Figure 3-3 shows the process for configuring route plans.

Figure 3-3 *Route Plan Configuration Process*

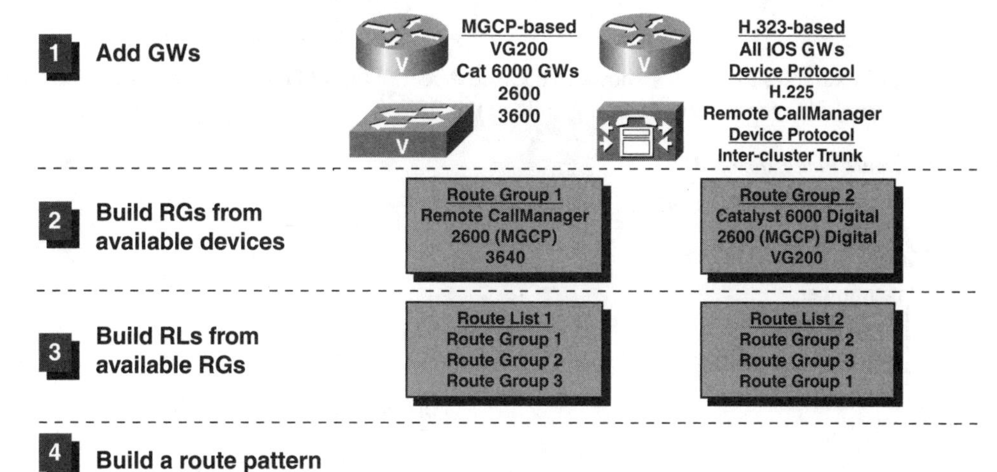

Now that you have a general understanding of a route plan, let's look at the general process for configuring a route plan.

The process of building a route plan can be generalized into four steps:

Step 1 **Add a gateway.** Gateways are defined in Chapter 6, "Cisco IP Telephony Devices." There are a variety of gateways to choose from that can support either analog or digital signaling.

Step 2 **Build route groups from available devices.** From the gateways in the Cisco CallManager system, select and place those gateways in an ordered list in a route group.

Step 3 **Build route lists from available route groups.** From the route groups in the Cisco CallManager system, select and order the route groups in a route list.

Step 4 **Build a route pattern from available route lists or gateways.** The route pattern is the key that connects the user to the appropriate available gateway to an external device. When Cisco CallManager matches this pattern, it routes the call to the assigned route list or gateway.

You can assign a route pattern directly to a gateway, or you can assign a route pattern to a route list for more flexibility.

NOTE If a gateway is assigned directly to a route pattern, that gateway cannot be used as part of a route group.

For example, Figure 3-4 shows that Cisco Access Digital Gateway 1 is designated as the first choice for routing outgoing calls to the PSTN.

Figure 3-4 *Route Plan Example 1*

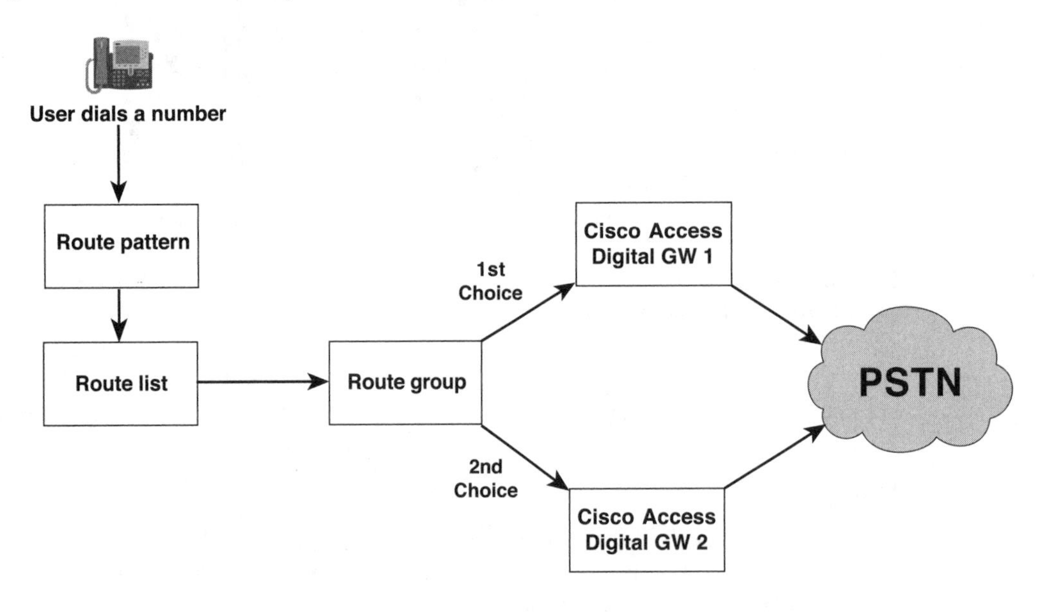

NOTE If a gateway is not assigned to a route pattern, it cannot place calls to the PSTN or PBX. To assign a route pattern to an individual port on a gateway, you must assign a route list and route group to that port. Figure 3-4 shows the effects of using route patterns with a single route group.

When using an H.323 gateway, Cisco CallManager has no control over which port gets used and the gateway has to have dial-peer information configured in it.

In this example, the route pattern is assigned to a route list, and that route list is associated with a single route group. The route group supports a list of devices that are selected based on availability. If all ports on the first-choice gateway are busy or out of service, the call is routed to the second-choice gateway.

NOTE It is important that after adding or changing route pattern information, you must reset the gateway for the new or updated information to be recognized.

Making a change to a route pattern configuration will automatically reset any gateways that are assigned directly to the route pattern. (This is a good reason why Cisco Systems recommends not assigning gateways directly to a route pattern.) Once a gateway has been configured and reset once, any subsequent route pattern changes that point to a route list should not require that the gateway be reset.

Figure 3-5 shows the effects of using route patterns with two route groups.

Figure 3-5 *Route Plan Example 2*

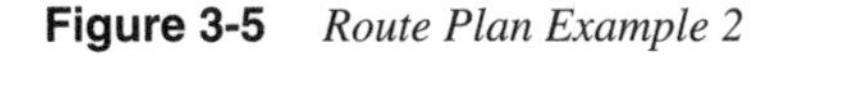

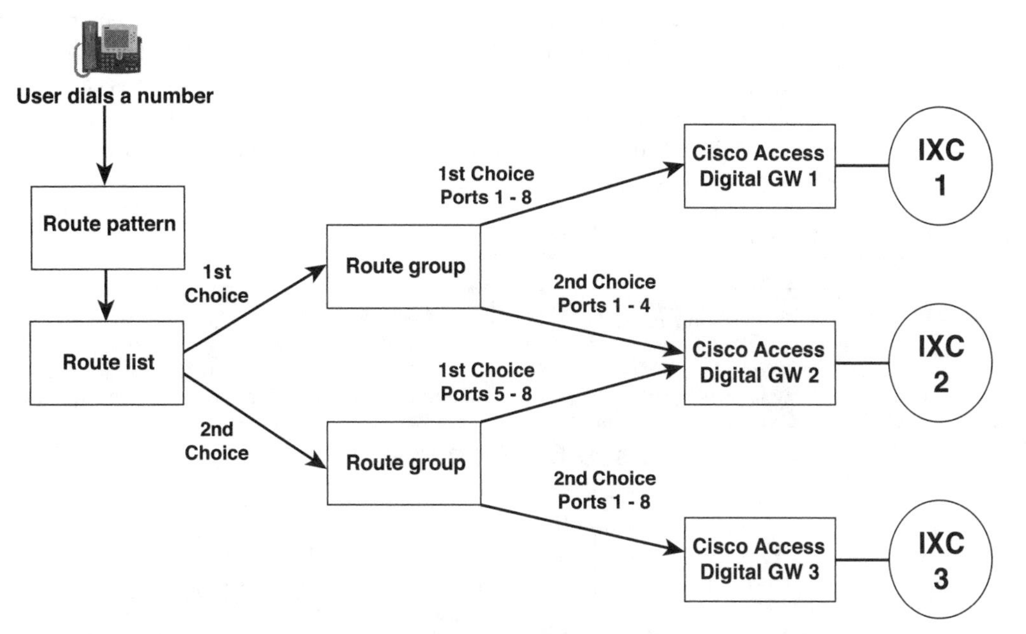

In this example, the route pattern is assigned to a route list, and that route list is associated with two route groups. RG 1 is associated with ports 1 through 8 on GW 1, which route all calls to Interexchange Carrier 1 (IXC 1). RG 1 is also associated with ports 1 through 4 on GW 2. RG 2 is associated with ports 5 through 8 on GW 2 and all ports on GW 3.

Each route group supports a list of devices that are selected based on availability. For RG 1, if ports 1 through 8 on the first-choice gateway are busy or out of service, calls are routed to ports 1 through 4 on the second-choice gateway. If all routes in RG 1 are unavailable, calls are routed to RG 2. For RG 2, if ports 5 through 8 on the first-choice gateway are busy or out of service, calls are routed to ports 1 through 8 on the second-choice gateway. If no ports on any gateway in either route group are available, the call is routed to an all trunks busy tone.

NOTE When using an H.323 gateway, Cisco CallManager has no control over which port gets used, and the gateway has to have dial-peer information configured in it.

Route Patterns

This section discusses route patterns in Cisco CallManager. Figure 3-6 illustrates how the terms directory number and route pattern relate to internal and external reachability.

Cisco CallManager uses route patterns to route or block external calls. A directory number is a type of specific route pattern that is applied to a Cisco IP Phone. Directory numbers are automatically entered into the Cisco CallManager route plan. Route patterns can also use more complex route patterns that can contain wildcards. Gateways can route ranges of numbers and manipulate directory numbers before Cisco CallManager passes them onto an adjacent system such as a CO or PBX.

Figure 3-6 *Internal Versus External Reachability*

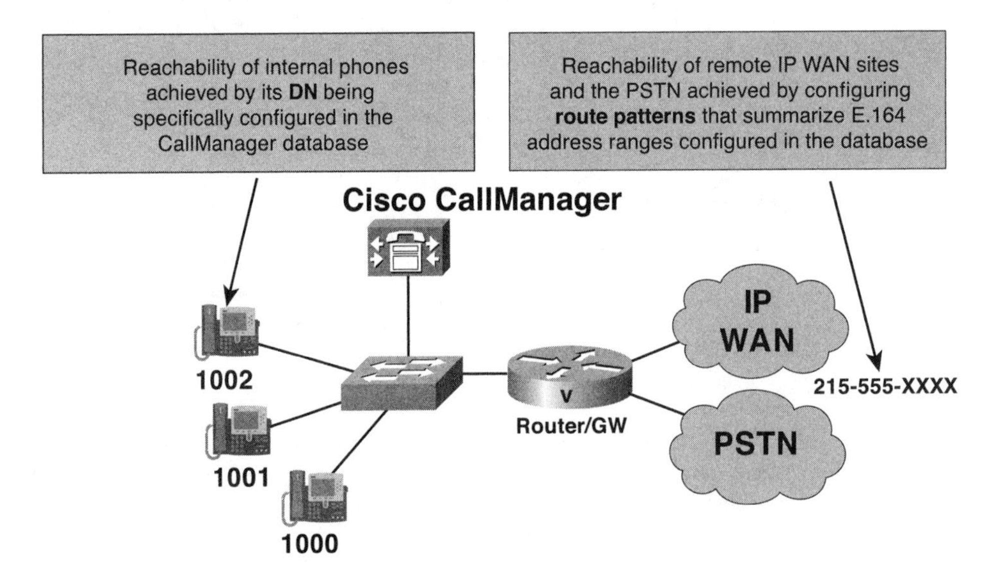

CAUTION If a gateway has no route pattern associated with it, or it doesn't belong to a route group, it cannot route or block any outbound calls. Based on the dial-peer configuration in the gateway, the gateway can still allow inbound calls to a CIPT network.

Figure 3-7 illustrates how the simplest route pattern is just a set of one or more digits.

Figure 3-7 *Simple Route Pattern*

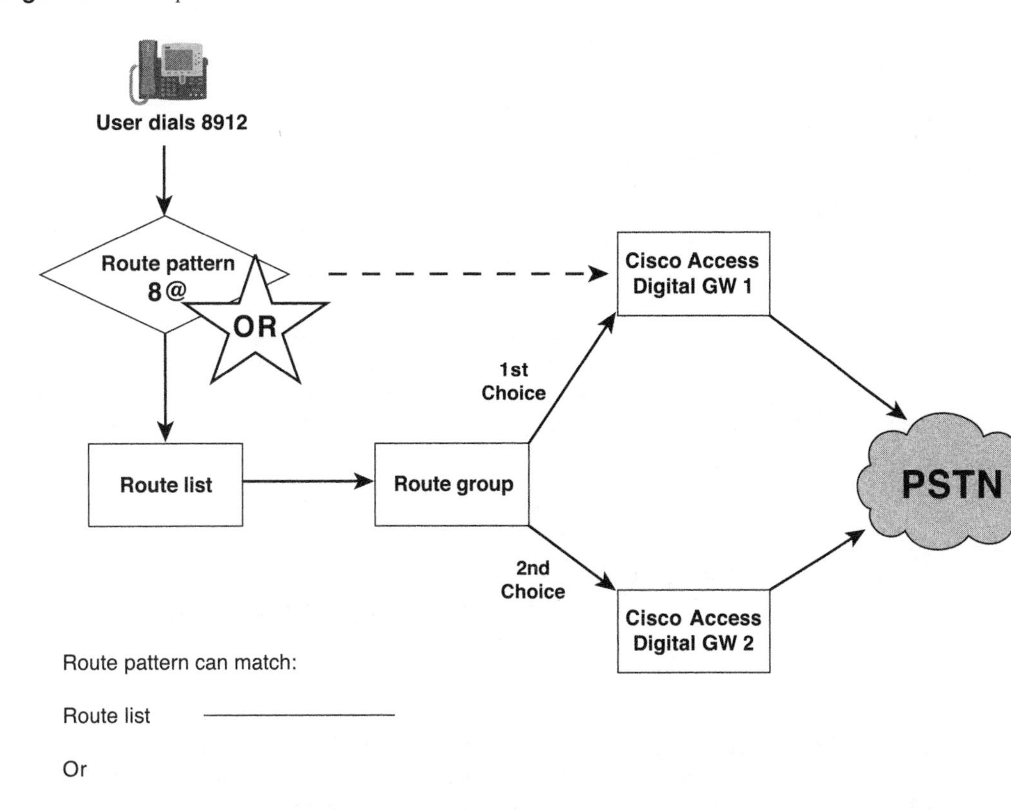

For example, the number 8@ is a route pattern. When assigned to a gateway or a route list, Cisco CallManager directs any calls to 8@ to the assigned device. If called party transformations are configured, Cisco CallManager manipulates the dialed address before passing the call to the route list or gateway.

If a gateway is assigned directly to a route pattern, you will not be able to assign that gateway to a route group and vice versa.

Route Pattern Wildcards and Special Characters

Table 3-2 lists the most commonly used route pattern wildcards. Route pattern wildcards and special characters allow a single route pattern to match a range of numbers (addresses). Use these wildcards and special characters also to build instructions that enable Cisco CallManager to match a number before sending it to an adjacent system.

Table 3-2 *Most Commonly Used Wildcards*

Wildcard	Description
X	Single digit (0-9)
@	NANP
!	One or more digits (0-9)
[x-y]	Generic range notation
[^x-y]	Exclusion range notation
.	Terminates access code
#	Terminates inter-digit timeout

TIP When @ is used in a routing pattern, the octothorpe, pound, or hash (#) is automatically recognized as an end-of-dialing character for international calls. For routing patterns that don't use @, you must include the # in the routing pattern to be able to use the # character to signal the end-of-dialing.

Table 3-3 describes all the wildcards and special characters supported by Cisco CallManager. Table 3-4 lists Cisco CallManager Administration fields that require route patterns and shows the valid entries for each field.

NOTE The NANP is the numbering plan for the PSTN in the United States and its territories, Canada, Bermuda, and many Caribbean nations. It includes any number that can be dialed and is recognized in North America. The @ wildcard represents the NANP, which represents anything that can be dialed from your home phone in North America. Route filters can be used to filter out some of those 300 route patterns.

Table 3-3 *Wildcards and Special Characters*

Character	Description	Examples
@	The at-symbol wildcard matches all NANP numbers. Each route pattern can have only one @ wildcard.	The route pattern 9.@ routes or blocks all numbers recognized by the NANP. The following route pattern examples show NANP numbers encompassed by the @ wildcard: 0 1411 19725551234 101028819725551234 01133123456789
X	The X wildcard matches any single digit in the range 0 through 9.	The route pattern 9XXX routes or blocks all numbers in the range 9000 through 9999.
!	The exclamation point wildcard matches one or more digits in the range 0 through 9.	The route pattern 91! routes or blocks all numbers in the range 910 through 91999999999999999999999.
?	The question mark wildcard matches zero or more occurrences of the preceding digit or wildcard value.	The route pattern 91X? routes or blocks all numbers in the range 91 through 91999999999999999999999.
+	The plus sign wildcard matches one or more occurrences of the preceding digit or wildcard value.	The route pattern 91X+ routes or blocks all numbers in the range 9100 through 91999999999999999999999.
[ ]	The bracket characters enclose a range of values. The values within the brackets can only represent a single digit.	The route pattern 813510[012345] routes or blocks all numbers in the range 8135100 through 8135105.
-	The hyphen character, used with the brackets, denotes a range of values.	The route pattern 813510[0-5] routes or blocks all numbers in the range 8135100 through 8135105.
^	The circumflex character, used with the brackets, negates a range of values. It must be the first character following the opening bracket ([). Each bracket can have only one ^ character, but a route pattern can have multiple brackets each containing one ^ character.	The route pattern 813510[^1-5] routes or blocks all numbers in the range 8135106 through 8135109.

continues

Table 3-3 *Wildcards and Special Characters (Continued)*

Character	Description	Examples
.	The dot character used as a delimiter separates the Cisco CallManager access code from the directory number. Use this special character with the DDIs to strip off the Cisco CallManager access code before sending the number to an adjacent system. Each route pattern can have only one . character. The dot character is not a wildcard that represents any digits and does not determine when outside dial tone is played.	The route pattern 9.@ identifies the initial 9 as the Cisco CallManager access code in an NANP call.
*	The asterisk character can provide an extra digit for special dialed numbers. This is not a wildcard; the asterisk matches the "*" on the phone button keypad.	You can configure the route pattern *411 to provide access to the internal operator for directory assistance.
#	The octothorpe character generally identifies the end of the dialing sequence. The # character must be the last character in the pattern.	The route pattern 9011819105555# routes or blocks an international number dialed from within the NANP. The # character after the 5 identifies this as the last digit in the sequence.

Table 3-4 *CallManager Administration Field Entries*

Field	Valid Entries
Call Park Number/Range	[^ 0 1 2 3 4 5 6 7 8 9 -] X * #
Calling Party Transform Mask	0 1 2 3 4 5 6 7 8 9 X * #
Called Party Transform Mask	0 1 2 3 4 5 6 7 8 9 X * #
Caller ID Directory Number (gateways)	0 1 2 3 4 5 6 7 8 9 X * #
Directory Number	[^ 0 1 2 3 4 5 6 7 8 9 -] + ? ! X * # +
Directory Number (Call Pickup Group)	0 1 2 3 4 5 6 7 8 9
External Phone Number Mask	0 1 2 3 4 5 6 7 8 9 X * #
Forward All	0 1 2 3 4 5 6 7 8 9 * #
Forward Busy	0 1 2 3 4 5 6 7 8 9 * #

Table 3-4 *CallManager Administration Field Entries (Continued)*

Field	Valid Entries
Forward No Answer	0 1 2 3 4 5 6 7 8 9 * #
Meet-Me Conference Number	[^ 0 1 2 3 4 5 6 7 8 9 -] + ? ! X * # +
Prefix Digits	0 1 2 3 4 5 6 7 8 9 * #
Prefix Directory Number (gateways)	0 1 2 3 4 5 6 7 8 9 * #
Route Filter Tag Values	[^ 0 1 2 3 4 5 6 7 8 9 -] X * #
Route Pattern	[^ 0 1 2 3 4 5 6 7 8 9 -] + ? ! X * # + . @
Translation Pattern	[^ 0 1 2 3 4 5 6 7 8 9 -] + ? ! X * # + . @

Table 3-5 shows examples of route patterns and the possible matches to the pattern examples.

Table 3-5 *Route Pattern Examples*

Pattern	Result
1234	Matches 1234.
1*1X	Matches numbers between 1*10 and 1*19.
12XX	Matches numbers between 1200 and 1299.
13[25-8]6	Matches 1326, 1356, 1366, 1376, 1386.
13[^3-9]6	Matches 1306, 1316, 1326, 13*6, 13#6.
13!#	Matches any number that begins with 13, is followed by one or more digits, and ends with #; 135# and 13579# are example matches.

Within Cisco CallManager, digit analysis uses the following to help route dialed digits:

- Digit collection
- Closest match routing
- Inter-digit timeout

Cisco CallManager collects every digit dialed and tries to match or find a potential match for each digit dialed in sequence. Every time a user enters a digit (by pressing the keypad on the phone) the Cisco CallManager collects those digits. After each digit dialed is collected, Cisco CallManager tries to match that digit to a pattern in the route plan table. For example, if phone A (DN-1000) calls phone B (1001), then when phone A dials 1, Cisco CallManager collects that digit and checks if there are any patterns that match or are potential matches. Phone A then dials; Cisco CallManager digit collection now has 10 and tries to find a match or potential match for 10. Phone A then dials 0; Cisco CallManager digit collection now has 100 and tries to find a match or potential match for 100. Phone A then dials 1; Cisco CallManager digit collection now has 1001 and tries to find a match or potential match for 1001. Phone B (DN-1001) is an exact match and so Cisco CallManager will route the call.

Closest match routing handles potential matches that overlap. For example, the patterns 1000 and 100X overlap. In this case, Cisco CallManager uses closest match routing to determine where to route the call. If both patterns 1000 and 100X exist and the digits dialed are 1000, Cisco CallManager will match the pattern 1000 because that is the closest match. In other words, the pattern 1000 has only one match, where the pattern 100X has a possible 10 matches.

Inter-digit timeout is mostly used for international dialing. Because of the multiple length dialing plans outside the NANP, Cisco CallManager is not sure when a user has completed dialing. The T302 Timer service parameter is set to 10 seconds (value=10,000 ms) and that is the inter-digit timeout. Inter-digit timeout provides 10 seconds for the user to complete dialing digits. Use the octothorpe wildcard to terminate this inter-digit timeout. For user flexibility, a pattern with the "#" and a pattern without the "#" should be configured as route patterns. This enables the user to terminate the inter-digit timeout or, if they are not aware, wait for the inter-digit timeout and still have the call routed.

Cisco CallManager Administration allows you to use wildcards, special characters, and settings to perform the following tasks:

- Enabling a single route pattern to match a range of numbers.
- Removing a portion of the dialed digit string.
- Manipulating the appearance of the calling party's number for outgoing calls.
- Manipulating the dialed digits, or called party's number, for outgoing calls.

Figure 3-8 shows an example using wildcards for digit matching and transformation masks.

The following steps you through the example in Figure 3-8.

There are four users with the following directory numbers:

- **A**—DN 5062
- **B**—DN 5063
- **C**—DN 5064
- **D**—DN 1000

Step 1 Users A, B, and C dial three different numbers:

— A dials 91234

— B dials 91324

— C dials 91432

Wildcards make it possible for those three different dialed numbers to match one route pattern.

Figure 3-8 *Wildcards and Special Characters*

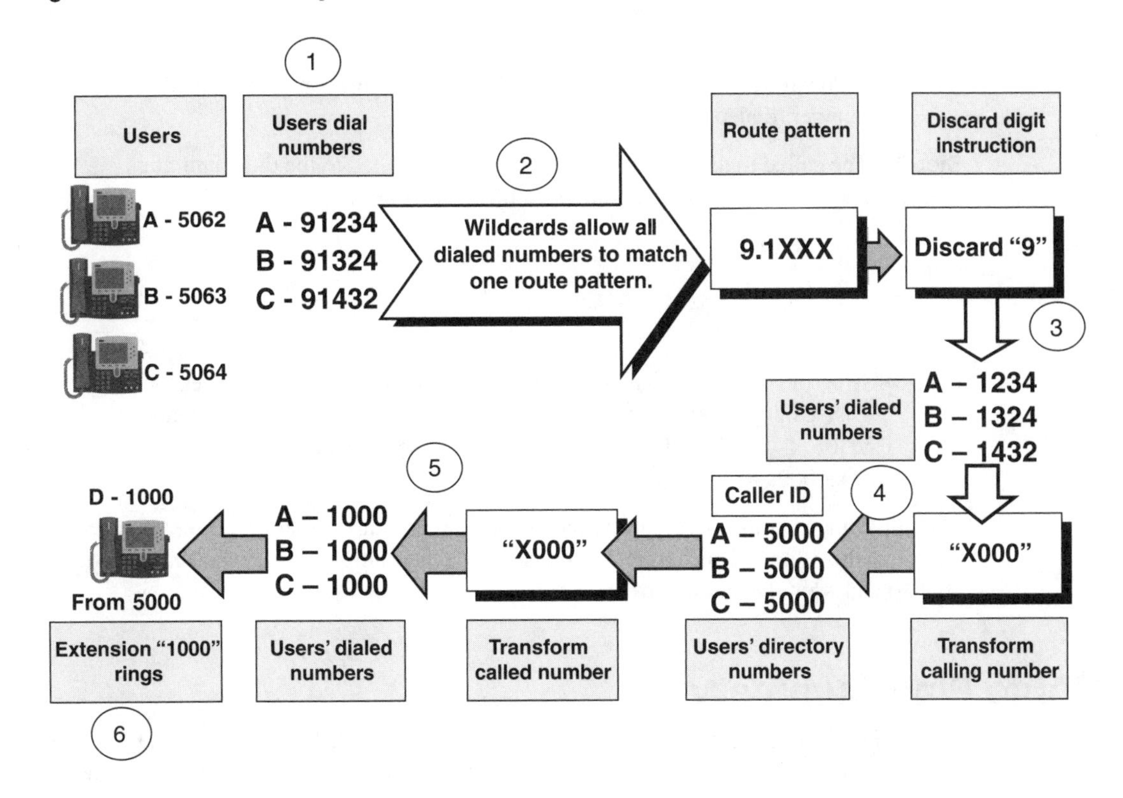

Step 2 Dialed numbers match the route pattern 9.1XXX.

Wildcards and special characters allow for DDIs.

Step 3 The DDI discards the 9 from the dialed digits. Now the dialed digits are as follows:

— **A**—1234

— **B**—1324

— **C**—1432

Wildcards and special characters allow transformation of the calling number (caller ID).

Step 4 The calling transformation number X000 is applied to the users' directory numbers and changes them to the following:

— **A**—5000

— **B**—5000

— **C**—5000

Wildcards and special characters allows transformation of the called number (dialed number).

Step 5 The called transformation number X000 is applied to the dialed number and changes the dialed number to the following:

— **A**—1000

— **B**—1000

— **C**—1000

Step 6 After the route pattern matches and after Cisco CallManager applies the DDI, calling transformation, and called transformation, the call is extended to user D who has directory number 1000. User D sees the call is from DN 5000.

The preceding example brings route pattern, DDI, and transformation masks together. Although this is not a typical real world example, this example shows when each feature is applied and what the feature affects.

Route Plan Summary

A route plan is essentially the front end that enables users to dial one another. The goal of a successful route plan is to provide diverse call routing and provide ease of dialing for users. The following are definitions of the functions in the route plan discussed in this section:

- **Route Patterns**—Match of an E.164 address range or specific address points to a single RLs or a device.
- **RLs**—How to reach a destination via prioritized route groups.
- **RG**—Forms a prioritized Trunk Group by pointing to devices.
- **Devices**—Gateways or remote Cisco CallManagers.

Figure 3-9 shows a summary of a route plan.

Table 3-6 summarizes the procedure for establishing a route plan through the steps highlighted in Figure 3-9:

Step 1 Add the gateway.

Step 2 Add a route group for the gateway.

Step 3 Add a route list for the route group.

Step 4 Add route patterns to the route list.

Figure 3-9 *Route Plan*

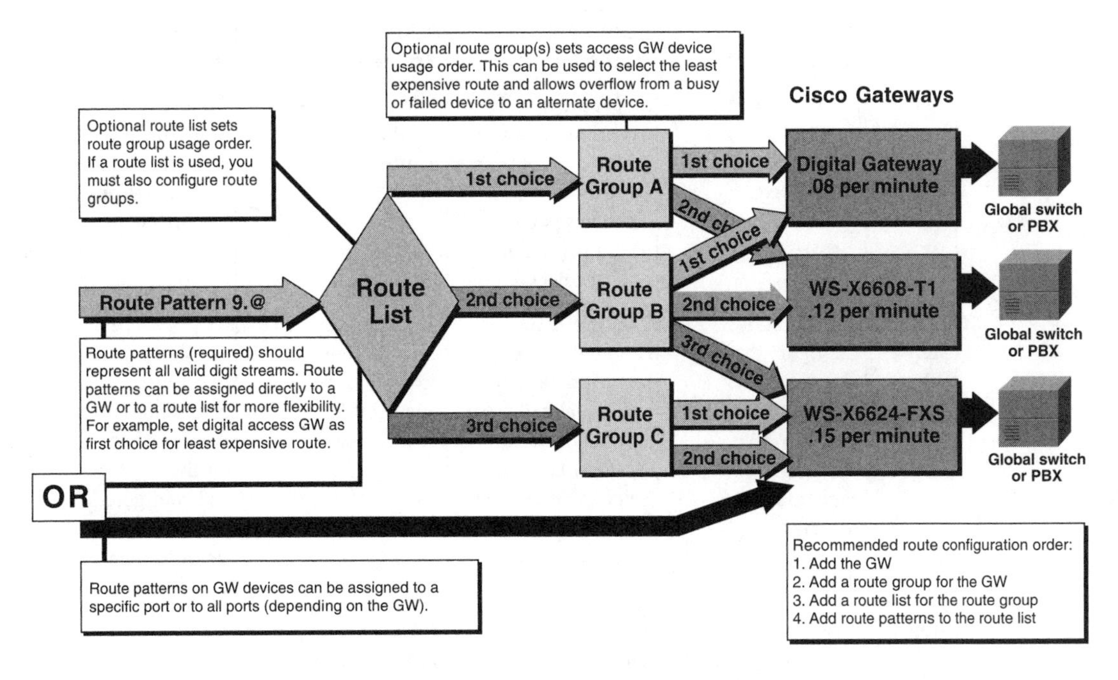

Table 3-6 *Creating a Route Plan*

	Adding a Gateway	Adding a Route Group	Adding a Route List	Adding Route Patterns
Step 1	Open Cisco CallManager Administration and add a new device (**Device > Add a New Device**).	Open the Route Group Configuration page (**Route Plan > Route Group**).	Open the Route List Configuration page (**Route Plan > Route List**).	Open the Route Pattern Configuration page (**Route Plan > Route Pattern**).
Step 2	Select a device type and then select **Next**.	Enter a Route Group Name and continue to the next step.	Enter a route list name and description and then select **Insert**.	Enter a route pattern. Use wildcards and special characters in the route pattern.

continues

Table 3-6 *Creating a Route Plan (Continued)*

	Adding a Gateway	Adding a Route Group	Adding a Route List	Adding Route Patterns
Step 3	Select gateway type and device protocol; then select **Next**.	Select the first device to add to this route group. For certain gateways you can pick a particular port or select **All**. Select **Insert**.	Add a route group by selecting **Add Route Group**.	Select a route list. Select the route list created from the **Gateway/Route List** menu.[1]
Step 4	Configure the gateway. Enter the parameters on the Gateway Configuration page; then select **Insert**.	If you want to add an additional gateway select **Add Another Gateway**. Select **Add Device**.	Select a route group and then select **Add**.	Finish the route pattern configuration. Select **Other Options** and configure transformation settings; then select **Insert**.
Step 5	Reset the gateway. After configuration changes have been made to a gateway, the gateway needs to be reset. Select **Reset**.	Add more gateways to the route group. Repeat Steps 3 and 4 to add other gateways to an route group.[2]	Select **Transformation Information**. (This is covered in a section later in this chapter.) Select **Insert**.[3]	
Step 6	Select **OK**.	Create a new route group. Select **Add New Route Group** from the left column.	Add another route group. Select the route list from the left column.	
Step 7	Repeat Steps 1 –6 to add more gateways.	Configure a route group. Repeat Steps 2–5 to configure a new route group.	Complete the adding of another route group. Repeat Steps 3–5.	

1 If a gateway is a member of a route group, it can no longer be assigned directly to a route pattern.

2 When creating a route group, the order of the gateways in the route group can be adjusted.

3 If transformations are applied at the route list, these transformation settings will override any transformation settings configured on the route pattern.

Discard Digits Instructions

A DDI removes a portion of the dialed digit string before passing the number on to the adjacent system. Portions of the digit string must be removed (for example, when an external access code is needed to route the call to the PSTN, but the PSTN switch does not expect that access code). Table 3-7 shows the different DDIs that can be assigned.

Table 3-7 *DDIs Assuming the Pattern is 9.5@*

Instructions	Discarded Digits	Used For
PreDot	**95** 1 214 555 1212	Access codes
PreAt	**95** 1 214 555 1212	Access codes
11D/10D→7D	95 **1 214** 555 1212	Toll bypass
11D→10D	95 **1** 214 555 1212	Toll bypass
IntlTollBypass	95 **011 33** 1234 #	Toll bypass
10-10-Dialing	95 **1010321** 1 214 555 1212	Suppressing carrier selection
Trailing-#	95 1010321 011 33 1234 **#**	PSTN compatibility

The different instructions can be combined and are selected from a drop-down menu. For example, you can select "PreDot Training-#," which discards anything to the left of the dot and the trailing "#." Most of the DDIs only apply to the route patterns that have the @ wildcard. The only DDIs that apply to route patterns without an @ wildcard are the PreDot and None.

Table 3-8 lists DDIs and describes the effects of applying each DDI to a dialed number.

Table 3-8 *DDIs*

DDI	Effect	Example
10-10-Dialing	This DDI removes IXC access code and carrier code.	Route pattern: 9.@ Dialed digit string: 910102889725555000 After applying DDI: 99725555000
10-10-Dialing Trailing-#	This DDI removes: IXC access code and carrier code End-of-dialing character for international calls	Route pattern: 9.@ Dialed digit string: 9101028801181910555# After applying DDI: 901181910555

continues

Table 3-8 *DDIs (Continued)*

DDI	Effect	Example
11/10D->7D	This DDI removes: Long distance direct-dialing code—the 1 before the area code in a long distance call. Long distance operator-assisted dialing code—the 0 before the long distance number. IXC access code Area code Local-area code This DDI creates a 7-digit local number from an 11- or 10-digit dialed number.	Route pattern: 9.@ Dialed digit string: 919725555000 or 99725555000 After applying DDI: 95555000
11/10D->7D Trailing-#	This DDI removes: Long distance direct-dialing code Long distance operator-assisted dialing code IXC access code Area code Local-area code End-of-dialing character for international calls This DDI creates a 7-digit local number from an 11- or 10-digit dialed number	Route pattern: 9.@ Dialed digit string: 919725555000 or 99725555000 After applying DDI: 95555000
11D->10D	This DDI removes: Long distance direct-dialing code Long distance operator-assisted dialing code IXC access code	Route pattern: 9.@ Dialed digit string: 919725555000 After applying DDI: 99725555000

Table 3-8 *DDIs (Continued)*

DDI	Effect	Example
11D->10D Trailing-#	This DDI removes: Long distance direct-dialing code Long distance operator-assisted dialing code End-of-dialing character for international calls IXC access code	Route pattern: 9.@ Dialed digit string: 919725555000 After applying DDI: 99725555000
Intl TollBypass	This DDI removes: International access code, which is 01. International direct-dialing code. This is the 1 after the 01. Country code IXC access code International operator-assisted dialing code. This is the 0 if you dial 010.	Route pattern: 9.@ Dialed digit string: 901181910555 After applying DDI: 9910555
Intl TollBypass Trailing-#	This DDI removes: International access code International direct-dialing code Country code IXC access code International operator-assisted dialing code End-of-dialing character	Route pattern: 9.@ Dialed digit string: 901181910555# After applying DDI: 9910555
NoDigits	This DDI removes no digits.	Route pattern: 9.@ Dialed digit string: 919725555000 After applying DDI: 919725555000

continues

Table 3-8 *DDIs (Continued)*

DDI	Effect	Example
Trailing-#	This DDI removes end-of-dialing character for international calls.	Route pattern: 9.@ Dialed digit string: 901181910555# After applying DDI: 901181910555
PreAt	This DDI removes all digits prior the @ wildcard.	Route pattern: 8.9@ Dialed digit string: 899725555000 After applying DDI: 9725555000
PreAt Trailing-#	This DDI removes: All digits prior to the @ wildcard End-of-dialing character for international calls	Route pattern: 8.9@ Dialed digit string: 8901181910555# After applying DDI: 01181910555
PreAt 10-10-Dialing	This DDI removes: All digits prior to the @ wildcard IXC access code	Route pattern: 8.9@ Dialed digit string: 89101028897255555000 After applying DDI: 9725555000
PreAt 10-10-Dialing Trailing-#	This DDI removes: All digits prior to the @ wildcard IXC access code and carrier code End-of-dialing character for international calls	Route pattern: 8.9@ Dialed digit string: 891010288011819105555# After applying DDI: 01181910555

Table 3-8 *DDIs (Continued)*

DDI	Effect	Example
PreAt 11/10D->7D	This DDI removes: All digits prior to the @ wildcard: Long distance direct-dialing code Long distance operator-assisted dialing code IXC access code and carrier code Area code Local-area code This DDI creates a 7-digit local number from an 11- or 10-digit dialed number.	Route pattern: 8.9@ Dialed digit string: 8919725555000 or 899725555000 After applying DDI: 5555000
PreAt 11/10D->7D Trailing-#	This DDI removes: All digits prior to the @ wildcard Long distance direct-dialing code Long distance operator-assisted dialing code IXC access code and carrier code Area code Local-area code End-of-dialing character for international calls This DDI creates a 7-digit local number from an 11- or 10-digit dialed number.	Route pattern: 8.9@ Dialed digit string: 8919725555000 or 899725555000 After applying DDI: 5555000
PreAt 11D->10D	This DDI removes: All digits prior to the @ wildcard Long distance direct-dialing code Long distance operator-assisted dialing code IXC access code	Route pattern: 8.9@ Dialed digit string: 8919725555000 After applying DDI: 9725555000

continues

Table 3-8 *DDIs (Continued)*

DDI	Effect	Example
PreAt 11D->10D Trailing-#	This DDI removes: All digits prior to the @ wildcard: Long distance direct-dialing code Long distance operator-assisted dialing code IXC access code and carrier code End-of-dialing character for international calls	Route pattern: 8.9@ Dialed digit string: 8919725555000 After applying DDI: 9725555000
PreAt Intl TollBypass	This DDI removes: All digits prior to the @ wildcard International access code International direct-dialing code Country code IXC access code and carrier code International operator-assisted dialing code	Route pattern: 8.9@ Dialed digit string: 8901181910555 After applying DDI: 910555
PreAt Intl TollBypass Trailing-#	This DDI removes: All digits prior to the @ wildcard International access code International direct-dialing code Country code IXC access code International operator-assisted dialing code End-of-dialing character	Route pattern: 8.9@ Dialed digit string: 8901181910555# After applying DDI: 910555

Table 3-8 *DDIs (Continued)*

DDI	Effect	Example
PreDot	This DDI removes all digits before the dot in the route pattern.	Route pattern: 8.9@ Dialed digit string: 899725555000 After applying DDI: 99725555000
PreDot Trailing-#	This DDI removes: All digits before the dot in the route pattern End-of-dialing character for international calls	Route pattern: 8.9@ Dialed digit string: 8901181910555# After applying DDI: 901181910555
PreDot 10-10-Dialing	This DDI removes: All digits before the dot in the route pattern IXC access code and carrier code	Route pattern: 8.9@ Dialed digit string: 89101028899725555000 After applying DDI: 99725555000
PreDot 10-10-Dialing Trailing-#	This DDI removes: All digits before the dot in the route pattern IXC access code and carrier code End-of-dialing character for international calls	Route pattern: 8.9@ Dialed digit string: 891010288011819105555# After applying DDI: 901181910555
PreDot 11/10D->7D	This DDI removes: All digits before the dot in the route pattern Long distance direct-dialing code Long distance operator-assisted dialing code IXC access code and carrier code Area code Local-area code This DDI creates a 7-digit local number from an 11- or 10-digit dialed number.	Route pattern: 8.9@ Dialed digit string: 8919725555000 or 899725555000 After applying DDI: 95555000

continues

Table 3-8 *DDIs (Continued)*

DDI	Effect	Example
PreDot 11/10D->7D Trailing-#	This DDI removes: All digits before the dot in the route pattern Long distance direct-dialing code Long distance operator-assisted dialing code IXC access code and carrier code Area code Local-area code End-of-dialing character for international calls This DDI creates a 7-digit local number from an 11- or 10-digit dialed number.	Route pattern: 8.9@ Dialed digit string: 8919725555000 or 899725555000 After applying DDI: 95555000
PreDot 11D->10D	This DDI removes: All digits before the dot in the route pattern Long distance direct-dialing code Long distance operator-assisted dialing code IXC access code and carrier code	Route pattern: 8.9@ Dialed digit string: 8919725555000 After applying DDI: 99725555000
PreDot 11D->10D Trailing-#	This DDI removes: All digits before the dot in the route pattern Long distance direct-dialing code Long distance operator-assisted dialing code IXC access code and carrier code End-of-dialing character for international calls	Route pattern: 8.9@ Dialed digit string: 8919725555000 After applying DDI: 99725555000

Table 3-8 *DDIs (Continued)*

DDI	Effect	Example
PreDot Intl TollBypass	This DDI removes: All digits before the dot in the route pattern International access code International direct-dialing code Country code IXC access code and carrier code International operator-assisted dialing code	Route pattern: 8.9@ Dialed digit string: 8901181910555 After applying DDI: 9910555
PreDot Intl TollBypass Trailing-#	This DDI removes: All digits before the dot in the route pattern International access code International direct-dialing code Country code IXC access code and carrier code International operator-assisted dialing code End-of-dialing character	Route pattern: 8.9@ Dialed digit string: 8901181910555# After applying DDI: 9910555

Transformation Settings Used in Route Plans

Figure 3-10 shows how transformations can be used in a route plan.

There are two types of transformation settings—calling party and called party. Transformations are used to change a directory number or a dialed number. Figure 3-10 reviews what was discussed in the "Route Pattern Wildcards and Special Characters" section, and the intent is to focus on the bottom flow where the directory numbers and dialed numbers are being transformed before extending a call to a device. The following are the two transformation settings and what they change:

- Calling party transformations change the caller ID.
- Called party transformations change the called party number.

Figure 3-10 *Transformations*

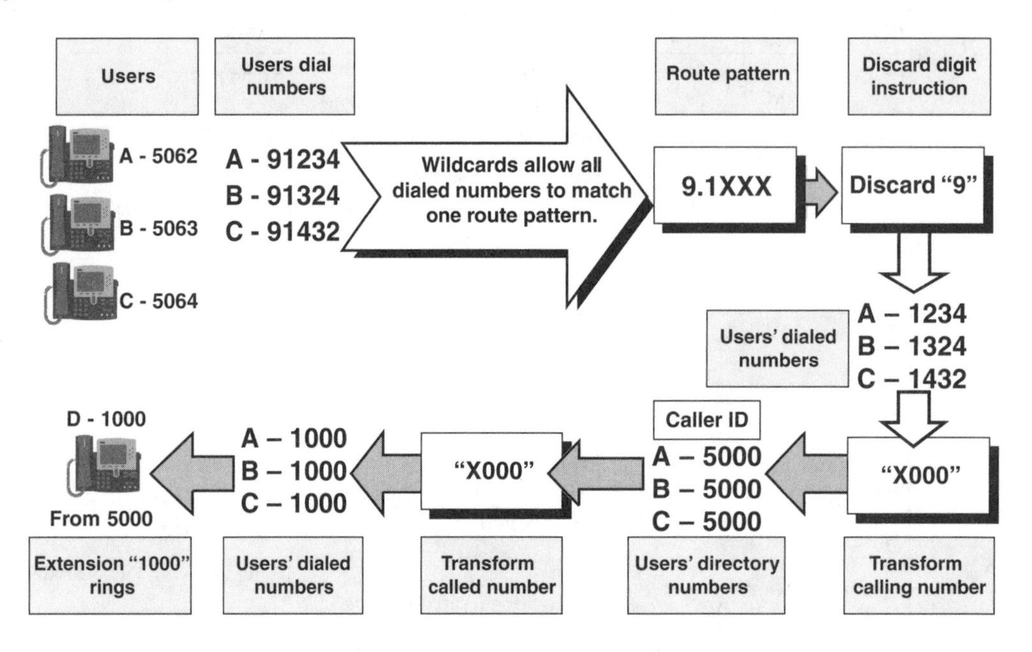

A mask is applied to a number in order to extend or truncate it. A mask can use digits 0–9, *, #, and X for transformations. Figure 3-11 shows how to extend or truncate numbers using masks.

Figure 3-11 *About Masks*

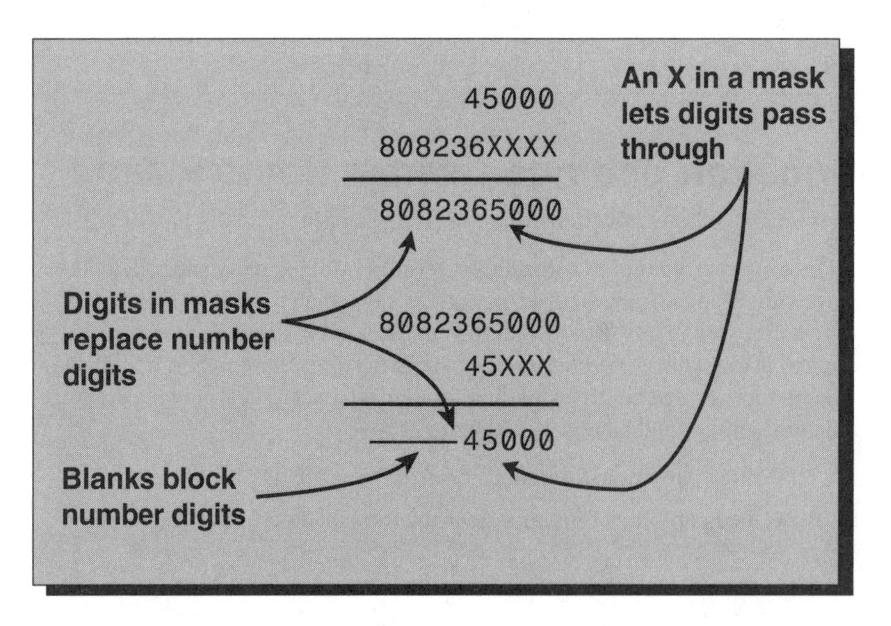

Using Figure 3-11, the following is a summary of how masks are applied:

- If there is an X in the mask, it enables digits in the original number to pass through.
- If there are digits in the mask, it changes or adds any digits that are in the original number.
- If there are blanks in the mask, it blocks any digits in the original number.
- The digits are always right-justified with the mask. This is always done before applying the transformation mask.

Calling Party Transformation Settings

Figure 3-12 shows where you can configure calling party transformation settings and the order that those settings are applied to the calling number.

Figure 3-12 *Calling Party Transformation Order*

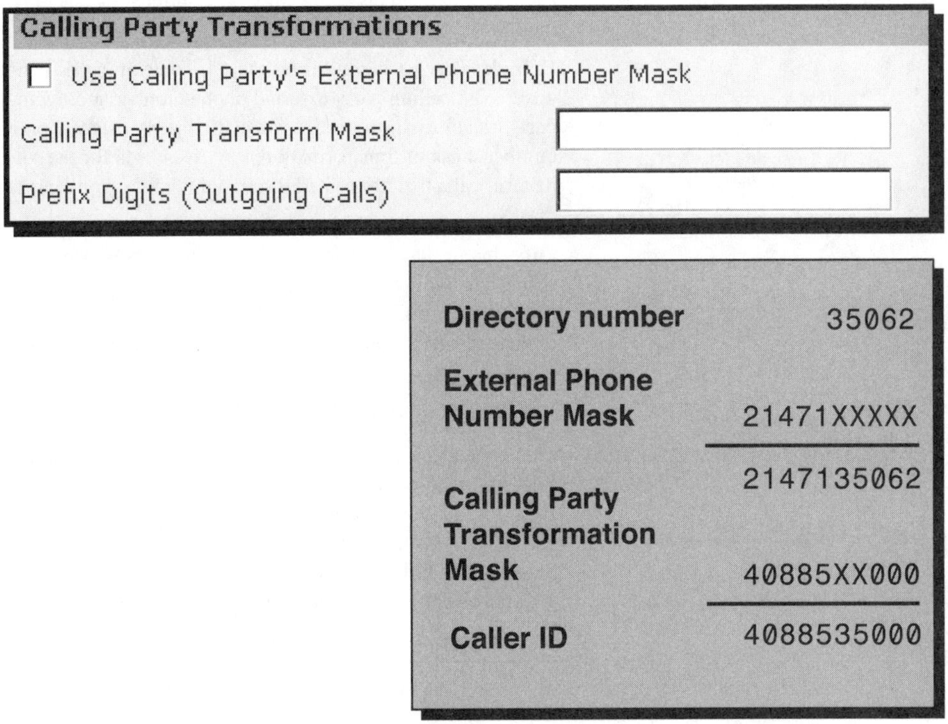

Calling Party Transformations
Use Calling Party's External Phone Number Mask
Calling Party Transform Mask
Prefix Digits (Outgoing Calls)

Directory number 35062
External Phone Number Mask 21471XXXXX
2147135062
Calling Party Transformation Mask 40885XX000
Caller ID 4088535000

Calling party transformation settings allow you to manipulate the appearance of the calling party's number for outgoing calls. The calling party's number is used for CLID. During an outgoing call, the CLID passes to each PBX, CO, and IXC as the call progresses. The CLID is also delivered to the called party when the call completes.

The calling party transformation settings used in route lists are assigned to the individual route groups composing the list, rather than the route list as a whole. The calling party transformation settings assigned to the route groups in a route list override any calling party transformation settings assigned to a route pattern associated with that route list.

Table 3-9 describes the fields, options, and values used to specify calling party transformation settings for a route group. This field can be found in the route details of a route list, route pattern, and translation pattern configuration pages.

Table 3-9 *Calling Party Transformation Settings*

Field Name	Description
Use Calling Party's External Phone Number Mask	This field determines whether the full external phone number is used for CLID on outgoing calls. The options for this field include Default, Off, and On. **Default**: This setting indicates that the route group does not govern the calling party external phone number and calling party transformation masks. If a calling party external phone number mask or transformation mask is chosen for the route pattern, calls that are routed through this route group use those masks. **Off**: This setting indicates that the calling party's external phone number is not used for CLID. If no transformation mask is entered for this route group, the CLID is the directory number or the external phone number mask configured on the line. **On**: This setting indicates that the calling party's full external number is used for CLID.
Calling Party Transform Mask	This field specifies the calling party transformation mask for all calls routed through this route group. Valid values for this field are the numbers 0 through 9 and the X wildcard character. You can also leave this field empty. If it is blank and the preceding field is set to Off, the CLID is the directory number or the external phone number mask configured on the line. The calling party transformation mask can contain up to 50 digits.

Called Party Transformation Settings

Figure 3-13 shows where you configure called party transformation settings and the order in which those settings are applied to the called number. This field can be found in the route details of a route list, route pattern, and translation pattern configuration pages.

Figure 3-13 *Called Party Transformation Order*

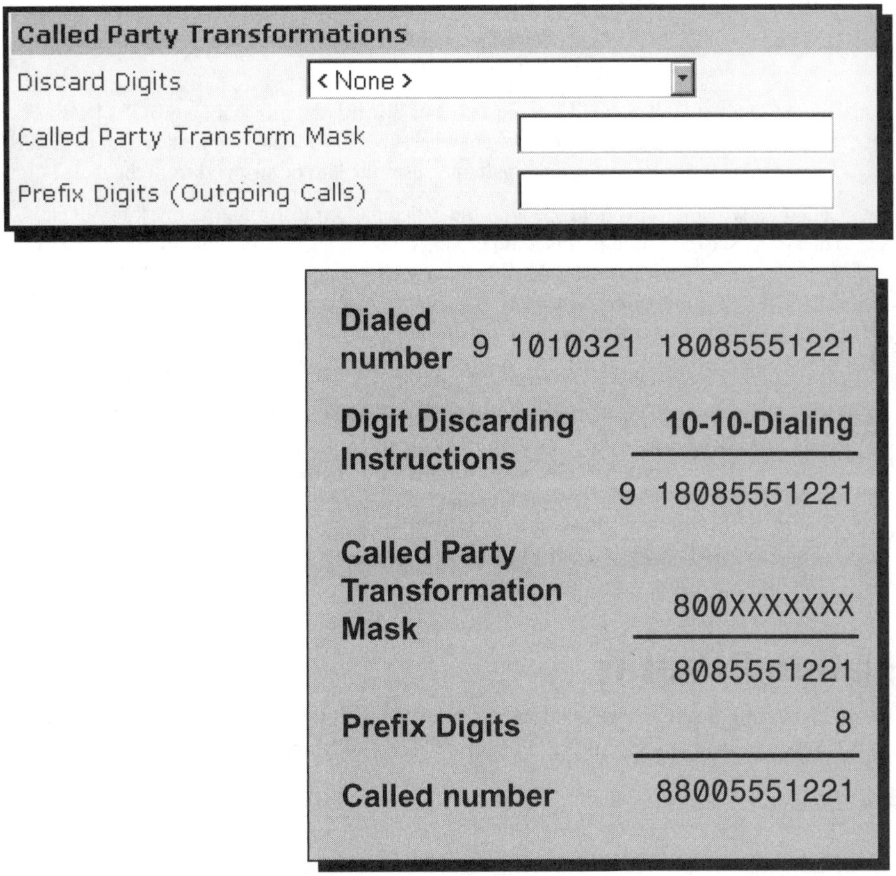

Called Party Transformations

Discard Digits < None >

Called Party Transform Mask

Prefix Digits (Outgoing Calls)

Dialed number	9 1010321 18085551221
Digit Discarding Instructions	**10-10-Dialing**
	9 18085551221
Called Party Transformation Mask	800XXXXXXX
	8085551221
Prefix Digits	8
Called number	88005551221

Called party transformation settings enable you to manipulate the dialed digits, or called party's number, for outgoing calls. Examples of manipulating called numbers include appending or removing prefix digits (outgoing calls), appending area codes to calls dialed as seven-digit numbers, appending area codes and office codes to interoffice calls dialed as four- or five-digit extensions, and suppressing carrier access codes for equal access calls.

The called party transformation settings used in route lists are assigned to the individual route groups composing the list, rather than the route list as a whole. The called party transformation settings assigned to the route groups in a route list override any called party transformation settings assigned to a route pattern associated with that route list.

Table 3-10 describes the fields, options, and values used to specify called party transformation settings for a route group.

Table 3-10 *Called Party Transformation Settings*

Field Name	Description
Dial Plan[1]	This field determines which dialing plan is used. If it is not already chosen, change this field to NANP.
Discard Digits	This field contains a list of discard patterns that control the DDIs. For example, in a system where users must dial 9 to make a call to the PSTN, the PreDot discard pattern causes the 9 to be stripped from the dialed digit string.
Called Party Transform Mask	This field specifies the called party transformation mask for all calls routed through this route group. Valid values for this field are the numbers 0 through 9 and the wildcard character X. You can also leave this field blank. If this field is blank, no transformation takes place; Cisco CallManager sends the dialed digits exactly as dialed. The calling party transformation mask can contain up to 50 digits.
Prefix Digits (Outgoing Calls)	This field contains a prefix digit or a set of prefix digits (outgoing calls) that are appended to the called party number on all calls routed through this route group. Valid values for this field are the numbers 0 through 9 and blank. Prefix Digits (Outgoing Calls) can contain up to 50 digits.

1 The Dial Plan field appears only when a route group is inserted in a route list. Once the route group is inserted, you cannot modify this field.

Translation Patterns

This section discusses translation patterns used in Cisco CallManager. Figure 3-14 shows that if a pattern matches a translation pattern, Cisco CallManager does digit analysis a second time.

Figure 3-14 *Translation Pattern*

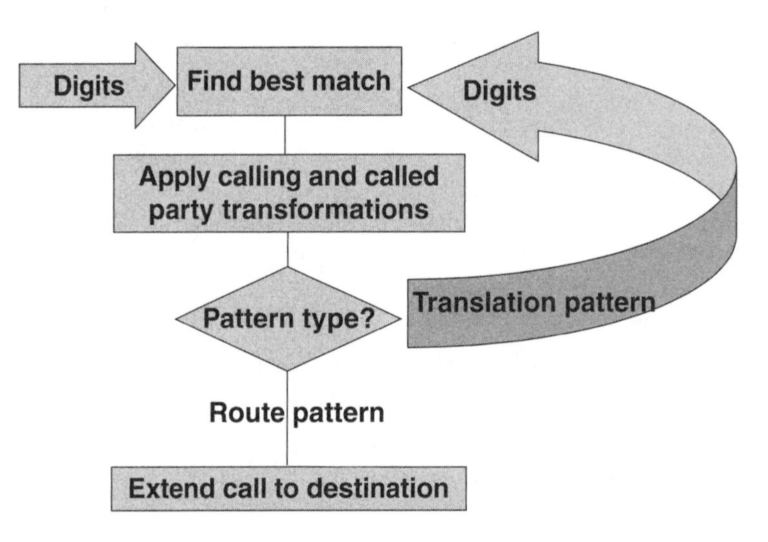

Cisco CallManager uses translation patterns to manipulate dialed digits before routing a call. In some cases, the system does not use the dialed number. In other cases, the PSTN does not recognize the dialed number.

Translation patterns are similar to route patterns in that they uses wildcards, transformations, and special characters. If digits match a translation pattern, Cisco CallManager applies any transformation, and after the pattern is translated, it will make another trip through digit analysis.

Translation patterns can be used to handle extension mapping and configuring Private Line Automatic Ring-Down (PLAR).

The CSS parameter on translation patterns provides a mechanism to handle overlapped dial plans such as are common in a multiple tenant situation.

Figure 3-15 shows two common examples to use translation patterns with.

Figure 3-15 *Common Uses of Digit Translation*

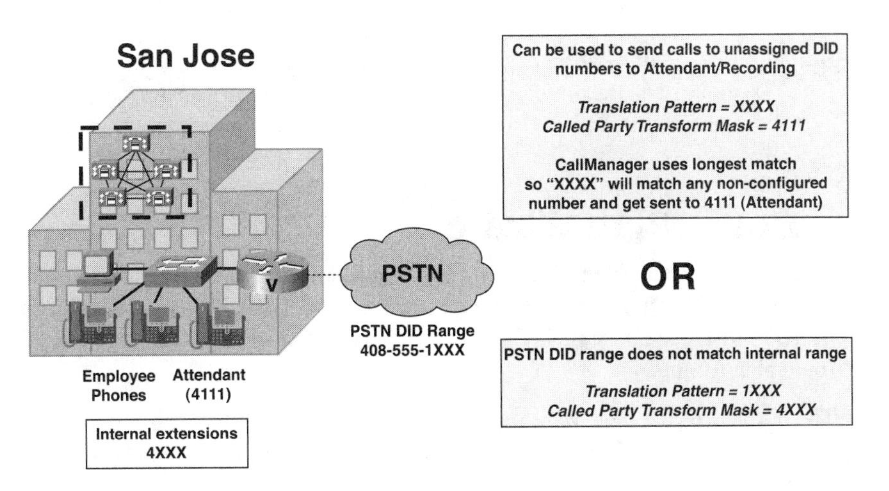

The two most common uses of translation patterns are the following:

- Use translation patterns to send calls of unassigned direct inward dial (DID) numbers to an attendant or recording.
- Use translation patterns when the PSTN DID range does not match the internal extension range.

Another use for digit translation is for PLAR. PLAR is a feature similar to the Batman phone, where the user does not have to dial digits to ring an extension. When the user goes off-hook, the translation pattern of no digits is transformed to a valid directory number and rings that extension.

Configuring Translation Patterns, DDI, and Calling/Called Party Transformations

The following steps are used to configure a translation pattern and apply DDIs, calling party transformation, and called party transformation settings:

Step 1 Open Cisco CallManager Administration and the Translation Pattern Configuration page (**Route Plan > Translation Pattern**).

Step 2 Enter a Translation Pattern. Enter **5.5XXXX** for the translation pattern.

Step 3 Enter Calling Party Transformation settings. Enter **88088** for the Calling Party Transform Mask.

Step 4 Enter Called Party Transformation settings. Select PreDot from the Discard Digit drop-down menu. Enter **<valid directory number in cluster>** for the Called Party Transform Mask.

Step 5 Click **Insert**.

Step 6 Test the translation pattern. From a phone, dial the translation pattern. It should ring the directory number in the called party transformation mask, and the called phone should display that the call is coming from 88088.

External Route Plan Wizard

The External Route Plan Wizard in Cisco CallManager enables Cisco CallManager administrators to quickly configure external routing to the PSTN, to PBXs, or to other Cisco CallManager systems. The External Route Plan Wizard can only be used to create a route plan for the NANP.

The purpose of the External Route Plan Wizard is to configure toll-bypass configurations for centralized CallManager configurations that span different local access and transport areas (LATAs).

The External Route Plan Wizard generates a single tenant, multi-location, partitioned route plan for the NANP area using information provided by the administrator through a series of prompts.

The route plan generated by the External Route Plan Wizard includes the following elements:

- Route filters
- Route groups
- Route lists
- Route patterns
- Partitions

- CSSs
- Calling party digit translations and transformations
- Access code manipulation

Figure 3-16 shows the generated naming convention from the External Route Plan Wizard.

Figure 3-16 *Generated Naming Conventions*

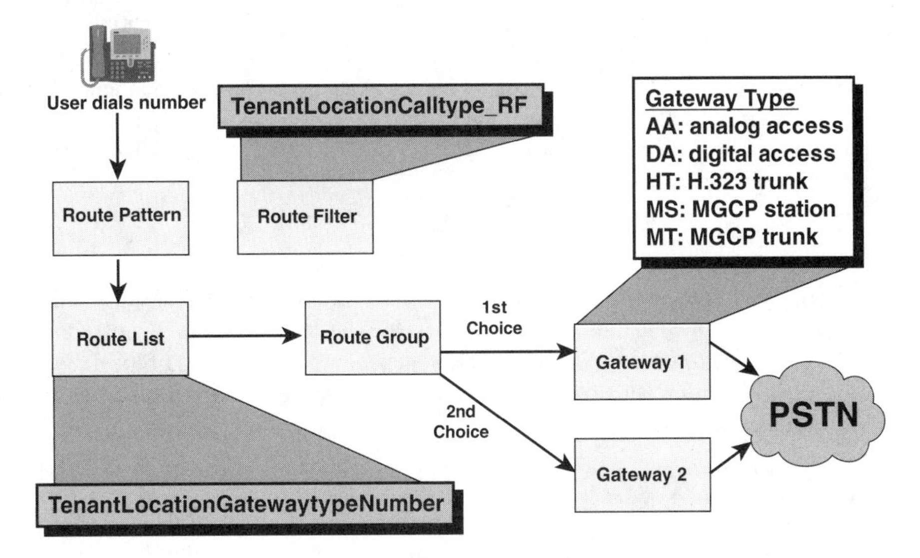

A generated route filter permits or restricts access through a route list using route patterns. The External Route Plan Wizard associates each route list with a particular route filter. It names route filters using the TenantLocationCalltype convention and appends the suffix RF to each route filter for easy identification.

A generated route group sets the order of preference for gateway and port usage. The External Route Plan Wizard assigns one gateway to each generated route group and uses all ports on the gateways. It does not support using partial resources for generated external route plans. For example, a gateway with many ports can only be added as one gateway, not a port that is part of a gateway as a resource.

The External Route Plan Wizard names route filters using the TenantLocationGateway-typeNumber convention for easy identification. The gateway type is abbreviated using AA, DA, HT, MS, and MT as shown in Table 3-1.

The External Route Plan Wizard identifies route groups associated with multiple gateways of the same type by attaching a number suffix to all route groups. For example, if there are three MT gateways at the Cisco Dallas location, the External Route Plan Wizard names the associated route groups CiscoDallasMT1, CiscoDallasMT2, and CiscoDallasMT3.

If a route list includes more than one route group and more than one gateway (with one gateway for each route group), the order in which the External Route Plan Wizard lists the route groups is arbitrary. The only order imposed is that route groups associated with the local gateways are listed before the route groups associated with remote gateways. If needed, change the order manually after the route plan is generated.

NOTE All gateways belonging to a location are shared resources for that location. However, you can go into the route lists and route groups generated by the External Route Plan Wizard and modify them. Or if you have reserved gateways, you can exclude them when entering gateways into the External Route Plan Wizard configuration screens, and then the External Route Plan Wizard won't incorporate them into the routing plan. You can manually add the reserved gateways later.

A generated route list sets the order of preference for route group usage and defines the route filters applied to those route groups. The External Route Plan Wizard creates between five and seven route lists for each location depending on the types of local dialing choices available. Therefore, the total number of route lists depends on the local dialing scheme and the number of locations served by the route plan.

Figure 3-17 shows the how the route patterns are generated using the External Route Plan Wizard.

A generated route pattern directs calls to specific devices and either includes or excludes specific dialed-digit strings. The External Route Plan Wizard generates only route patterns that require an access code prefix. The typical route pattern for routing a call to the PSTN has the prefix construction 9.@. The typical route pattern for routing a call to the PBX has the prefix construction 9.9@.

The External Route Plan Wizard associates a route list, a route filter, and a partition with each route pattern. The route pattern provides the appropriate calling party transformation mask, called party transformation mask, DDIs, and prefix digits for the associated route list.

The External Route Plan Wizard bases route patterns for calls to an adjacent PBX on the access code and the range of directory numbers served by that PBX. For example, if the access code used to direct calls to the adjacent PBX is 9 and the range of directory numbers served by that PBX is 1000 through 1999, then the External Route Plan Wizard generates the route pattern 9.1XXX for enterprise calls.

To use the External Route Plan Wizard to quickly configure external routing to the PSTN, PBXs, or to other Cisco CallManager systems, do the following:

Step 1 Open Cisco CallManager Administration and display the External Route Plan Wizard page (**Route Plan > External Route Plan Wizard**).

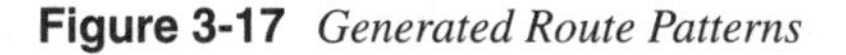

Figure 3-17 *Generated Route Patterns*

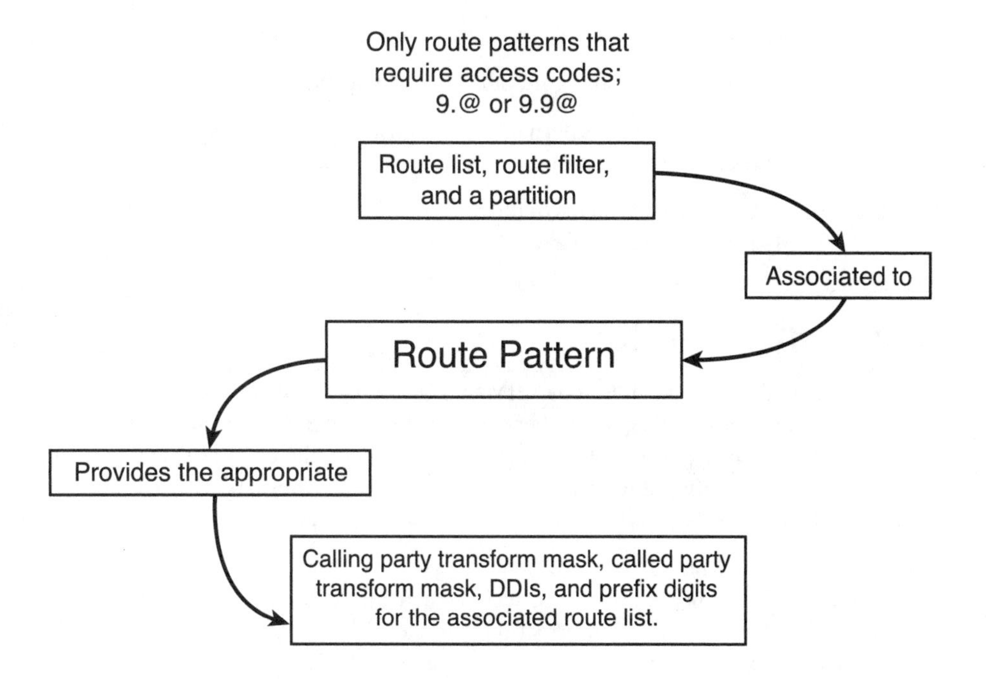

Take note of the prerequisites at the bottom of the page and stated below:

— All gateways must be already defined in the system.

— The External Route Plan Wizard can only be used to create a route plan for the NANP.

Step 2 Select **Next**.

Step 3 Routing Options. Choose criteria for routing options. Select **Next**.

Step 4 Tenant Information. This is used if the Cisco CallManager cluster is going to support multiple tenants. Enter **1** for the number of physical locations. Select **Next**.

Step 5 Location Entry. Configure location name, area codes, and main switchboard number, and then select how many digits it takes for a local call. Select **Next**.

Step 6 Gateway List. Select the gateways to be configured in the route plan. Select **Next**.

Step 7 Gateway Information. Select the location and type of carrier the gateways are connected to. If connected to a PBX, are there digits that need to be discarded? What extensions does the PBX serve? Select **Next**.

Step 8 Repeat Step 7 to configure other gateway information.

Step 9 Confirmation. The External Route Plan Wizard generates a summary of the route plan. Select **Next**.

Step 10 Status. Check the status of the route plan generated by the External Route Plan Wizard. Select **Next**.

Route Plan Report

The Route Plan Report in Cisco CallManager is a listing of all call park numbers, call pickup numbers, conference numbers, route patterns, and translation patterns in the system. The Route Plan Report enables you to view either a partial or full list and to go directly to the associated configuration pages by selecting a route pattern, partition, route group, route list, call park number, call pickup number, conference number, or gateway.

In addition, the Route Plan Report enables you to save report data into a .csv file that you can import into other applications, such as the Bulk Administration Tool (BAT). The .csv file contains more detailed information than the Web pages, including directory numbers for phones, route patterns, and translation patterns.

The External Route Plan Wizard enables Cisco CallManager administrators to quickly configure external routing to the PSTN, to PBXs, or to other Cisco CallManager systems.

To view the route plan report, do the following:

Step 1 Open the Route Plan Report page (**Route Plan > Route Plan Report**).

Step 2 Scroll down to view and identify the naming conventions of the route plan generated by the External Route Plan Wizard.

Step 3 At the top right of the page, select **View in File**. Select **Open This File from Its Current Location**. Select **Notepad** to view the file.

Step 4 View the .csv file. Scroll across and identify route plan components. Close Notepad.

Partitions and Calling Search Spaces

This section discusses partitions and CSSs in Cisco CallManager. Partitions and CSSs can also be referred to as class of service, which can mean different things depending on the technology you are working with. Figure 3-18 shows how class of service is defined internal to a PBX.

Figure 3-18 *Class of Service*

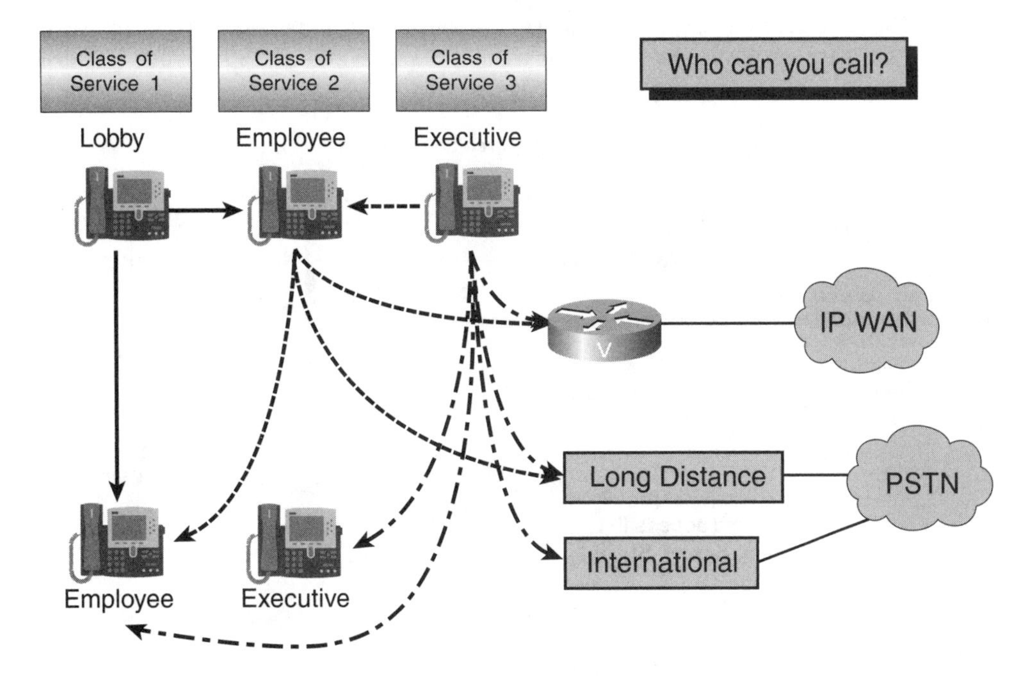

Newton's Telecom Dictionary, 17th Updated and Expanded Edition (pp. 149), has three definitions for class of service:

1) Internal to a PBX, 2) On the public switched network, and 3) On a packet switched network, courtesy of Cisco Systems, Inc.

Class of service internal to a PBX, excerpted from *Newton's Telecom Dictionary*:

"Each phone in a corporation telephone system may have a different collection of privileges and features assigned to it, such as access to long distance, international calls, 900 area code calls, 976 local calls, etc. Let's say you are afraid that your people will waste the company's money by frivolously calling some expensive numbers, you might wish to define "Class of Service" assignments in your PBX. You could have one that's called " ability to dial everywhere except 900 area code, international calls and all 976 numbers." That could be Class of Service Assignment B. When you give a phone to an employee, you could give that person COS B. Big bosses, on the other hand might need to call internationally, but not 900 area code or 976 calls. That could be called Class of Service Assignment A. Class of Service assignments, if properly organized, can become an important tool in controlling telephone abuse."

Cisco CallManager has the capability to apply class of service to devices by configuring partitions and CSSs. Partitions and CSSs are configured to provide call routing restrictions, but not necessarily restrictions on accessing features.

Understanding Partitions and Calling Search Spaces

This section discusses partitions and CSSs that are used to provide class of service in a CIPT network. Figure 3-19 shows an analogy of partitions and CSSs to routers with access lists.

Figure 3-19 *Understanding Partitions and CSSs—Compared to Access Lists*

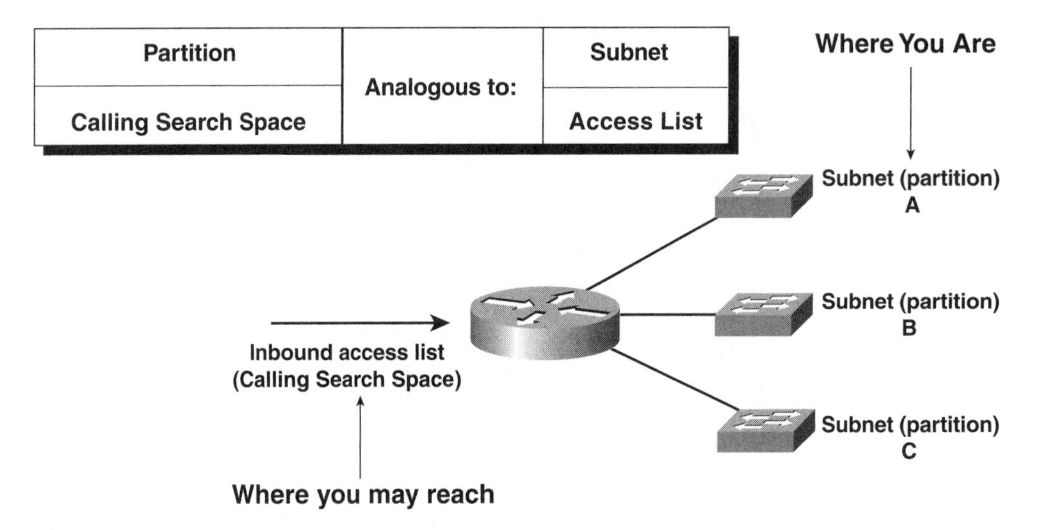

Think of a partition as an IP subnet where you place users. A CSS is likened to an inbound access list that dictates which subnet you can reach.

Another analogy to describe partitions and CSSs is with a phone book as shown in Figure 3-20.

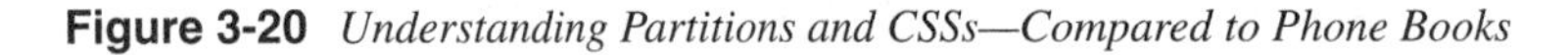

Figure 3-20 *Understanding Partitions and CSSs—Compared to Phone Books*

Every city, county, or state has a phone book published that lists phone numbers. The phone numbers listed in the phone book are the same as directory numbers or route patterns listed in a partition. A CSS is the bookshelf that holds the phone books from the different cities, counties, and states.

A device is assigned a CSS and wants to dial a directory number or route pattern. The device looks in its CSS (on its bookshelf) to see if it has the correct partition (phone book) that has the directory number or route pattern that it wants to call. If the device has the partition (phone book) that contains the directory number or route pattern in its CSS (on its bookshelf), it is able to place the call. If it does not, that device cannot call that directory number or route pattern.

Defining Partitions and Calling Search Spaces

Figure 3-21 shows a way to group directory numbers and route patterns into partitions.

Figure 3-21 *Partition Definition*

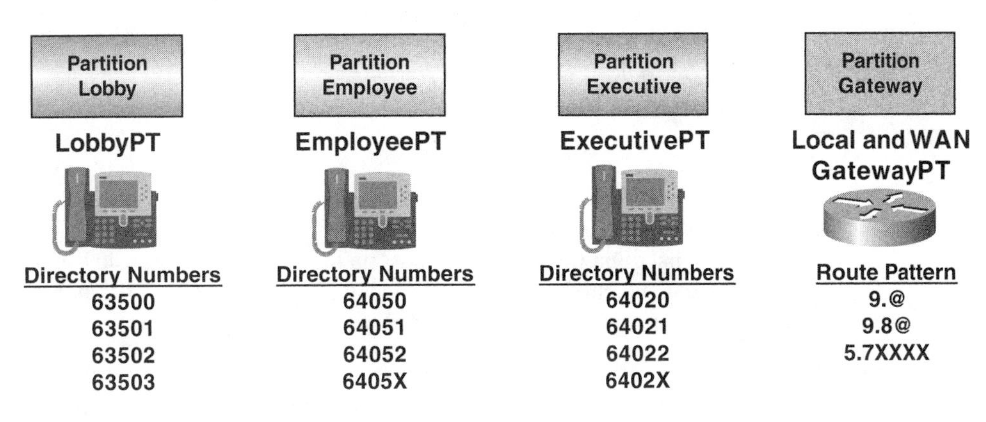

- A logical grouping of patterns.
- All patterns in a partition are equally reachable.
- Assigned to DNs and route patterns.

A partition is a logical grouping of directory numbers and route patterns with similar reachability characteristics. These are entities associated with directory numbers that users dial. For simplicity, partitions are usually named for their characteristics, such as "NYLongDistancePT," "NY911PT," and so on. When a directory number or route pattern is placed into a certain partition, this creates a rule together with a CSS that specifies what devices can call that directory number or route pattern.

A directory number needs to be unique within a partition, but can be duplicated if they are in different partitions. This enables Cisco CallManager to support a multi-tenant solution. Using the phone book analogy, the same phone number can be listed in different phone books on different book shelves.

Figure 3-22 shows how available partitions are listed in a CSS and assigned to devices.

Figure 3-22 *CSS Definition*

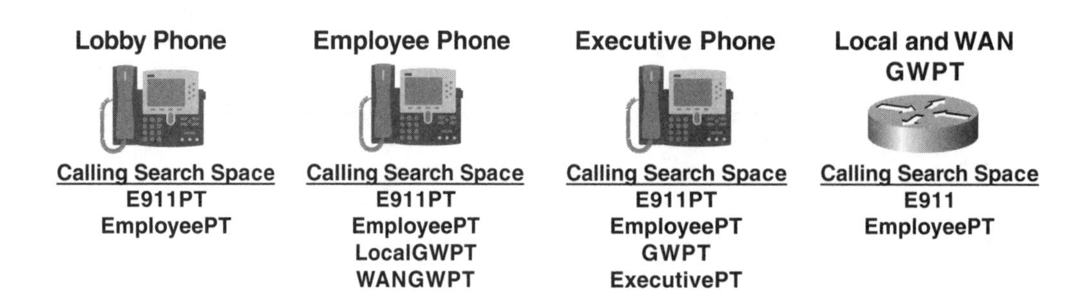

- An ordered list of partitions.
- Digit analysis looks through the caller's list of partitions when searching for the closest match for the caller's dialed number.
- Assigned to phones and GWs.

A CSS is an ordered list of partitions that devices can look at before being allowed to place a call. CSSs determine which partition's calling devices, including Cisco IP Phones, Cisco IP SoftPhones, and gateways, can search when attempting to complete a call.

When a CSS is assigned to a device, the list of partitions in the CSS is the only list of partitions that the device is allowed to reach. All other directory numbers in partitions not in the device's CSS are given a busy signal.

Partitions do not significantly impact the performance of digit analysis, but every partition specified in a calling device's search space does require that an additional analysis pass through the analysis data structures. Digit analysis looks through every partition in a CSS for the best match. The order of the partitions listed in the CSS is used only to break ties when there are equally good matches in two different partitions. If no partition is specified for a pattern, the pattern is listed in the null PT. Digit analysis always looks through the null partition last.

Partitions and CSSs are designed to address three specific problems:

- Routing by geographical location
- Routing by tenant
- Routing by class of user

Partitions and CSSs provide a way to segregate the system-wide dialable address space. The system-wide dialable address space is the complete set of dialing patterns to which Cisco CallManager can extend a call.

Partition/Calling Search Space Examples

This section discusses examples in using partitions and CSSs to apply a class of service to devices. Figure 3-23 illustrates a basic example of how partitions and CSSs may be used to provide dialing restrictions.

Figure 3-23 *Partitions/CSSs* for a Single Site

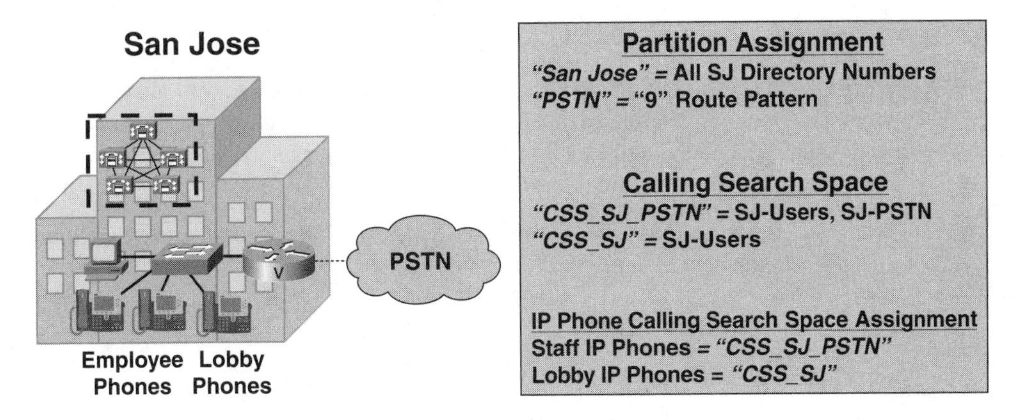

Employees—may dial each other and access the PSTN
Lobby phones—can dial SJ employee directory numbers only

In this example, staff employees have unrestricted dialing; whereas, the lobby phones have the ability to dial people only within the local site. As noted in the diagram, all IP phones are placed in the San Jose partition, and the route pattern 9 associated with the PSTN is placed in the PSTN partition. Two CSSs are created that denote two different dialing characteristics. A CSS called CSS_SJ_PSTN is created that has both San Jose and PSTN partitions in it. A second CSS called CSS_SJ is created with only the San Jose users in it. San Jose staff IP phones are assigned the CSS_SJ_PSTN CSS denoting that they may dial anywhere. The lobby phones are assigned the CSS_SJ CSS meaning they can only dial local phones in the building.

Tables 3-11 and 3-12 show the PT and CSS configuration for Figure 3-23.

Table 3-11 *Partition Configuration*

Partition	Directory Numbers and Route Patterns
<Company Name> "SanJose"	Directory numbers of San Jose employees.
PSTN	All external route patterns (Local PSTN).

Table 3-12 *CSS Configuration*

CSS	Partition (Ordered List)	Assigned To
CSS_SJ_PSTN	SanJose PSTN	Devices that can make internal and external calls.
CSS_SJ	SanJose	Devices that can only make internal calls.

Partitions and CSSs can accommodate different types of calls in a centralized call processing model. A centralized call processing model is one cluster that services multiple locations. Users may go to the different locations yet want to use the same type of dialing habits at each location. Partitions and CSSs accommodate for this type of end user behavior.

Figure 3-24 illustrates the types of calls partitions and CSSs accommodate in a multiple-site WAN deployment using centralized call processing.

Figure 3-24 *Multiple-Sites: Centralized Call Processing*

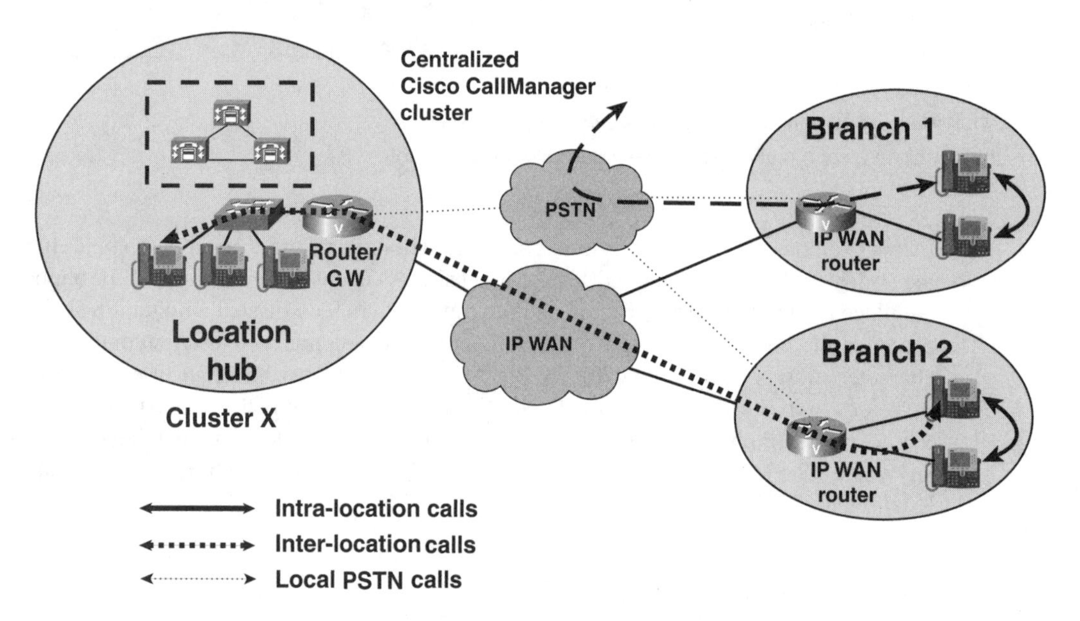

Intra-location calls are generally made between IP phones and other devices such as analog phones connected to gateway devices based on MGCP, H.323, or the Skinny gateway protocol. As within a cluster, all devices register with a single Cisco CallManager so that the availability of all devices is known.

When a call is attempted, the outcome is one of the following:

- The call succeeds.
- A busy tone is issued due to the remote device being active.
- A busy tone is issued due to insufficient WAN resources; a message might also appear on the device.

No configuration of a dial plan is required for intra-cluster calls in the majority of cases.

Each site uses the same dialed number to access the local PSTN. Based upon CSS, a local gateway is selected.

Tables 3-13 and 3-14 show the partition and CSS configuration for Figure 3-24.

Table 3-13 *Pattern Configuration*

Partition	Directory Numbers and Route Patterns
Cluster-X Users	Directory numbers of cluster X users.
Cluster-X Hub PSTN Access	PSTN gateway(s) at hub location.
Cluster-X Branch 1 PSTN Access	PSTN gateway at Branch 1.
Cluster-X Branch 2 PSTN Access	PSTN gateway at Branch 2.

Table 3-14 *CSS Configuration*

CSS	Partition (Ordered List)	Assigned To
Cluster-X Internal Only	Cluster-X Users	Devices that can only make internal calls.
Cluster-X Hub Unrestricted	Cluster-X Users Cluster-X Hub PSTN Access	Internal calls and PSTN calls through hub location gateways.
Cluster-X Branch 1 Unrestricted	Cluster-X Users Cluster-X Branch 1 PSTN Access	Internal calls and PSTN calls through Branch 1 location gateways.
Cluster-X Branch 2 Unrestricted	Cluster-X Users Cluster-X Branch 2 PSTN Access	Internal calls and PSTN calls through Branch 2 location gateways.

Table 3-15 shows some suggested partition names and what they should be assigned to.

Table 3-15 *Suggested Partition Names and Examples*

Partition Name	Example	Devices Applied To/Function
[Company's Name]	Cisco	Assigned to end-device directory numbers
Block	Block	Block certain numbers
Auto-Registration	Auto-Registration	Unadministered phone installation
[Location]_Local	RCH_Local	Local access
[Location]_Long Distance	RCH_LongDistance	LD access
[Location]_International	RCH_International	International access
Translations	Translations	Translation patterns
E911_[Physical Location of Gateway]	E911_RCH3	911 access per building

The name of a CSS should reflect the partitions listed in that CSS. For example,

- CSS_E911_Company
- CSS_E911_Company_Local
- CSS_E911_C_LD_Intl

A CSS provides the class of service for the device. The CSS includes the partitions the device is enabled to call. The following is a format for the naming of a CSS:

[Location]_[List of Partitions in Calling Search Space]

Table 3-16 provides suggested names of CSS names and their definitions.

Table 3-16 *Suggested CSS Names with Definitions*

CSS Names	Probable Devices Assigned To	Where Can Those Devices Call and What About 911?
RCH3_E911Cisco	Phones outside building 3.	These devices can only call **Cisco** devices and when **911** is dialed, it will use the gateway physically located in **Richardson**, building **3**.
RCH2_E911CiscoLocal	Lobby Phones, Admin Phones located in building 2.	This device can call **Cisco** devices and **local** calls. When **911** is dialed, it will use the gateway physically located in **Richardson**, building **2**.

Table 3-16 *Suggested CSS Names with Definitions (Continued)*

CSS Names	Probable Devices Assigned To	Where Can Those Devices Call and What About 911?
RCH5_E911 CiscoLocalLongDistance	Guest cube, admin, or restricted employee phones located in building 5.	This device can call **Cisco** devices, **local** calls and **long distance**. When **911** is dialed, it will use the gateway physically located in **Richardson**, building **5**.
RCH6_E911 CiscoLocalLongDistanceInternational	Employee phones located in building 6.	This device can call **Cisco** devices, **local** calls, **long distance** and **international**. When **911** is dialed, it will use the gateway physically located in **Richardson**, building **6**.
RCH8_E911	Emergency phones in building 8.	This device can only call **911**. When 911 is dialed, it will use the gateway physically located in **Richardson**, building **8**.

CSS names do not need to be cryptic. Use what works best for your deployment. Descriptive names assist during troubleshooting.

CAUTION CSSs have a character limitation in a colon-delimited list of the partitions in the CSS. The limit is 1024 characters.

On a Cisco IP Phone configuration, CSSs can be configured at the device or directory number level. Cisco CallManager digit analysis searches through the directory number level first then the device level. If there are any ties, the directory number level CSSs is the priority. Being able to configure the CSS at the device and directory number level allows for shared line appearances in multiple locations the ability to access the local PSTN (for 911 access) based on the device-level CSS.

CSSs can be applied to the call forward settings of a phone. This is covered in more detail in Chapter 6.

Using partitions and CSSs, you can provide a class of service for users and devices.

The following steps are used to configure partitions and CSSs:

To configure partitions, do the following:

Step 1 Open Cisco CallManager Administration and display the Partition Configuration page (**Route Plan > Partition**).

Step 2 Partition name and description. Enter a partition name and description. Select **Insert**.

Step 3 Add two more PTs. Select **Add New Partition** from the left menu and repeat Step 2, twice.

Step 4 Confirm entry. In the left column you will see the partitions you created.

To configure CSSs, do the following:

Step 1 Open the Calling Search Space Configuration page (**Route Plan > Calling Search Space**).

Step 2 CSS name and description. Enter a CSS name and description.

Step 3 Add partitions. Select partitions to add to the CSS from the available partition box. Highlight a partition and use the arrows between the available partition box and the selected partition box to move the selected partition to the selected partition box.

Step 4 Order partitions. Highlight a partition in the selected partition box and use the arrows keys to the right of the selected partitions box to place them in the order you would like. Select **Insert**.

Step 5 Create another CSS. Select **Copy** and repeat steps 2–4.

To apply partitions and CSSs to a device, do the following:

Step 1 Open the Find and List Phones page (**Device > Phone**).

Step 2 Find your phone. Use the search criteria or select **Find** to list all phones. In the Device Pool column select your phone.

Step 3 Assign a CSS. At the Phone Configuration page select a CSS from the drop-down menu. Select **Update**. Select **OK** in the dialog box.

By adding a CSS to a phone, you have just determined which partitions that phone can call, which also determines the directory numbers and route patterns that phone can call.

Step 4 Assign a partition to the directory number. From the left column select Line 1. Select a partition from the drop-down menu. Select **Update**.

By assigning a partition to the directory number, you have restricted other devices that can call that directory number if they have the CSS that includes the partition that the directory number is assigned to.

Step 5 Reset Phone. In the top right corner of the page, select **Configure Device** (SEP<*MAC Address*>). Select **Reset Phone**. Select **Reset**; then select **OK** from the dialog box or select **Restart** from the dialog box.

Summary

This section summarizes topics discussed in this chapter. After completing this chapter, you should have learned the following:

- Identify and describe the components of a route plan.
- Construct a flow chart of a basic route plan.
- Identify route pattern wildcards.
- Identify the characteristics of partitions and CSSs.
- Configure partitions and CSSs in Cisco CallManager Administration.
- A route plan is essentially the front end that allows users to dial one another. The goal of a successful route plan is to provide diverse call routing and provide ease of dialing for users. The following are definitions of the functions in the route plan discussed in this section:
  - **Route Patterns**—Match of an E.164 address range or specific address points to a single route lists or a device.
  - **Route Lists**—How to reach a destination via prioritized route groups.
  - **Route Group**—Forms a prioritized Trunk Group by pointing to devices.
  - **Devices**—Gateways or remote Cisco CallManagers.

The dial plan architecture in Cisco CallManager generally can handle two types of calls:

- Internal calls to Cisco IP Phones registered to the CallManager cluster itself.
- External calls via a PSTN gateway or to another CallManager cluster via the IP WAN.

Partitions and CSSs draw an analogy to routers with access lists. Think of a partition as an IP subnet where you will place users. A CSS is likened to an inbound access list that dictates which subnet you can reach.

By configuring partitions and CSSs, you can provide a class of service for users and devices.

Post-Test

Did you get all that? Test yourself by taking this post-test to assess or reinforce your knowledge on this topic. The results of the quiz also help you determine whether you need to read this chapter again. You can find the answers to the post-test for this chapter in Appendix B, "Answers to Chapter Pre-Test and Post-Test Questions."

1 Write or draw the process used for configuring a route plan.

____

____

____

2 Match the most commonly used route pattern wildcards to its definition.

Route Pattern Wildcards						
X	!	#	[x-y]	.	@	[^x-y]

a. One or more digits (0 – 9)

b. Generic range notation

c. Terminates access code

d. Exclusion range notation

e. North American Numbering Plan

f. Terminates inter-digit timeout

g. Single digit (0 – 9)

3 What are the two things a route pattern can be assigned to?

____

____

____

4 Write or draw the flow of a call when a user dials digits.

____

____

____

5 If you are using an "@" in a route pattern, what is automatically recognized as an end-of-dialing character for international calls?

____

____

____

6 What removes a portion of the dialed digit string before passing the number on to the adjacent system?

____

____

____

7 Which one of the transformation settings changes the dialed number?

____

____

____

8 If dialed digits match a translation pattern, what does Cisco CallManager do with those digits?

____

____

____

9 What are the prerequisites for using the External Route Plan Wizard?

____

____

____

10 List five of the route plan components generated by the External Route Plan Wizard.

____

____

____

11 Match the gateway-type naming convention used by the External Route Plan Wizard to the gateway type.

Gateway types naming convention				
AA	DA	HT	MS	MT

a. Digital Access

b. MGCP Trunk

c. MGCP Station

d. Analog Access

e. H.323 Trunk

12 The Route Plan Report can be saved as what type of file?

____

____

____

13 What does Cisco CallManager use to apply class of service to devices?

____

____

____

14 What are directory numbers and route patterns assigned to?

____

____

____

15 What does a CSS provide for the device it is assigned to?

____

____

____

16 What are the three specific problems that partitions and CSSs are designed to address?

____

____

____

Upon completing this chapter you will be able to do the following tasks:

- Given a Cisco CallManager cluster, Cisco IP Voice Media Streaming Application, and hardware digital signal processor (DSP) resources, configure software and hardware conference bridge resources.
- Given a Cisco CallManager cluster and hardware DSP resources, configure transcoding resources.
- Given a Cisco CallManager cluster, Cisco IP Voice Media Streaming Application, and hardware DSP resources, configure MTP resources.
- Given a Cisco CallManager cluster and Cisco IP Voice Media Streaming Application, configure MOH resources.
- Given a Cisco CallManager cluster and configured media resources, configure and assign media resource groups and lists.
- Given a list of the services, provide a brief description of the function of that service in a Cisco CallManager cluster.

CHAPTER 4

Cisco CallManager Administration Service Menu

This chapter discusses the configurable attributes in the Service menu of Cisco CallManager Administration. The Service menu provides access to media resources that can be shared among all Cisco CallManagers in the cluster. Using the menu items in the Service menu, you can configure and adjust service parameters and Cisco TFTP settings. The following topics are discussed in this chapter.

- Abbreviations
- Media Resources
- Conference Bridge (ConfBr)
- Media Termination Point (MTP)
- Music on Hold (MOH)
- Transcode (XCODE)
- Media Resource Management (MRM)
- Services

Pre-Test

Do you already know this information? The pre-test is designed to help you gauge your knowledge about this chapter. Of the nine questions, if you answer one to three questions correctly, we recommend that you read this chapter. If you answer four to six questions correctly, we recommend that you skim through this chapter, reading those sections that you need to know more about. If you answer seven to nine questions correctly, you probably understand this information well enough to skip this chapter. You can find the answers to the pre-test for this chapter in Appendix B, “Answers to Chapter Pre-Test and Post-Test Questions.”

1 Name the two types of conferences and briefly describe each type.

____

____

____

2 What is the supported codec for ConfBrs?

____

____

____

3 Supplementary services on H.323v1 gateways are enabled through which media resource?

____

____

____

4 What is the name of the application that provides for software conferencing and MTP?

____

____

____

5 Name the two types of hold that can use MOH.

____

____

____

6 Describe an XCODE. What does it do or provide?

____

____

____

7 Media resource groups and lists have the same type of relationship and manner of configuration as what two other components in CCMAdmin?

____

____

____

8 If you have software ConfBr resources listed before hardware ConfBr resources in a MRG, in what order does the media resource manager allocate ConfBr resources?

____

____

____

9 Where in Cisco CallManager Administration do you go to assign a bulk number of devices to a MRGL?

____

____

____

Abbreviations

This section defines the abbreviations used in this chapter. For more information about terms and abbreviations used in this chapter refer to the *IP Telephony Network Glossary* at the following URL:

www.cisco.com/univercd/cc/td/doc/product/voice/evbugl4.htm

Table 4-1 provides the abbreviation and the complete term.

Table 4-1 *Abbreviation with Definition*

Abbreviation	Complete Term
ConfBr	conference bridge
CLI	command-line interface
CMI	Cisco Messaging Interface
.cnf	configuration file
GSM	Global System for Mobile communications (aka, Groupe Speciale Mobile)
GW	gateway
HW	hardware
IP	Internet Protocol
MAC	media access control
MOH	Music on Hold
MTP	media termination point
MRG	media resource group

continues

Table 4-1 *Abbreviation with Definition (Continued)*

Abbreviation	Complete Term
MRGL	media resource group list
PC	personal computer
RTP	Real-Time Transport Protocol
SMDI	station message desk interface or simplified message desk interface
SW	software
TAC	Technical Assistance Center
TCD	Telephony Call Dispatcher
TFTP	Trivial File Transfer Protocol
TCP	Transport Control Protocol
UconfBr	Unicast conference bridge
XCODE	Transcoder
XML	extensible markup language

Understanding Media Resources

This chapter provides an overview of media resources in a Cisco CallManager cluster. Media resources in Cisco CallManager include the following types:

- MOH server
- UconfBr
- Media streaming application server (software MTP)
- XCODE

The following reasons explain why resources are shared:

- To enable both hardware and software devices to coexist within a Cisco CallManager.
- To enable Cisco CallManager to share and access resources available in the cluster.
- To enable Cisco CallManager to perform load distribution within a group of similar resources.
- To enable Cisco CallManager to allocate resources based on user preferences.

Conference Bridge (ConfBr)

ConfBrs in Cisco CallManager are available both as a *software application* and a *hardware solution* designed to enable both Ad Hoc and Meet-Me voice conferencing. Each ConfBr is capable of hosting several simultaneous, multiple-party conferences.

- **Ad Hoc conference**—In an Ad Hoc conference, a user, known as the conference controller, adds participants to a conference by calling the new participant and (once connected) pressing the **Conference** button or **Confrn** soft key. Only a conference controller can add participants to an Ad Hoc conference. An Ad Hoc conference continues even if the conference controller hangs up, but no new participants can be added.
- **Meet-Me conference**—In a Meet-Me conference, a user, known as the conference controller, presses the **MeetMe** button or soft key and establishes the conference to which participants dial into. A range of Meet-Me directory numbers must be configured in Cisco CallManager Administration. The conference controller advertises the directory number to participants, and at the appointed time, participants simply dial the specified directory number to join the conference. A tone plays each time a participant joins the conference. Participants can leave the conference by hanging up the conference call. As long as at least two participants remain on the bridge, a conference continues even if the conference controller hangs up.

NOTE Conference devices configured only for software support G.711 codecs. Some hardware conference devices, such as Catalyst 6608, support G.711, G.723, and G.729, while others, such as the Catalyst 4000, support only G.711. Any of the G.711-only conference devices can support G.723 and G.729 using a separate transcoding resource.

Configuring ConfBrs

ConfBrs are configured individually in Cisco CallManager Administration and can be shared among the cluster. Use the following quick steps for configuring ConfBrs.

Step 1 Open the Conference Bridge Configuration page in Cisco CallManager Administration (click **Service > Conference Bridge**). Figure 4-1 shows the Conference Bridge Configuration page.

Step 2 Examine the Conference Bridge Type. The following are three ConfBr types:

— Cisco Conference Bridge Software

— Cisco Conference Bridge Hardware

— Cisco IOS Software Conference Bridge

Figure 4-1 *Conference Bridge Configuration*

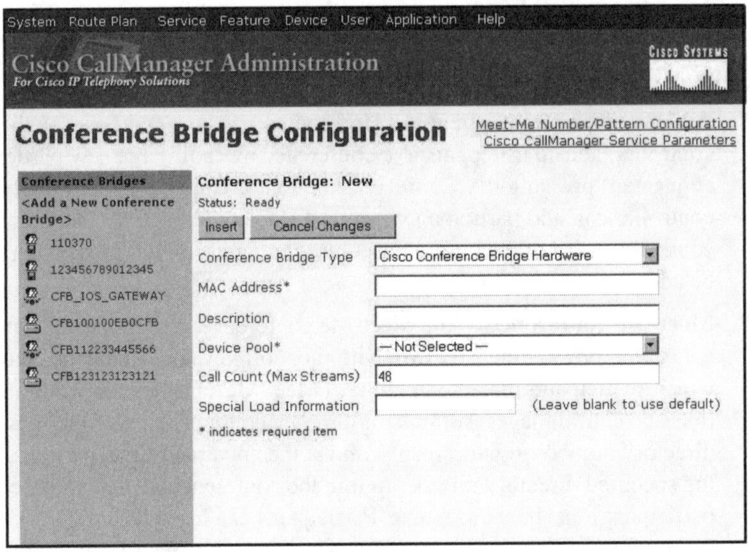

Step 3 View existing ConfBrs. Select a ConfBr device from the left column and view the settings of this ConfBr resource.

Step 4 Add a new ConfBr. Select **Add a New Conference Bridge** from the left column. By default, the ConfBr type will be "software." Select **hardware** for the ConfBr type.

Step 5 Enter the MAC Address. If you do not know the MAC address and need to determine it, you may need to access the CLI of the resource. For a Catalyst 6000 Voice T1 (or E1) and Services module, type **show port** *mod/port* from the global configuration mode. For a Catalyst 4000 Access Gateway module, the command is **show voicecard conference**. Determine the MAC address of the port you wish to utilize and enter that address into the MAC Address field.

Step 6 In the Description field, enter a description, such as the location of the hardware or software (for example, San Jose or Dallas2).

Step 7 In the Device Pool field, select a device pool from the drop-down menu.

Step 8 In the Call Count (Max Streams) field, enter the number of calls the ConfBr can handle.

Step 9 Select **Insert**. The new HW ConfBr has been inserted into the Cisco CallManager database.

Media Termination Point (MTP)

An MTP is a software device that enables supplementary services to calls routed through an H.323v1 gateway. Supplementary services are features, such as hold, transfer, call park, and conferencing, that are otherwise not available when a call is routed to an H.323v1 endpoint. H.323v1 endpoints do not support empty capability sets and, as such, require an MTP to provide supplementary services. H.323v2 endpoints support empty capability sets, enabling Cisco CallManager to extend supplementary services without the use of an MTP.

TIP If you are interested in learning detailed information about the way Cisco CallManager processes calls, read the book, *Cisco CallManager Fundamentals* (Cisco Press, ISBN: 1-58705-008-0)

Figure 4-2 shows how an MTP works with an H.323v1 gateway.

Figure 4-2 *MTP*

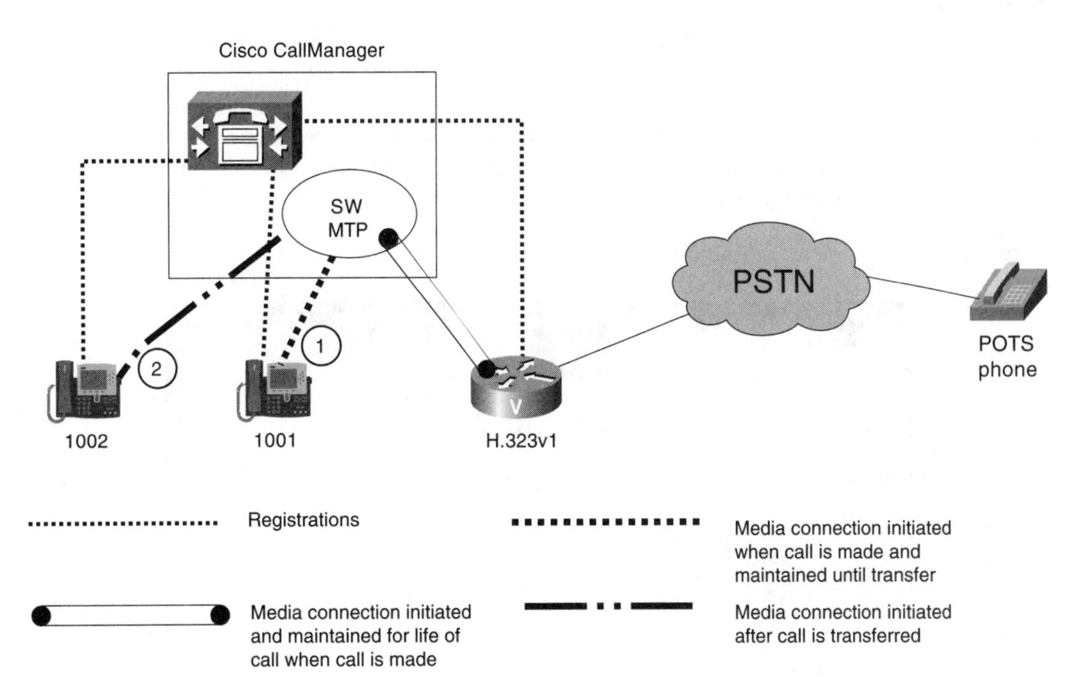

The Cisco IP Voice Media Streaming Application must be installed and running on the server on which the MTP is configured. This application is common to both the MTP and ConfBr applications. Under Windows 2000, the application runs as a service.

There are two ways to add an MTP. Both methods utilize a software resource to provide the functionality.

- An MTP is automatically added if you choose to install the optional component, Cisco IP Voice Media Streaming Application, during the automated installation of Cisco CallManager.
- You can manually install the Cisco IP Voice Media Streaming Application on a networked server and configure an MTP on that server through Cisco CallManager Administration. Cisco IP Voice Media Streaming Application is part of the Cisco CallManager installation process.

NOTE The Cisco IP Voice Media Streaming Application automatically adds an MTP device and software ConfBr device for that server.

Configuring MTP

Step 1 MTPs are configured individually in Cisco CallManager Administration and can be shared among the cluster. Use the following quick steps for configuring MTPs. Open the Media Termination Point Configuration page in Cisco CallManager Administration (click **Service > Media Termination Point**). Figure 4-3 shows the Media Termination Point Configuration page.

Figure 4-3 *Media Termination Point Configuration*

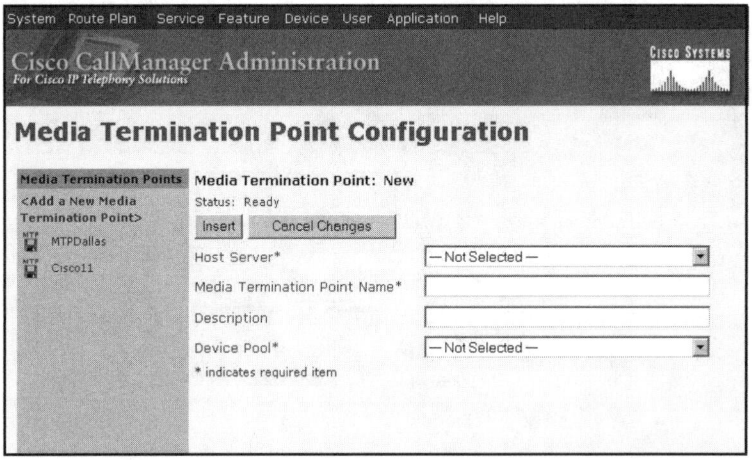

View existing MTP resources. If the Cisco IP Voice Media Streaming Application was installed with the server, there are MTP devices already configured. Select an MTP device from the left column and take note of the Host Server.

Step 2 In the left column, click **Add a New Media Termination Point**.

Step 3 In the **Host Server** field, choose a server.

Step 4 In the **Media Termination Point** field, enter a descriptive name for the MTP.

Step 5 In the **Description** field, enter a description that in future will help you recall which MTP resource this is.

Step 6 In the **Device Pool** field, choose the device pool to which you want the MTP device to belong.

Step 7 Click **Insert**. The MTP has been inserted into the Cisco CallManager database.

Music on Hold (MOH)

The integrated MOH feature allows users to place OnNet and OffNet users on hold with music provided from a streaming source. The MOH feature allows two types of hold:

- User hold
- Network hold, which includes hold that is automatically invoked during a transfer operation, conference hold, and hold that is automatically invoked during a call park operation

You can customize MOH so that specific recordings are played based on the directory number from which callers were placed on hold or based on the line number the caller dialed.

In user hold, MOH takes effect when, during an active call between phone A and phone B, phone A places phone B on hold. If an MOH resource is available, music is streamed from an MOH server to phone B.

An example of network hold is when phone A presses the **Call Park** button or **Park** soft key. The caller on phone B is automatically placed on hold at a specified call park directory. Another example occurs when a call is transferred. Phone A presses the **Transfer** button or **Trnsfer** soft key which causes the caller on phone B to be placed on hold while the call is being transferred. As with user hold, if an MOH resource is available, music is streamed from an MOH server to phone B while phone B is on hold as a result of being parked or transferred.

For Cisco CallManager Release 3.1(1), there are four levels of prioritized audio source IDs. Level four has the highest priority and level one has the lowest priority:

- Level four is directory number/line-based. It's for devices that have no line definition, such as gateways. The system selects the audio source IDs at this level.
- If no priority is defined in level four, the system searches any selected audio source IDs in level three. Level three is device-based.
- If no level four or level three audio source IDs are selected, the system selects audio source IDs defined in level two, which is device pool-based.
- If all the preceding levels have no audio source IDs selected, the system searches for audio source IDs in the final level, level one, which are service-wide service parameters.

An MOH server handles up to 500 simplex, Unicast audio streams. An MRG includes one or more MOH servers. An MOH server supports 51 audio sources, with one audio source sourced from a fixed device using the local computer audio driver and the rest sourced from files on the local MOH server.

You can use a single file for multiple MOH servers, but the fixed device can be used as a source for only one MOH server. The MOH audio source files are stored in the proper format for streaming. Cisco CallManager allocates the simplex Unicast streams among the MOH servers within a cluster.

After the discussion on media resource management, there is an example on the functionality of the MOH feature.

Configuring MOH

MOH services are configured individually in Cisco CallManager Administration and can be shared among the cluster. Use the following quick steps for configuring MOH.

Step 1 Open the Music On Hold (MOH) Audio Source Configuration page in Cisco CallManager Administration (click **Service > Music on Hold**). Figure 4-4 shows the Music On Hold (MOH) Audio Source Configuration page.

Step 2 Explore the Sample Audio Source. Select **SampleAudioSouce** from the left column or the MOH Audio Source File drop-down list. View the settings for this audio source.

Step 3 Working outside of Cisco CallManager Administration, copy music files. Most standard .wav and .mp3 file formats are valid audio source files. Get a .wav or .mp3 file and copy it to the directory C:\Cisco\DropMOHAudioSourceFilesHere.

Figure 4-4 *Music On Hold (MOH) Audio Source Configuration*

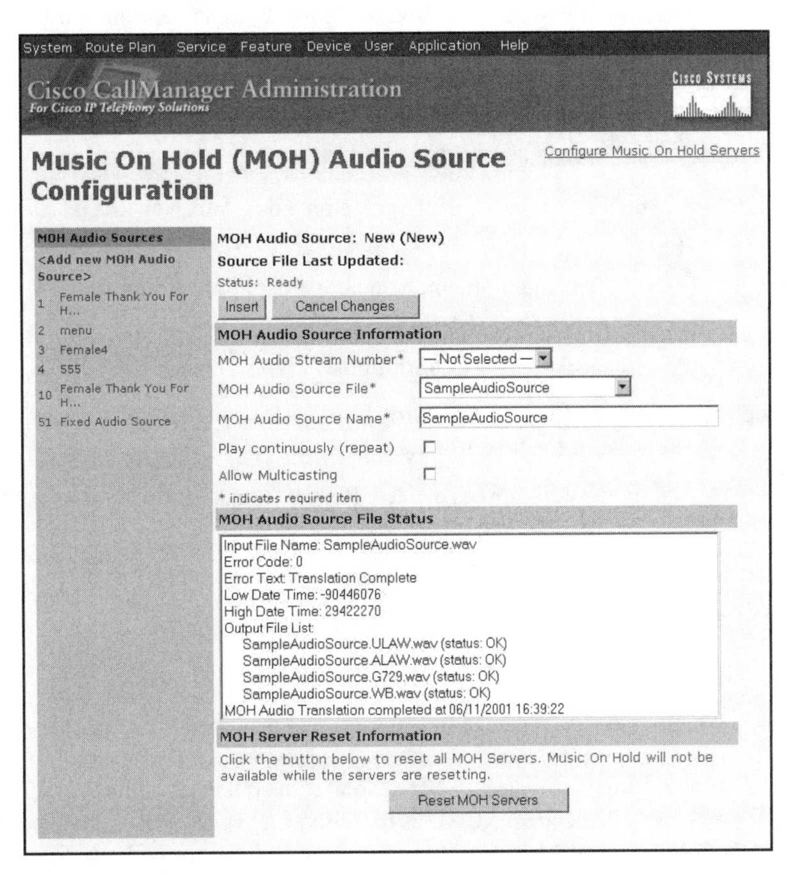

CAUTION Step 4 should not be performed during business hours. Verifying audio source translation consumes 100 percent of the CPU resources and could affect call processing.

Step 4 Wait a few moments for the translation to complete. Then verify the translation of the audio source file. Repeat Step 1 to redisplay the MOH Audio Source Configuration page.

Step 5 In the column on the left, select the file you copied or select it from the MOH Audio Source File drop-down list. You can determine whether the audio file was translated correctly by checking the File Status. The File Status should be listed as (status: OK).

Step 6 Click **Update**. The audio source has been inserted into the Cisco CallManager database.

Step 7 Add audio sources. In the left column, click **Add New MOH Audio Source**.

Step 8 In the MOH Audio Stream Number field, choose a number.

Step 9 In the MOH Audio Source File field, choose the name of the audio source file and check the **Play continuously (repeat)** box.

Step 10 Check the **Allow Multicasting** box if you want this audio source to be Multicast. If the box is not checked, the audio stream is Unicast only.

Step 11 Add more audio files as needed. Repeat steps 7–10 to add more audio files.

Step 12 Click **Insert**. The new audio source is now available.

XCODEs

An XCODE is a device that takes the output stream of one codec and transcodes (converts) it from one compression type to another compression type. For example, an XCODE can take an output stream from a G.711 codec and convert it in real time to a G.729 input stream accepted by a G.729 codec. XCODEs for Cisco CallManager transcode between G.711, G.723, G.729, and GSM codecs. In addition, an XCODE provides MTP capabilities and may be used to enable supplementary services for H.323 endpoints when required.

Figure 4-5 shows how an XCODE enables two different codecs to communicate and how it can also provide an MTP for H.323 endpoints.

Cisco CallManager invokes an XCODE on behalf of endpoint devices when the two devices are using different codecs and are normally not able to communicate. When inserted into a call, the XCODE converts the streams between the two disparate codecs to enable communications between them.

An XCODE requires specific hardware to run. For Catalyst 6608, the same hardware can support ConfBrs, XCODEs, or PRI. The Catalyst 4000 Access Gateway Module, in gateway mode, allocates some DSPs for transcoding and others for conferencing.

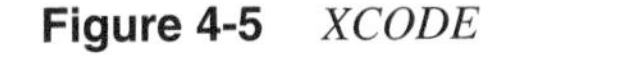

Figure 4-5 *XCODE*

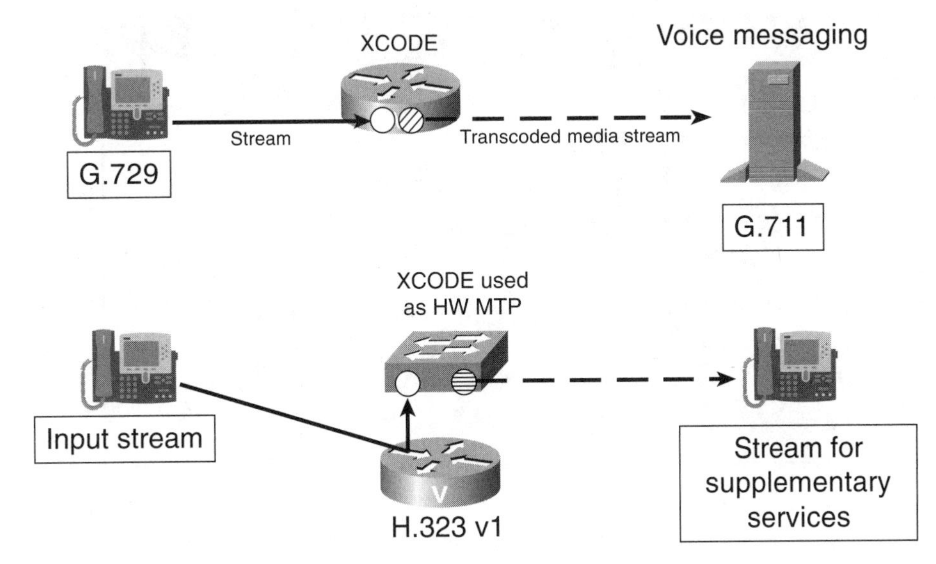

Configuring the XCODE

XCODEs are configured individually in Cisco CallManager Administration and can be shared among the cluster. Use the following quick steps for configuring an XCODE:

Step 1 Open the Transcoder Configuration page in Cisco CallManager Administration (click **Service > Transcoder**). Figure 4-6 shows the Transcoder Configuration page.

Step 2 In the Transcoder Type field, use the default, **Cisco Media Termination Point Hardware**.

Step 3 In the MAC Address field, enter the MAC address. If you do not know the MAC address and need to determine it, you may need to access the CLI of the resource. For the MAC address of the port on a Catalyst 6608 module, type **show port** [*module number/port number*] from the global configuration mode. For the Catalyst 4000 Access Gateway Module, the command is **show voicecard transcode**. Determine the MAC address of the port you wish to utilize and enter that address into the MAC Address field.

Figure 4-6 *Transcoder Configuration*

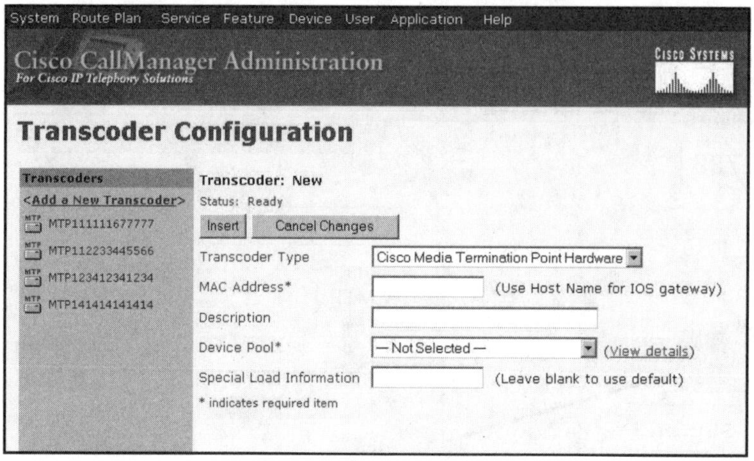

Step 4 In the Description field, enter a description that will help you identify the XCODE.

Step 5 In the Device Pool field, choose a device pool from the drop-down menu.

Step 6 Select **Insert**. The XCODE has been inserted into the Cisco CallManager database.

Media Resource Manager (MRM)

Media resources provide services such as transcoding, conferencing, MOH, and media termination. The Media Resource Manager (MRM) is an integral component of Cisco CallManager. The MRM controls and manages the media resources within a cluster, allowing all Cisco CallManagers within the cluster to share these media resources.

Figure 4-7 shows how the MRM controls all the media resources that are shared among a Cisco CallManager cluster.

The MRM enhances Cisco CallManager features by making Cisco CallManager more readily able to deploy transcoding, conferencing, and MOH services. Distribution throughout the cluster uses resources to their full potential, making them more efficient and more economical.

The following reasons explain why resources are shared:

- To enable both hardware and software devices to coexist within a Cisco CallManager.
- To enable Cisco CallManager to share and access resources available in the cluster.

- To enable Cisco CallManager to perform load distribution within a group of similar resources.
- To enable Cisco CallManager to allocate resources based on user preferences.

The MRM controls the media resources using MRGs and lists. These groups and lists are configured in a similar manner to route groups and route lists. Although they are configured the same way, the way devices use and interact with the resources is different. The following sections provide more information about MRGs and lists.

Figure 4-7 *Media Resource Management*

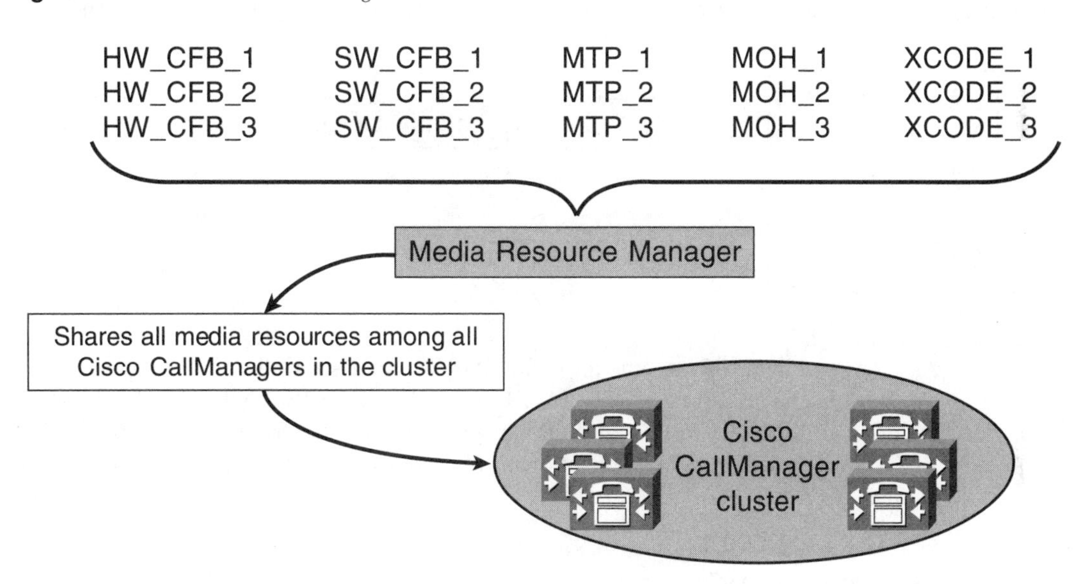

Media Resource Groups (MRGs)

Media resource groups define logical groupings of media servers. You can associate an MRG with a geographical location or a site. You can also form MRGs to control the usage of servers or the type of service (Unicast or Multicast) desired.

Figure 4-8 shows how media resources are allocated to devices when they are listed in an MRG.

Cisco CallManager provides a default list of media resources. The default list of media resources includes all media resources that have not been assigned to an MRG. Once a resource is assigned to an MRG, it is removed from the default list. If media resources have been configured but no MRGs have been defined, all media resources belong to the default list, and as such, all media resources are available to all Cisco CallManagers within a given cluster.

Figure 4-8 *Media Resource Group*

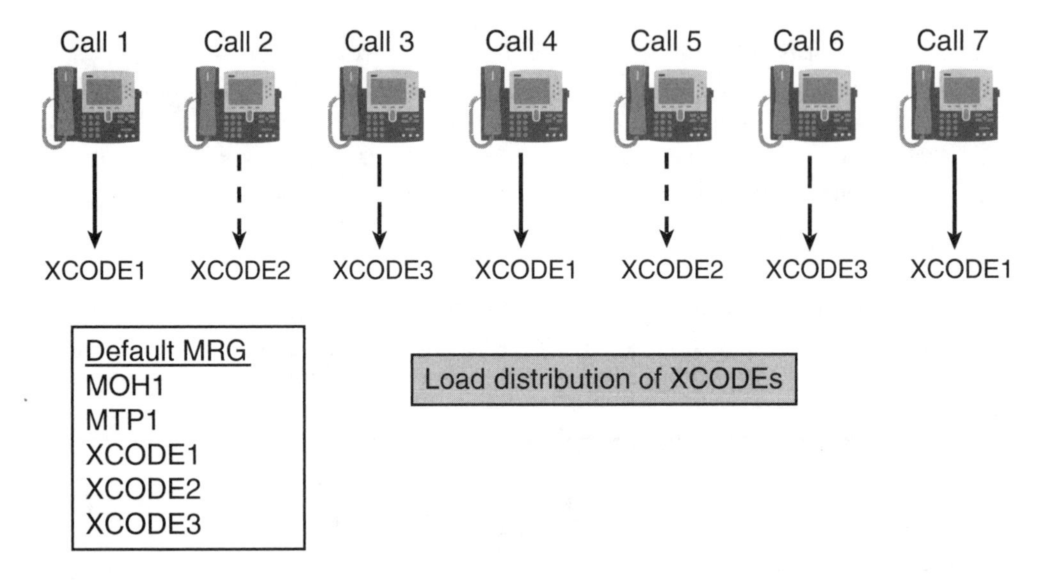

As shown in Figure 4-8, assume the default MRG for a Cisco CallManager comprises the following media resources: MOH1, MTP1, XCODE1, XCODE2, and XCODE3. For calls requiring an XCODE , this Cisco CallManager distributes the load evenly among the XCODEs in its default MRG. The following allocation order occurs for incoming calls that require XCODEs:

- Call 1 - XCODE1
- Call 2 - XCODE2
- Call 3 - XCODE3
- Call 4 - XCODE1
- Call 5 - XCODE2
- Call 6 - XCODE3
- Call 7 - XCODE1

This assumes all XCODEs use the same type of hardware.

Media Resource Group Lists (MRGLs)

Figure 4-9 shows the hierarchical ordering of media resources. It also illustrates that MRGs and MRGLs are similar to route groups and route lists.

Figure 4-9 *Media Resource Group*

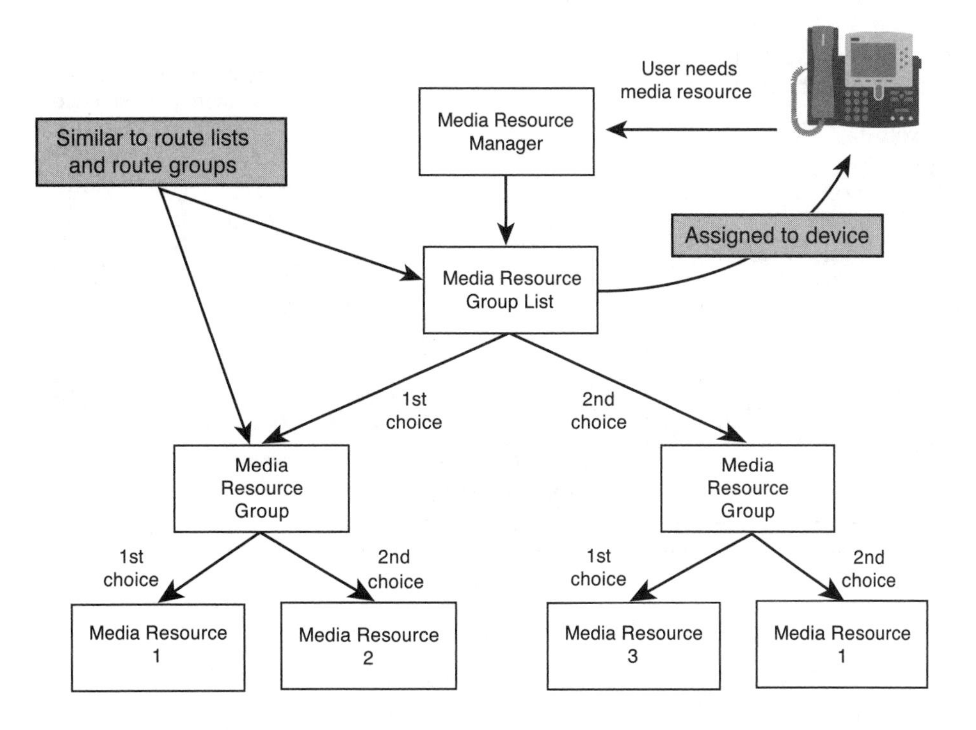

MRGLs specify a list of prioritized MRGs. An application can select required media resources among the available resources according to the priority order defined in the MRGL. MRGLs, which are associated with devices, provide MRG redundancy.

Cisco CallManager provides a default list of media resources. When a device needs a media resource, it searches its own MRGL first. If none are available, the device searches the MRGL for the device pool to which the device is assigned. If still none are available, the device tries to allocate a resource from the default list. The default list of media resources includes all media resources that have not been assigned to an MRG. Once a resource is assigned to an MRG, it is removed from the default list.

Example: Using an MRGL to Group Resources by Type

Figure 4-10 shows how conference resources are allocated if the resources are grouped by type and the software conference resource group is listed before the hardware conference resource group in the MRGL.

Figure 4-10 *Group Resources by Type*

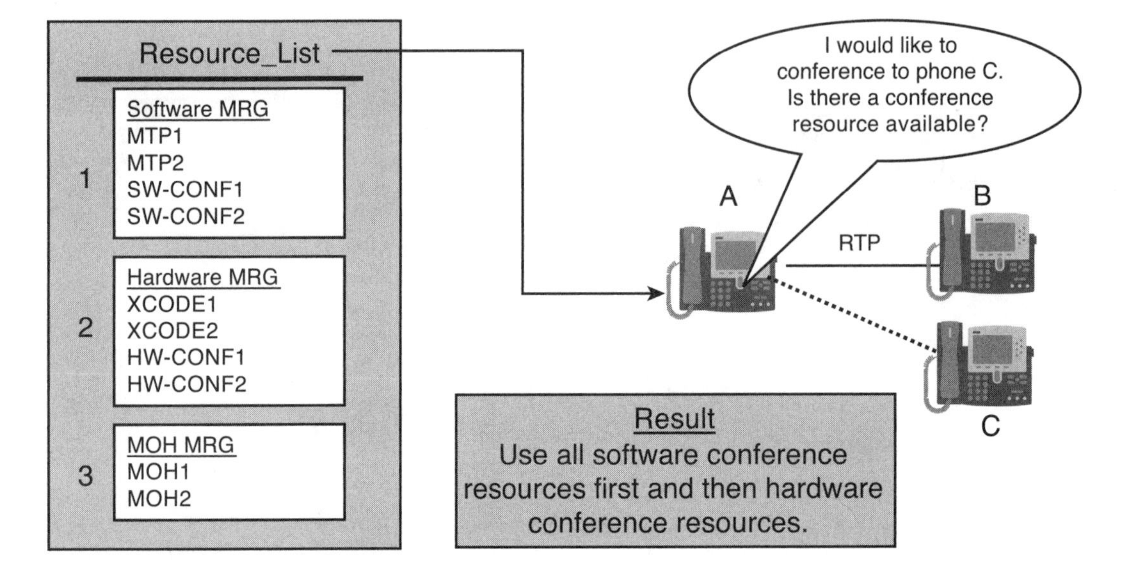

Assign all resources to three MRGs as listed:

- Software MRG: MTP1, MTP2, SW-CONF1, and SW-CONF2
- Hardware MRG: XCODE1, XCODE2, HW-CONF1, and HW-CONF2
- MOH MRG: MOH1 and MOH2

Create an MRGL called RESOURCE_LIST and assign the MRGs in this order: Software MRG, Hardware MRG, MOH MRG.

Result: With this arrangement, when a conference is needed, Cisco CallManager allocates the software conference resource first; the hardware conference is not used until all software conference resources are exhausted.

Example: Using an MRGL to Group Resources by Location

Figure 4-11 shows media resources grouped by location and how devices in those locations use the media resources in their location before using the media resources at the central site (Hub).

This example is for a multiple-site WAN deployment using centralized call processing. All Cisco CallManagers are at the central site, Hub, so all software resources are also at the central site. For devices at the Dallas and San Jose locations, it is more efficient to use media resources physically located at their location rather than using a resource across the WAN.

Figure 4-11 *Group Resources by Location*

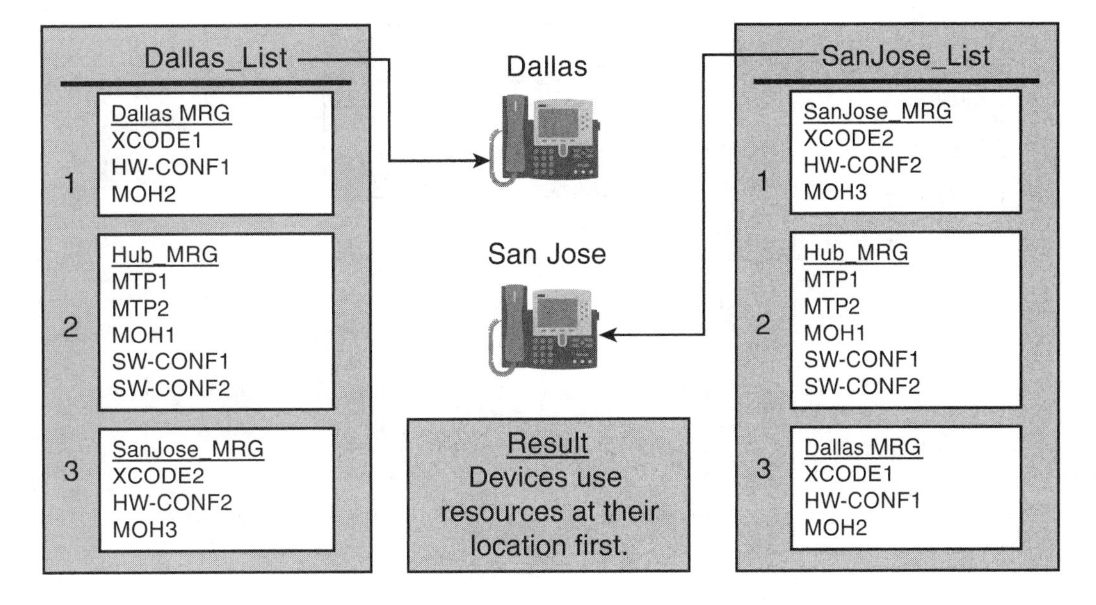

Assign resources to three MRGs as listed:

- Hub MRG: MTP1, MTP2, MOH1, SW-CONF1, and SW-CONF2
- Dallas MRG: XCODE1, HW-CONF1, and MOH2
- San Jose MRG: XCODE2, HW-CONF2, and MOH3

Create a Dallas_List MRGL and assign MRGs so that resources are available in this order: local hardware resources first (Dallas MRG); software resources second (Hub MRG); and distant hardware resources third (San Jose MRG).

Create a SanJose_List MRGL and assign MRGs so that resources are available in this order: local hardware resources first (SanJose MRG), software resources second (Hub MRG), and distant hardware resources third (Dallas MRG).

Assign a phone in Dallas to use Dallas_List and a phone in San Jose to use SanJose_List.

Result: With this arrangement, phones in Dallas use the Dallas_List resources before using the SanJose_List resources.

Example: Using an MRGL to Restrict Access to All Media Resources

Figure 4-12 shows how you can restrict media resources for a device by assigning an MRGL that has no media resources.

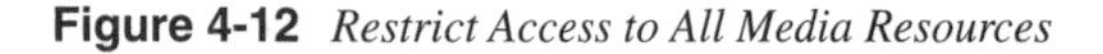

Figure 4-12 *Restrict Access to All Media Resources*

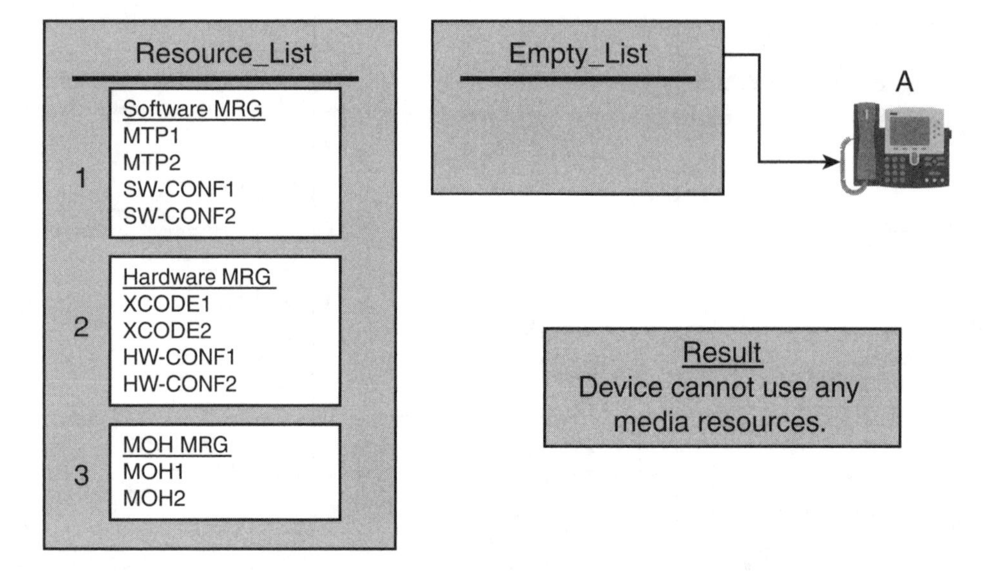

Assign resources to three MRGs as listed, assigning no resources to the MRGL that will contain the devices for which you want to restrict resource access:

- Software MRG: MTP1, MTP2, SW-CONF1, and SW-CONF2
- Hardware MRG: XCODE1, XCODE2, HW-CONF1, and HW-CONF2
- MOH MRG: MOH1 and MOH2

Create a Resource_List MRGL and assign MRGs so that resources are available in this order: Software MRG, Hardware MRG, and MOH MRG.

Create an Empty_List MRGL but do not assign any MRGs to it.

Configure a phone by assigning Empty_List as its MRGL.

Result: The phone cannot use any media resources when configured this way because none have been made available to the phone.

Example: Using an MRGL to Restrict Access to Conference Resources

Figure 4-13 shows how to restrict conference resources from devices by the way you configure your MRGs and lists.

Figure 4-13 *Restrict Access to Conference Resources*

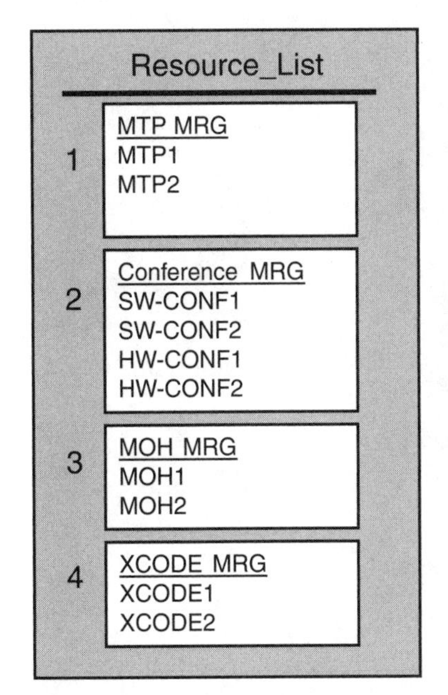

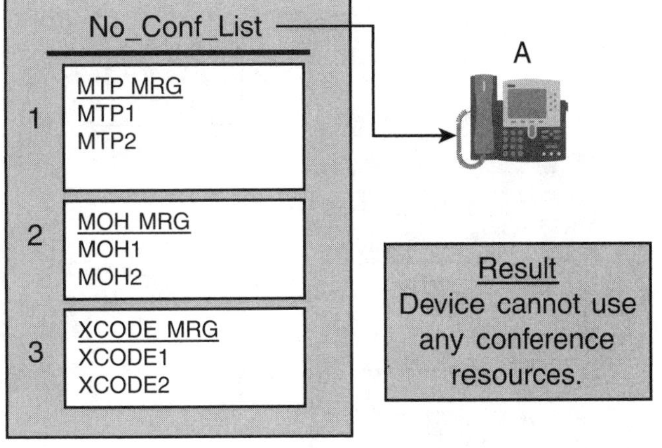

Assign all resources to four groups as listed, assigning no resources to the MRGL that will contain the devices for which you want resource access restricted:

- MTP MRG: MTP1 and MTP2
- Conf MRG: SW-CONF1, SW-CONF2, HW-CONF1, and HW-CONF2
- MOH MRG: MOH1 and MOH2
- XCODE MRG: XCODE1 and XCODE2

Create an MRGL called No_Conf_List and assign MRGs in this order: MTP MRG, XCODE MRG, MOH MRG.

In the device configuration, assign No_Conf_List as the device's MRGL.

Result: The device cannot use conference resources. Only the MTP, XCODE, and MOH resources are available to the device.

MOH Functionality

Now that we know about MRGs and lists, how does this affect the two types of MOH: user and network? Figure 4-14 shows how the MRGs and list are configured and the following sections show which music the held party hears based on the type of hold and the server to which the held device is assigned based on its MRGs and MRGLs.

Figure 4-14 *MOH Functionality*

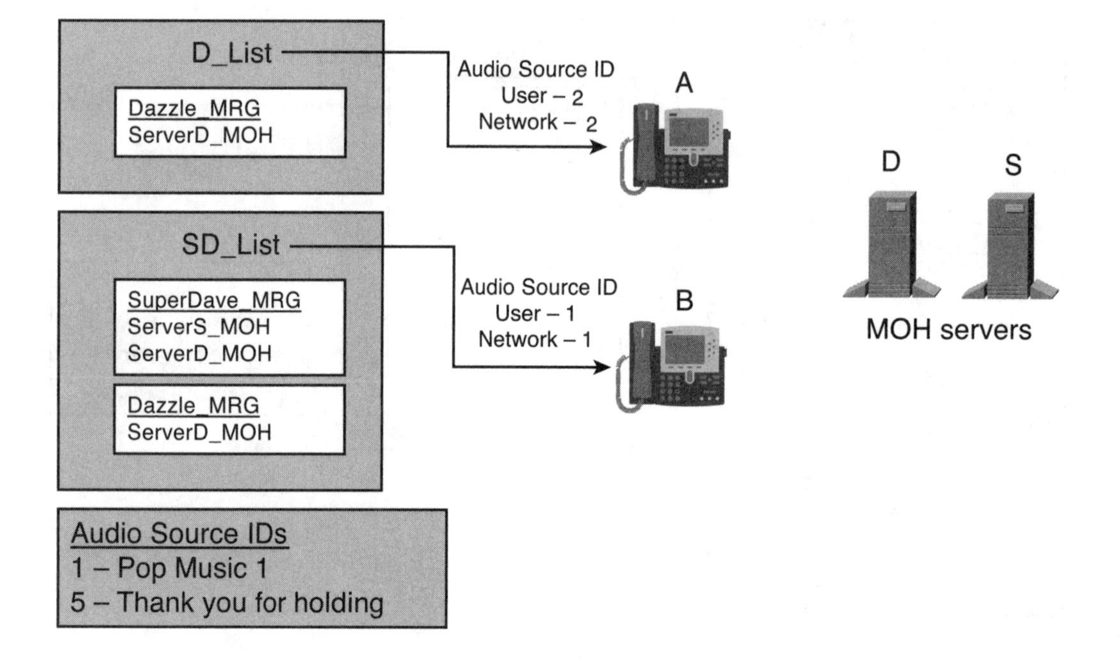

To demonstrate a simple example of MOH functionality, this figure shows the following actions:

1 Media resource groups have been configured.

— Dazzle_MRG comprises Server_D.

— SuperDave_MRG comprises Server_S and Server_D.

2 Media resource group lists have been configured.

— D_List comprises Dazzle_MRG.

— SD_List comprises SuperDave_MRG and Dazzle_MRG (prioritized order).

3 Configure a level one audio source ID.

4 Configure a level two audio source ID and level one MRGL.

5 Configure a phone with level two MRGLs and level three audio source IDs.

— Assign phone A audio source ID 5, "Thank you for holding" (for both user and network hold), and D_List.

— Assign phone B audio source ID 1, Pop Music 1 (for both user and network hold), and SD_List.

6 Configure a gateway with level two MRGLs and level three audio source IDs.

User Hold Example

This section shows how MOH functions when the user initiates the hold. Figure 4-15 shows a user placing a call on hold and which audio source the held device hears, based on the MRGL to which it is assigned.

Figure 4-15 *User Hold Example*

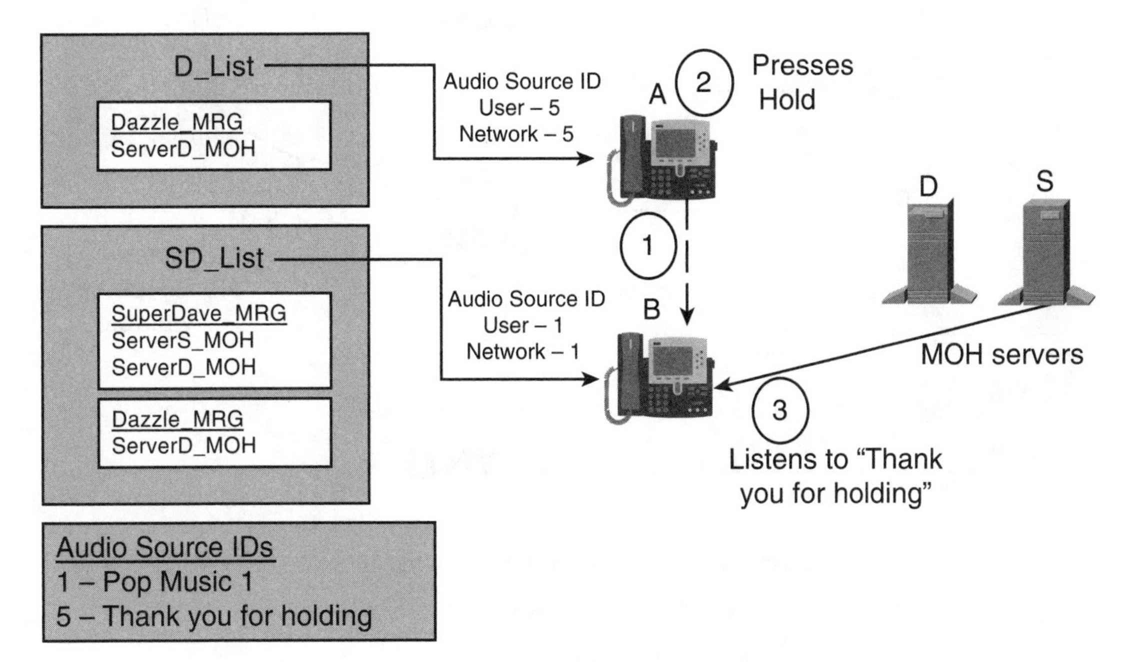

The sequence of actions illustrated in Figure 4-15 is as follows:

1 Phone A calls phone B, and phone B answers.

2 Phone A caller presses the **Hold** soft key.

The RTP stream stops and phone B refers to the MRM for its MRGL.

3 Phone B hears "Thank you for holding" streaming from MOH server S.

This happens because phone A dictated which audio source ID (5) to listen to and phone B goes to MOH server S because that is the MRGL to which phone B is assigned.

Network Hold Example 1

This section shows how MOH functions when the network initiates the hold. Figure 4-16 shows a user transferring a call and from which MOH server the held device gets the audio, based on the MRGL to which it is assigned.

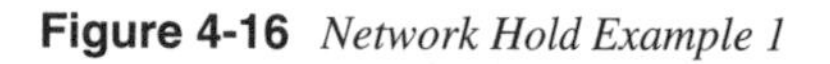

Figure 4-16 *Network Hold Example 1*

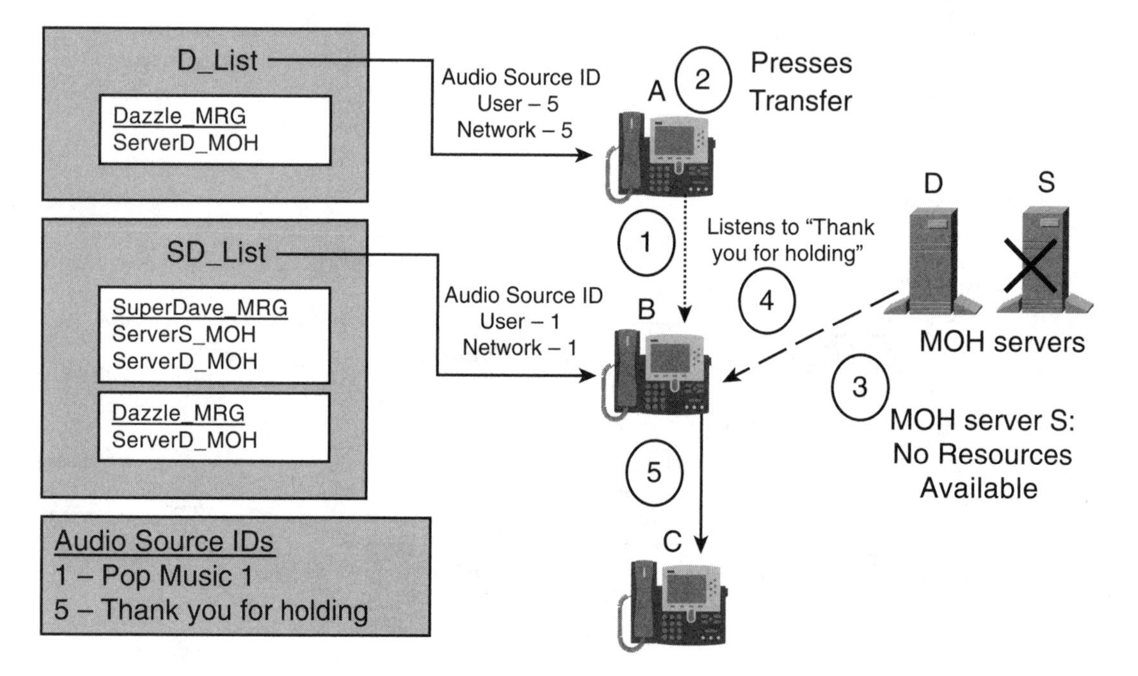

The sequence of actions illustrated in Figure 4-16 is as follows:

1. Phone A calls phone B, and phone B answers.
2. Phone A caller presses the **Trnsfer** soft key.

   The RTP stream stops and phone B refers to the media resource manager for its MRGL.
3. MOH server S has no available streams, but MOH server D does.
4. Phone B hears "Thank you for holding" streaming from MOH server D.
5. After phone A completes the transfer action, phone B disconnects from the music stream and is redirected to phone C, the transfer destination.

Network Hold Example 2

This is another example of network hold. Figure 4-17 shows a user parking a call and from which MOH server the held device gets the audio based on the MRGL to which it is assigned.

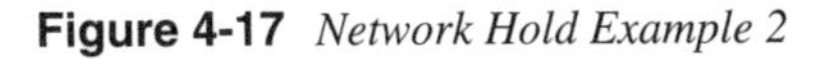

Figure 4-17 *Network Hold Example 2*

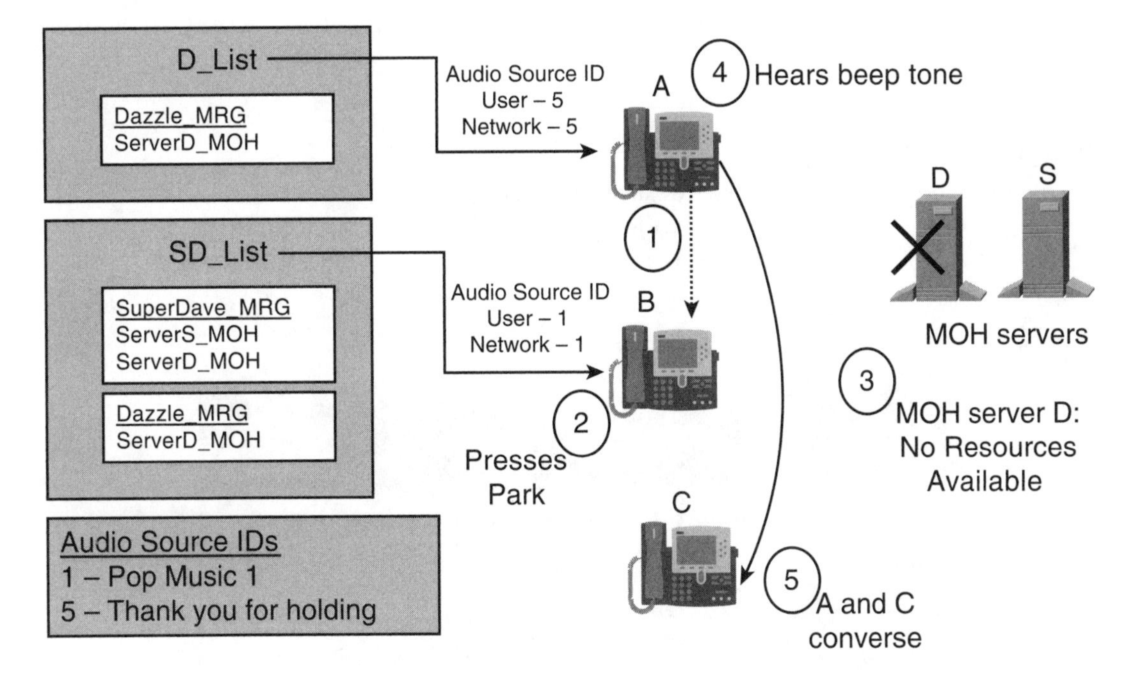

The sequence of actions illustrated in Figure 4-17 is as follows:

1. Phone A calls phone B, and phone B answers.
2. Phone B caller presses the **Park** soft key.
3. MOH server D has no available streams.
4. Phone A hears a beep tone.
5. Phone C retrieves the parked call. The call from phone A is redirected to phone C and phone A and phone C are conversing.

Configuring MRGs and MRGLs

The MRM controls media resources by way of MRGs and MRGLs.

The MRM makes Cisco CallManager more readily capable of deploying transcoding, conferencing, and MOH services. Distribution throughout the cluster uses resources to their full potential, making them more efficient and more economical.

The following are quick steps to configure ConfBrs, MTPs, MOH services, and XCODEs into MRGs and MRGLs.

Configuring MRGs

Step 1 Open the Media Resource Group Configuration page in Cisco CallManager Administration (click **Service > Media Resource Group**). Figure 4-18 shows the Media Resource Group Configuration page.

Figure 4-18 *Media Resource Group Configuration*

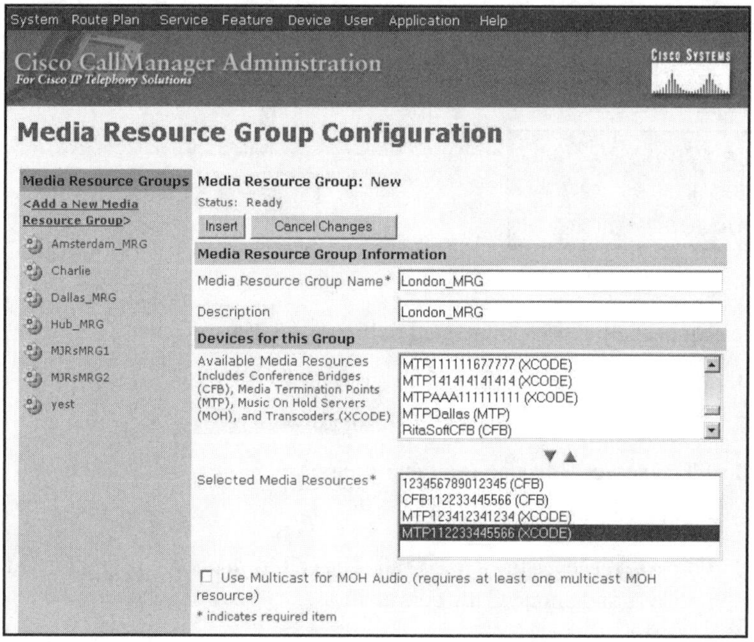

Step 2 Configure an MRG. Enter a name, provide a description, and then select available media resources and add them to the Selected Media Resources box in the desired priority.

Step 3 To use Multicast for MOH audio, check the **Use Multicast for MOH Audio** box. If you check this box, make sure that at least one of the Selected Media Resources is a Multicast MOH server.

Step 4 Click **Insert**.

Step 5 You can add another MRG by selecting **Add a New Media Resource Group** in the left column or by copying an existing MRG and modifying the properties.

Configuring MRGLs

Step 1 Open the Media Resource Group List Configuration page in Cisco CallManager Administration (click **Service > Media Resource Group List**). Figure 4-19 shows the Media Resource Group List Configuration page.

Figure 4-19 *Media Resource Group List Configuration*

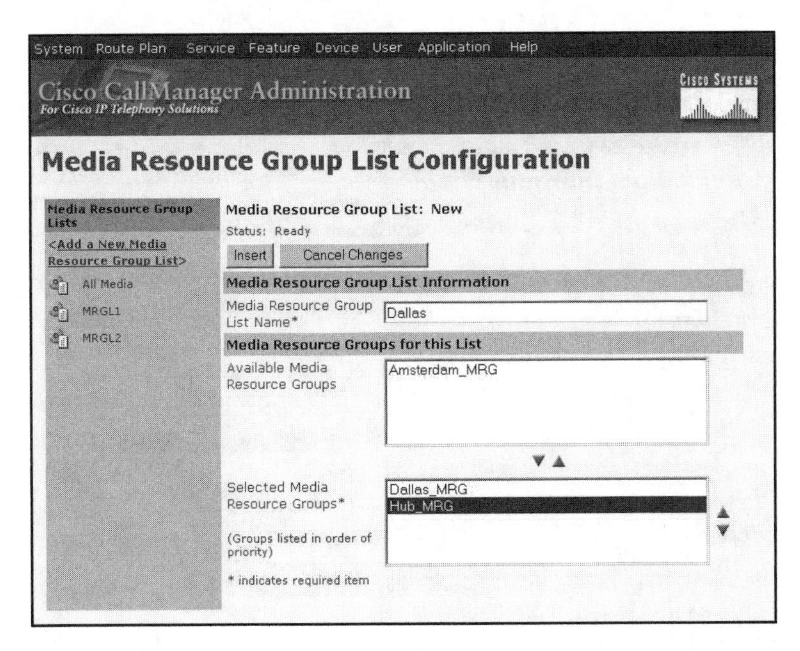

Step 2 Configure a MRGL. Enter a name, and then add available MRGs to the Selected Media Resource Group box and prioritize your groups.

Step 3 Click **Insert**.

Step 4 You can add another MRGL by selecting **Add a New Media Resource Group List** in the left column or by copying an existing MRGL and modifying the properties.

Specifying an MRGL for Devices

You can assign MRGLs to devices in bulk by assigning the MRGL to a device pool used by many devices or by assigning the MRGL on a device-by-device basis using the configuration page for each device.

The following steps describe how to assign an MRGL to a group of devices by defining the MRGL in a device pool used by those devices.

Step 1 Open the Device Pool Configuration page in Cisco CallManager Administration (click **System > Device Pool**). Figure 4-20 shows the Device Pool Configuration page.

Figure 4-20 *Device Pool Configuration*

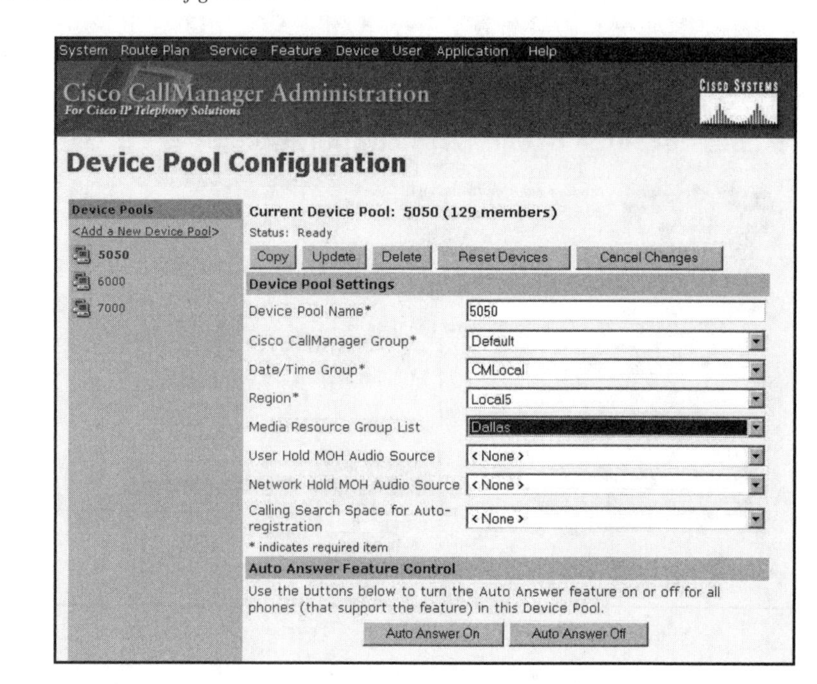

Step 2 In the column on the left, select the device pool used by Cisco IP Phones to which you want to assign a specific MRGL.

Step 3 In the Media Resource Group List field, choose an MRGL from the drop-down list.

Step 4 Click **Update**.

The following steps describe how to specify an MRGL for a single device.

Step 1 Open the Phone Configuration page in Cisco CallManager Administration (click **Device > Phone**). Figure 4-21 shows the Phone Configuration page.

Figure 4-21 *Phone Configuration*

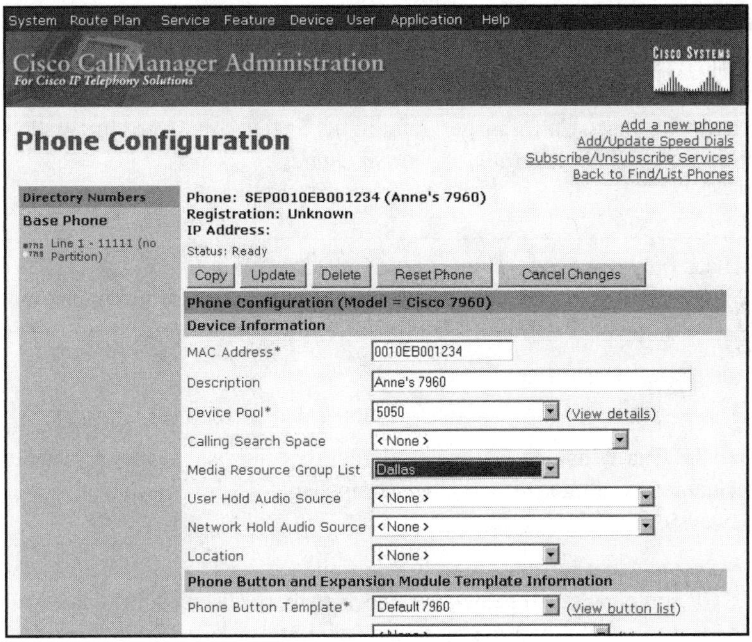

Step 2 Find a Cisco IP Phone and, in the Phone Configuration page, select an MRGL from the drop-down list for Media Resource Group List.

Step 3 Click **Update**.

Step 4 Click **Reset Phone** for the change to take effect.

Services That Require Advanced Configuration

This section describes the following services in the Services menu:

- CMI
- Cisco TFTP
- Cisco WebAttendant
- Service Parameters

These services are out of the scope of basic configuration and are discussed in Part III, "Advanced Configuration," of this book. The following sections provide some basic details about these services. Refer to Part III for more information.

Cisco Messaging Interface (CMI)

The CMI service enables you to interface Cisco CallManager Release 3.0 and later with an SMDI-compliant voice mail system. For detailed information on integrating a voice mail system with Cisco CallManager, refer to the SMDI Voice Mail Integration section in the online help, *Cisco CallManager System Guide*.

NOTE CMI can only reside on a single Cisco CallManager within a cluster. You cannot have two CMI services connected to the same voice mail system running at the same time.

Cisco TFTP

Cisco TFTP, a Windows NT service, builds .cnf files and serves embedded component executables and ringer files to CIPT components. These components include phones, gateways, and DSP resources.

A .cnf contains a prioritized list of Cisco CallManagers for a device (phones and gateways), the TCP port on which the device connects to those Cisco CallManagers and an executable load identifier. Configuration files for Cisco IP Phone 7960 and 7940 models also contain URLs for the messages, directories, services, and information buttons. Configuration files for gateways contain all the configuration information for the gateway.

Configuration files may be in a .cnf format or a .cnf.xml format, depending on the device type and your TFTP service parameter settings. When you set the BuildCNFFlag service parameter to True, the TFTP server builds both .cnf.xml and .cnf format .cnf files for devices. When you set the parameter to False, the TFTP server builds only .cnf.xml files for devices.

Cisco WebAttendant

Cisco WebAttendant, a plug-in, client-server application installed to a PC, enables a user, administrative assistant, or a receptionist to manage incoming and outgoing phone calls using a Web-based interface. Cisco WebAttendant provides quick directory access to look up phone numbers, the ability to monitor line states, and direct calls. Used in conjunction with a Cisco IP Phone, Cisco WebAttendant enables you to set up a Microsoft Windows-based PC as an attendant console.

The Cisco WebAttendant client can be installed on a PC with IP connectivity to the Cisco CallManager system. The client works with a Cisco IP Phone that is registered to a Cisco CallManager system (one client for each phone that will be used as an attendant console). Multiple clients can connect to a single Cisco CallManager system.

The Cisco WebAttendant client registers with and receives call dispatching services from the TCD services on Cisco CallManager.

Service Parameters

Service parameters for Cisco CallManager enable you to configure different services on selected servers. You can insert and delete services on a selected server, as well as modify the service parameters for those services. When a service is deleted, the associated NT service is also removed from the system. However, if a service is added, the associated NT service must be added separately.

CAUTION Some changes to service parameters may cause system failure. Cisco recommends that you do not make any changes to service parameters unless you fully understand the feature that you are changing or unless the Cisco TAC specifies the changes.

Summary

In this chapter you learned about the components of the Service menu in Cisco CallManager Administration and how to configure media resources into groups and lists.

Media resources are configured individually in Cisco CallManager Administration and can be shared among the cluster. The following are the media resources that can be configured in a Cisco CallManager:

- ConfBrs
- MTP
- MOH
- XCODE

The MRM controls resources using MRGs and MRGLs.

The MRM makes Cisco CallManager more readily able to deploy transcoding, conferencing, and MOH services. Distribution throughout the cluster uses resources to their full potential, making them more efficient and more economical.

Post-Test

Use this section to test yourself on how well you learned the concepts discussed in this chapter. Use the following scale to decide whether to re-read, skim this chapter, or go on to the next chapter. If you answer one to three questions correctly, we recommend that you re-read the chapter. If you answer four to six questions correctly, we recommend you skim the headers and read the sections related to the questions you missed. If you answer seven to nine questions correctly, go on to the next chapter. You can find the answers to the post-test for this chapter in Appendix B, "Answers to Chapter Pre-Test and Post-Test Questions."

1. Name the two types of conferences and briefly describe each type.

   ____

   ____

   ____

2. What is the supported codec for ConfBrs?

   ____

   ____

   ____

3. Supplementary services on H.323v1 gateways are enabled through which media resource?

   ____

   ____

   ____

4. What is the name of the application that provides for software conferencing and MTP?

   ____

   ____

   ____

5 Name the two types of hold that can use MOH.

____

____

____

6 Describe an XCODE. What does it do or provide?

____

____

____

7 Media resource groups and lists have the same type of relationship and manner of configuration as what two other components in CCMAdmin?

____

____

____

8 If you have software ConfBr resources listed before hardware ConfBr resources in a MRG, in what order does the media resource manager allocate ConfBr resources?

____

____

____

9 Where in Cisco CallManager Administration do you go to assign a bulk number of devices to a MRGL?

____

____

____

Upon completing this chapter you will be able to perform the following tasks:

- Given a Cisco CallManager cluster and Cisco IP Phones, configure the call park feature to park a call from one phone and retrieve the same call from another phone.
- Given the two types of call pickup, identify and define the characteristics of each type of call pickup.
- Given a Cisco CallManager cluster and Cisco IP Phones, configure the call pickup feature and pickup calls within the same group and from another group.
- Given a Cisco CallManager cluster, Cisco IP Phones, and Cisco IP Phone services, configure Cisco IP Phones services and access those services from the Cisco IP Phones.
- Given a Cisco CallManager cluster, add users to the embedded directory and access the Cisco IP Phone User Options Web page.

CHAPTER 5

Cisco CallManager Administration Feature and User Menus

This chapter discusses the Feature and User menus in CCMAdmin. The menu items on the Feature menu are call park, call pickup, Cisco IP Phone services, and Meet-Me number/pattern. The menu items on the User menu are add a user and global directory. This chapter addresses the following topics:

- Abbreviations
- Feature Menu
- Call Park
- Call Pickup
- Cisco IP Phones Services
- Meet-Me Number/Pattern
- User Menu
- Add a New User
- Global Directory

Pre-Test

Do you already know this information? The pre-test is designed to help you gauge your knowledge about this chapter. Of the 10 questions, if you answer one to three questions correctly, we recommend that you read through this chapter. If you answer four to seven questions correctly, we recommend that you skim through this chapter, reading those sections that you need to know more about. If you answer 8 to 10 questions correctly, you probably understand this information well enough to skip this chapter. You can find the answers to the pre-test for this chapter in Appendix B, "Answers to Chapter Pre-Test and Post-Test Questions."

1 What is the benefit of call park?

____

____

____

2 After configuring a call park directory number or range of directory numbers, how many calls can be parked at a given call park extension?

____

____

____

3 Describe the two types of call pickup.

____

____

____

4 Describe what Cisco IP Phone services provide.

____

____

____

5 Which two Cisco IP Phone models support phone services?

____

____

____

6 Name three of the Cisco AVVID IP Telephony components that access user information you configure in CCMAdmin?

____

____

____

7 Assuming you are using a Cisco IP Phone model 7940, list four of the seven items a user can select from the Cisco IP Phone User Options Web page.

____

____

____

8 What do you need to do in CCMAdmin to ensure that users can access the Cisco IP Phone User Options Web page?

____

____

____

9 Once phone services have been subscribed to a given phone, how does the user access those services?

____

____

____

10 Name the two Web pages through which a system administrator can subscribe to Cisco IP Phone services.

____

____

____

Abbreviations

This section defines the abbreviations used in this chapter. For more information about terms and abbreviations used in this chapter refer to the *IP Telephony Network Glossary* at the following URL:

www.cisco.com/univercd/cc/td/doc/product/voice/evbugl4.htm

Table 5-1 provides the abbreviation and the complete term.

Table 5-1 *Abbreviations and Complete Terms Used in the Chapter*

Acronym	Complete Term
CCMAdmin	Cisco CallManager Administration
HTTP	Hypertext Transfer Protocol
IIS	Microsoft Internet Information Service
LCD	liquid crystal display
LDAP	Lightweight Directory Access Protocol
PIN	personal identification number
TCP/IP	Transmission Control Protocol/Internet Protocol
XML	extensible markup language

Feature Menu

Use the Feature menu in CCMAdmin to configure the following features:

- Call park
- Call pickup/call pickup groups
- Cisco IP Phone services
- Meet-Me number/patterns

The following sections discuss these features in detail.

Call Park

The call park feature enables you to place a call at a specified directory number so that it can be retrieved from another phone in the same cluster in the system. Call park requires either a single directory number or a range of directory numbers to be configured as call park directory numbers in CCMAdmin. Only one call can be parked at each call park extension number.

Call park can be used in many ways. A good example of implementing the call park feature is in a department store with an overhead system. A call comes in to a cashier desk for an employee on the floor. The cashier can park the call, announce the call park code on the overhead and the employee on the floor can pick up the call using a nearby phone.

Figures 5-1 and 5-2 show an example of using the call park feature. Figure 5-1 indicates what happens when the call is parked.

Figure 5-1 *Call Park, Steps 1–3*

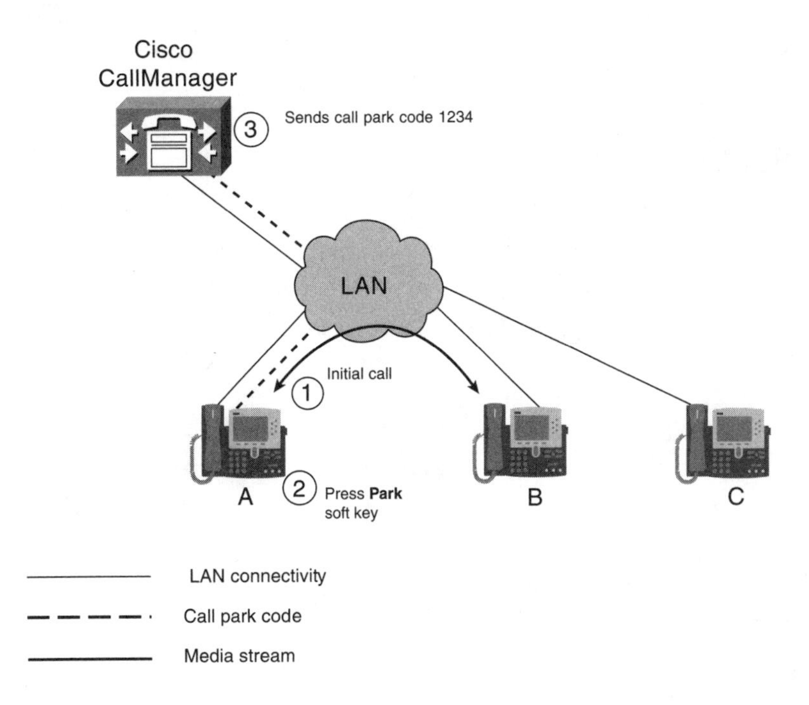

Based on the process shown in Figure 5-1, the following are the steps for using the call park feature:

1 User on phone A calls phone B.

2 User on phone A wants to take the call in a conference room for privacy. Phone A presses the **Park** soft key.

3 The Cisco CallManager to which phone A is registered sends the first available call park directory number, 1234, to be displayed on phone A. The user on phone A watches the display for the call park directory number.

The call park example concludes in Figure 5-2, which indicates what happens when the call is retrieved from park.

4 The user on phone A leaves her office and walks to an available conference room. The phone in the conference room is designated phone C. The user goes off-hook on phone C and dials 1234 to retrieve the call.

5 The call is established between phones C and B.

Figure 5-2 *Call Park, Steps 4 and 5*

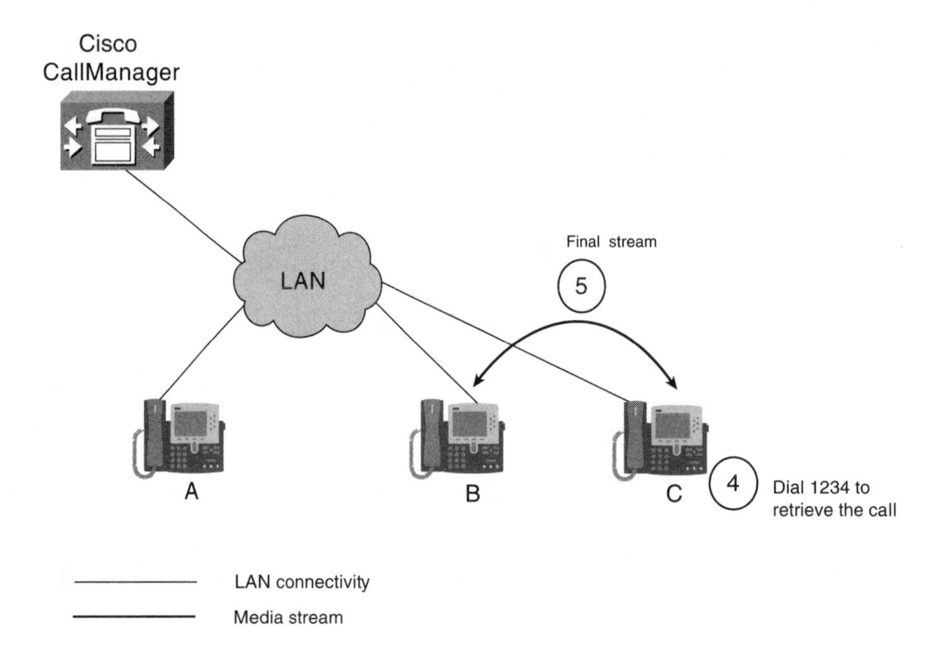

Configuring Call Park

The following are quick steps to configure call pickup:

Step 1 Open the Call Park Configuration page in CCMAdmin (click **Feature > Call Park**). Figure 5-3 shows the Call Park Configuration page.

Step 2 Enter a call park number or range.

Step 3 Choose a partition. If you want to use a route partition to restrict access to the call park numbers, choose the desired route partition from the drop-down list. If you do not want to restrict access to the call park numbers, choose **None** for the route partition. Call park only works for phones that are both in the same partition (the phone parking the call and the phone retrieving the call must both have the same calling search space).

The combination of the call park extension number and route partition are unique within the Cisco CallManager cluster.

Figure 5-3 *Call Park Configuration*

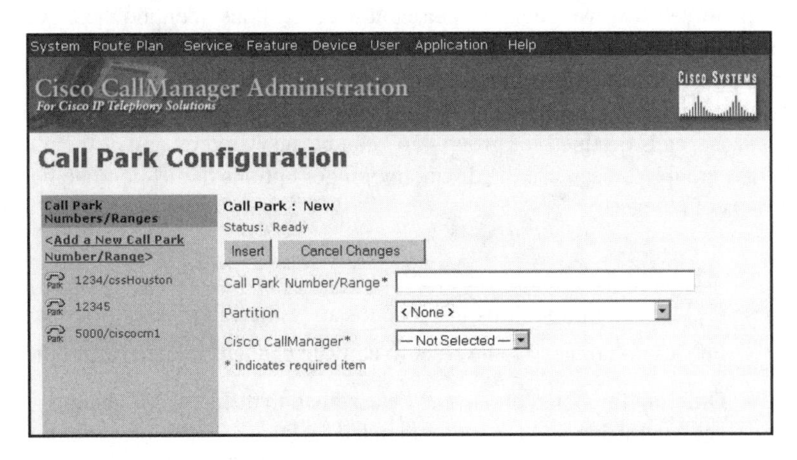

Step 4 Choose the Cisco CallManager to which this number or range will apply.

Step 5 Click **Insert**.

Step 6 Configure another call park number range and assign it to a different Cisco CallManager. Select **Add a New Call Park Number/Range** from the left column and repeat Step 2.

Call Pickup

Call groups enable users who have been configured as part of a group to answer calls that come in on a directory number other than their own. The features that use these call groups are as follows:

- **Call pickup**—Enables users to pick up incoming calls on any phone within their own group. When the user presses the **Call Pickup** button or **PickUp** soft key, Cisco CallManager automatically dials the appropriate call pickup group number.
- **Group call pickup**—Enables users to pick up incoming calls within their own group or in another group. Users press the **Group Call Pickup** button or **GPickUp** soft key and dial the appropriate call pickup group number.

The same procedures apply for configuring both of these features. Group call pickup numbers apply to lines or directory numbers.

The purpose of call pickup is to enable a group of users who are seated near each other (within hearing of the ringing phone) to handle incoming calls as a group. Whoever is available to take the next call does so. Only phones that are configured in a pickup group can use these features.

A good example of call pickup is for a sales group in an organization. For example, a company sells two widgets: one geared toward the consumer and the other for enterprises. The sales staff is broken into two call pickup groups: 1234 for consumers and 4567 for enterprises. When a call comes into the bank of phones in call pickup group 1234, any one of the phones assigned to that group can answer it by pressing the **PickUp** soft key. Likewise, calls destined for group 4567 can be answered by any of the phones configured in that group. You can use the distinctive ringer options to differentiate between the two groups of phones.

TIP Some useful configuration tips when configuring the call pickup group are as follows:

- Phones do not need to be reset to activate changes related to call pickup groups.
- Different lines on a phone can be assigned to different call pickup groups but that is very confusing and not recommended.
- Design efficient groups by assigning only phones that are located near each other to one or more call pickup groups.

Call Pickup Example

Figure 5-4 shows how the call pickup feature works.

Figure 5-4 *Call Pickup*

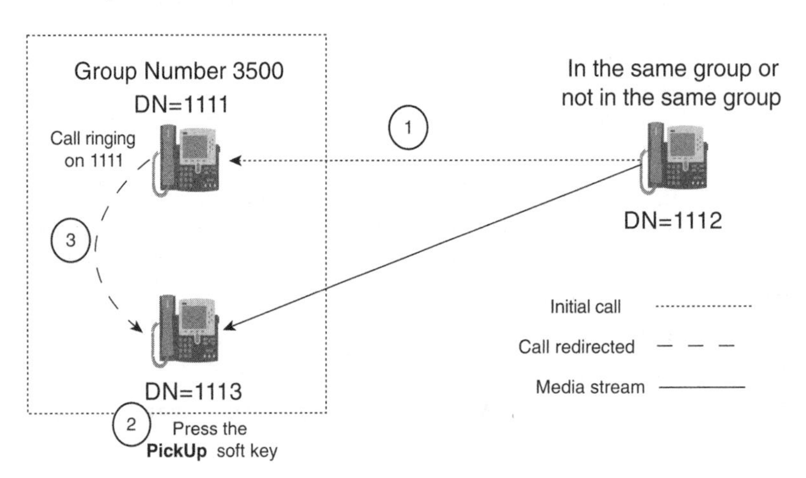

There is one call pickup group, 3500, and two directory numbers, 1111 and 1113, belong to that call pickup group. The sequence of events illustrated in Figure 5-4 is described as follows:

1. The initial call is for directory number 1111 from 1112. Whether 1112 is in the same group is unimportant.
2. Hearing the call ring on 1111 and realizing it is not being answered, the user at directory number 1113 goes off-hook and presses the **PickUp** soft key.
3. The call is redirected from 1111 to 1113.

Group Call Pickup Example

Figure 5-5 shows how the group call pickup feature works.

Figure 5-5 *Group Call Pickup*

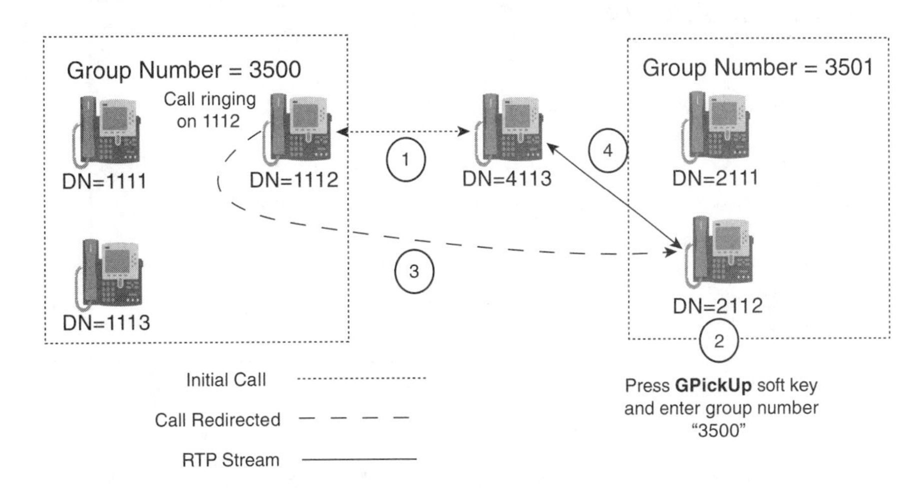

There are two call pickup groups, 3500 and 3501. Directory numbers 1111, 1112, and 1113 belong to the 3500 call pickup group and directory numbers 2111 and 2112 belong to the 3501 call pickup group. Directory number 4113 does not belong to either call pickup group. The sequence of transactions illustrated in Figure 5-5 is as follows:

1. The initial call is from directory number 4113 to 1112.
2. The user at directory number 2112 knows the user at 1112 is unavailable and plans to answer the incoming call. She goes off-hook, presses the **GPickUp** soft key, and enters the group number to which directory number 1112 belongs (3500).
3. The call is redirected to 2112.
4. The call is established between 4113 and 2112.

Call pickup groups can be configured to accommodate groups that have similar tasks. If an employee takes a break and their extension is ringing, another employee in their call pickup group can redirect the customer calls.

Call Pickup Configuration

The following are quick steps to configure call pickup:

Step 1 Open the Call Pickup Configuration page in CCMAdmin (click **Feature > Call Pickup**). Figure 5-6 shows the Call Pickup Configuration page.

Figure 5-6 *Call Pickup Configuration*

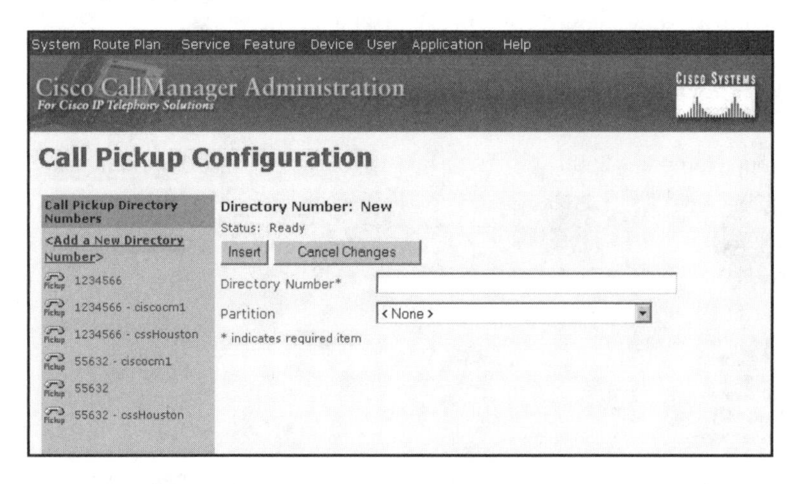

Step 2 In the Directory Number field, enter a directory number.

Step 3 In the Partition field, select a partition. Members of the same call pickup group must use the same partition. You can prohibit unauthorized phones from being able to pickup a call from your pickup group by excluding them from the partition used by the call pickup group.

Step 4 Click **Insert**.

Step 5 To configure another call pickup number, select **Add New Directory Number** from the left column and repeat Steps 2 through 4.

Step 6 Assign call pickup group numbers to a phone. To do so, find a phone (**Device > Phone**) and open the Phone Configuration page. Select a line number for that phone and scroll down to the Call Forward and Pickup Settings area on the Directory Number Configuration page. Figure 5-7 shows the Call Forward and Pickup Settings area on the Directory Number Configuration page.

Figure 5-7 *Designating a Call Pickup Group for a Phone*

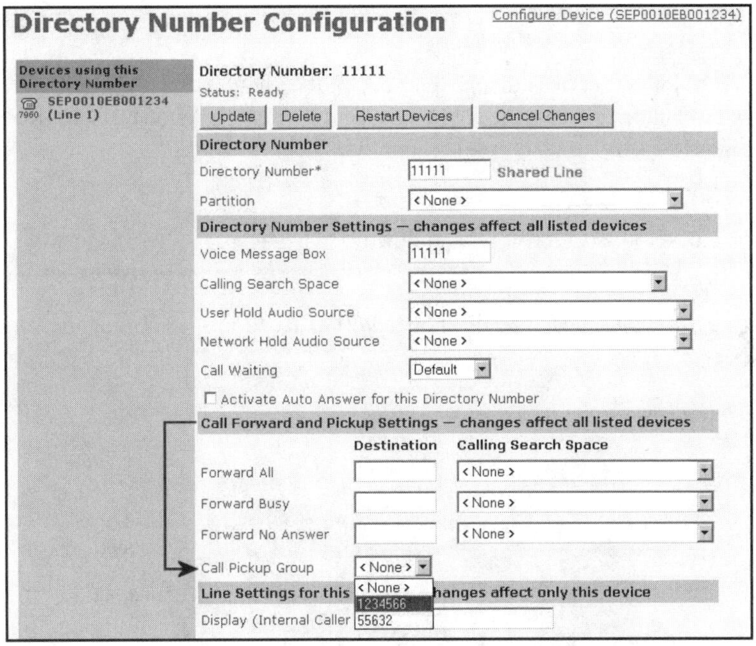

Step 7 Select a call pickup group from the **Call Pickup Group** drop-down menu.

Step 8 Click **Update**.

Step 9 Click **Restart Devices** for the change to take effect.

Cisco IP Phone Services

Use the Cisco IP Phone Services Configuration page in CCMAdmin (**Feature > Cisco IP Phone Services**) to define and maintain the list of Cisco IP Phone services to which users can subscribe. Cisco IP Phone services are XML applications that enable the display of interactive content with text and graphics on Cisco IP Phone models 7960 and 7940. You can create customized Cisco IP Phone applications for your site using the Cisco IP Phone Services Software Developer's Kit (Cisco XML SDK).

The SDK is available to registered and unregistered users at

www.cisco.com/warp/public/570/avvid/voice_ip/cm_xml/cm_xmldown.html

For information about the Cisco XML SDK, refer to the following links:

www.cisco.com/go/developersupport/
www.cisco.com/warp/public/cc/pd/unco/ippps/

NOTE Currently, only Cisco IP Phone models 7960 and 7940 support Cisco IP Phone Services.

Cisco IP Phone services display content on the phone's 133 x 65 LCD. Pressing the **services** button displays the list of services to which the phone is subscribed. The user can navigate the services and provide input via the following:

- Soft keys
- Rocker key (to scroll up and down)
- Keypad

Figure 5-8 shows the Cisco IP Phone 7960 with callouts describing the various parts of the phone.

Figure 5-8 *Cisco IP Phone 7960 with Callouts*

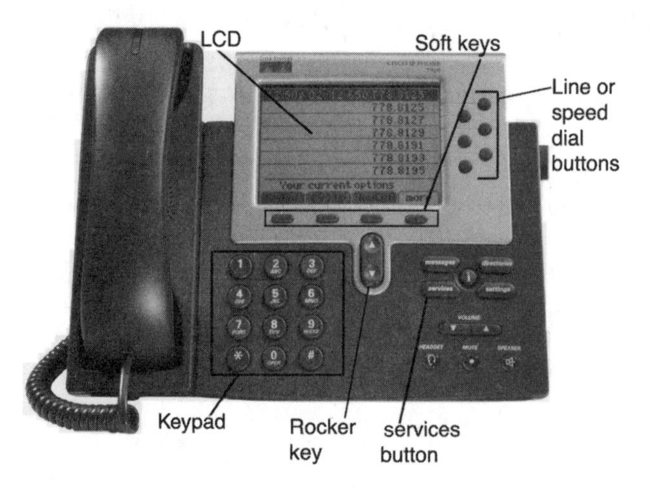

After you have configured the services in CCMAdmin, users can log on to the Cisco IP Phone User Options Web page and subscribe to services. This is discussed in the next section. Users can subscribe only to services that have already been configured in CCMAdmin. The following list describes the information that must be configured for each service:

- URL of the server that provides the content.
- Service name and description—this information helps users decide whether they want to subscribe to the service.
- *(Optional)* A list of parameters that are appended to the URL when it is sent to the server.

Parameters serve to customize a service. Examples of parameters include stock ticker symbols, city names, zip codes, or user IDs.

The following list describes some of the kinds of services that can be offered using Cisco IP Phone services.

- Weather check
- Yellow pages phone number lookup
- Mass transit schedules
- Stock ticker check
- Flight status
- Meeting room scheduler

Figure 5-9 shows examples of services displayed on the phone.

The following lists the examples in Figure 5-9, moving clockwise from the top left:

- **Menu**—A menu enables the user to scroll through the list of menu items and make selections.
- **Text**—The text from a Web page can be delivered for the user to view.
- **Input**—The input enables the user to enter information using the keypad.
- **Graphical**—The service can display graphics as well as text.
- **Directory**—The service can display directory information pulled from a database.
- **Image**—The services can deliver images in black, white, and shades of gray. Color images will be available when Cisco releases a Cisco IP Phone with a color display.

Configuring Cisco IP Phone Services

Once the service has been created using the Cisco XML SDK, you must configure it in CCMAdmin. Doing so makes it available for user subscription in the Cisco IP Phone User Options Web page.

TIP You can subscribe Cisco IP Phones to services in CCMAdmin. To do so, click **Device > Phone**, find and click on the phone you want to subscribe services to and then from the Phone Configuration page, click the link to Subscribe/Unsubscribe Services.

Figure 5-9 *Examples of Services as Shown on the Phone Display*

Figure 5-10 shows a service in the Cisco IP Phone Services Configuration page.

Figure 5-10 *Cisco IP Phone Services Configuration Page*

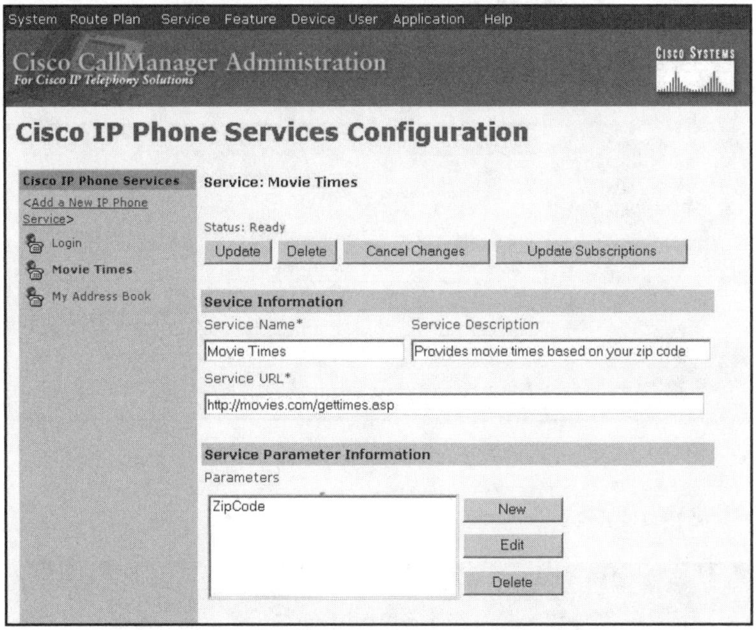

The following are quick steps to configure Cisco IP Phone services:

Step 1 Open the Cisco IP Phone Services Configuration page in CCMAdmin (click **Feature > Cisco IP Phone Service**). Figure 5-11 shows a new service being configured on the Cisco IP Phone Services Configuration page.

Step 2 Configure the service. Provide a service name and for the URL enter: **http://**<*ip_address*>**/ccmuser/**<*filename*>.<*file extension*> and provide a description. The description displays on the Cisco IP Phone User Options Web page to help users decide whether or not to subscribe to the service, so provide a useful and concise description of the service.

Step 3 Click **Insert**. When you insert the service, the second part of the configuration displays. Figure 5-12 shows the new service being configured with parameters.

Figure 5-11 *Step 1 of Configuring a Service in CCMAdmin*

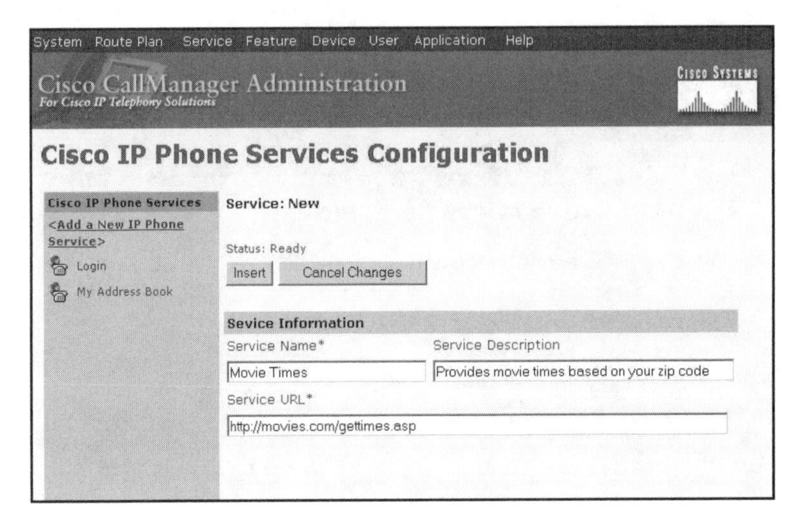

Figure 5-12 *Inserting a Service in CCMAdmin*

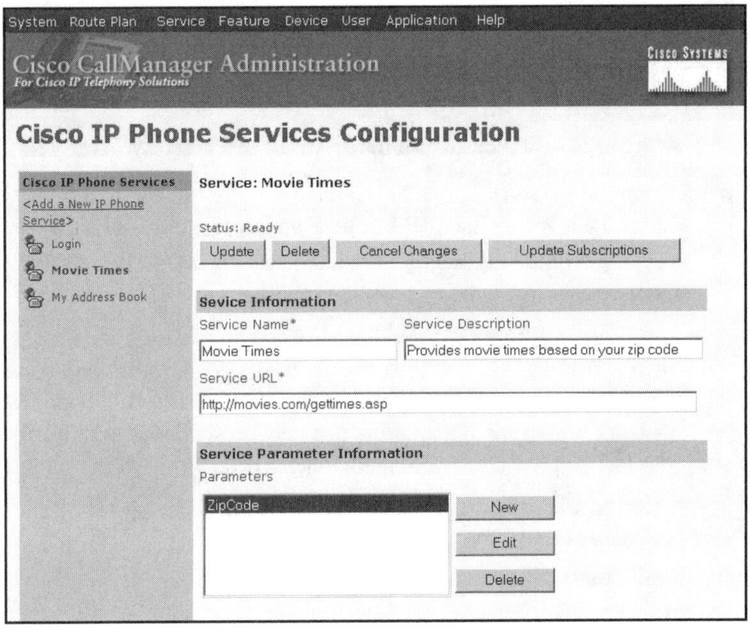

Step 4 If the service has parameters that need to be defined, such as zip codes for the fictional Movie Times service, then define those by clicking **New** in the Service Parameter Information area. Figure 5-13 shows the popup window that displays when you click **New** or **Edit**.

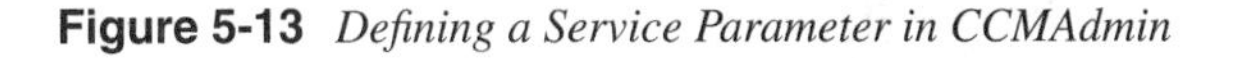

Figure 5-13 *Defining a Service Parameter in CCMAdmin*

Step 5 Define the parameter name, display name, and description. You can also provide a default value. For example, if the Movie Times service was available for employees in Mitchell, Indiana, then you would supply 47446 as the default zip code.

Step 6 You can also check the boxes to indicate whether the parameter is required and to mask the input. Masking the input displays asterisks on the Cisco IP Phone when the user enters the digits for the service parameter.

Step 7 Click **Insert** to add the parameter. Once fully configured, the service looks like the one shown in Figure 5-10.

Providing User Access to Cisco IP Phone Settings

Users can access a Web page called the Cisco IP Phone User Options to set speed dials, change their password or PIN, configure the Cisco Personal Address Book, manage subscriptions to Cisco IP Phone services, and more. To make this feature available, you must first configure the Cisco IP Phone User Options Web page and then broadcast the address and authentication credentials to users.

Figure 5-14 shows the Cisco IP Phone User Options Web page after login.

Figure 5-14 *Cisco IP Phone User Options Web Page*

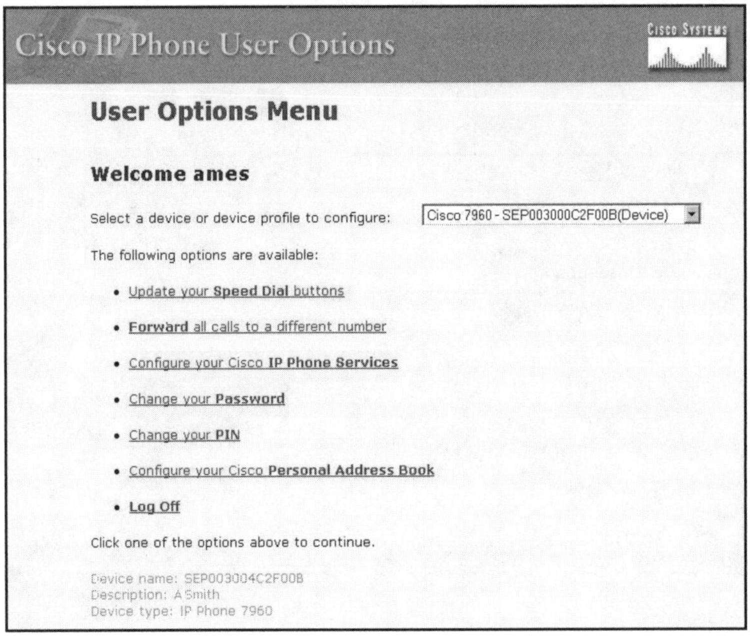

Users can access the Cisco IP Phone User Options Web page using the following URL:

http://<*server_name_OR_IP_address*>/CCMUser

At the Log On page, users need to enter their user name and password as specified in the User Configuration page in CCMAdmin. Once logged in, users can configure the following settings:

- Select a device or device profile to configure.

  The administrator is able to associate multiple devices or device profiles to a user. After logging into the Cisco IP Phone User Options Web page, the user is able to select a device or device profile to configure.

This provides the user the ability to customize the features of each device or device profile that is associated to them.

- Change their password.

  The user can change their password any time from the Cisco IP Phone User Options Web page by entering in their current password and then entering a new password.

  The password must contain at least four characters, but no more than 20.

  The administrator configures the initial password for users to log on.

- Change their PIN.

  Users can change their PIN anytime from the Cisco IP Phone User Options Web page by entering in their current PIN and then entering in their new PIN.

  The PIN must have at least five characters, but no more than 20.

  The administrator configures the initial PIN for users.

- Configure their Cisco Personal Address Book.

  Users can add other users into the Cisco Personal Address Book. Address book entry information includes: Name, Nickname, Email, Home Phone, Work Phone, and Mobile Phone.

  Fast dials, which are a combination of digits that represents an entry in the address book, can also be configured.

  To use the Cisco Personal Address Book from the Cisco IP Phone 7940 or 7960, the user will need to subscribe to the Cisco IP Phone service My Address Book.

On the selected device, users can configure the following settings.

- Update speed dial buttons.

  Users can configure the speed dial buttons for their Cisco IP Phone.

  Enter the phone number and a name for that speed dial number.

  The phone number entered for speed dial must follow dialing rules in Cisco CallManager. For example, if you dial "9" for access to the PSTN and then dial the 10-digit number, (972) 555-2487, the speed dial number entered must be "99725552487."

- Forward all calls to a different number.

  The user can forward calls to a number by checking the box and then entering the phone number to which they want all calls forwarded.

You can limit where users can forward their calls by assigning a calling search space to the Forward All setting on the Directory Number Configuration page in CCMAdmin.

The Forward All number must follow the dialing rules in Cisco CallManager. For example, if you dial "9" for access to the PSTN and then dial the 10-digit number, (972) 555-2487, the Forward All phone number entered by the user must be "99725552487."

- Subscribe to or unsubscribe from Cisco IP Phone services.

  The user can manage subscriptions for Cisco IP Phone services. This includes viewing a list of available services, subscribing to services, or unsubscribing from services.

Figure 5-15 shows the Cisco IP Phone Services page in the Cisco IP Phone User Options Web page.

Figure 5-15 *Subscribing to a Service in the Cisco IP Phone User Options Web Page*

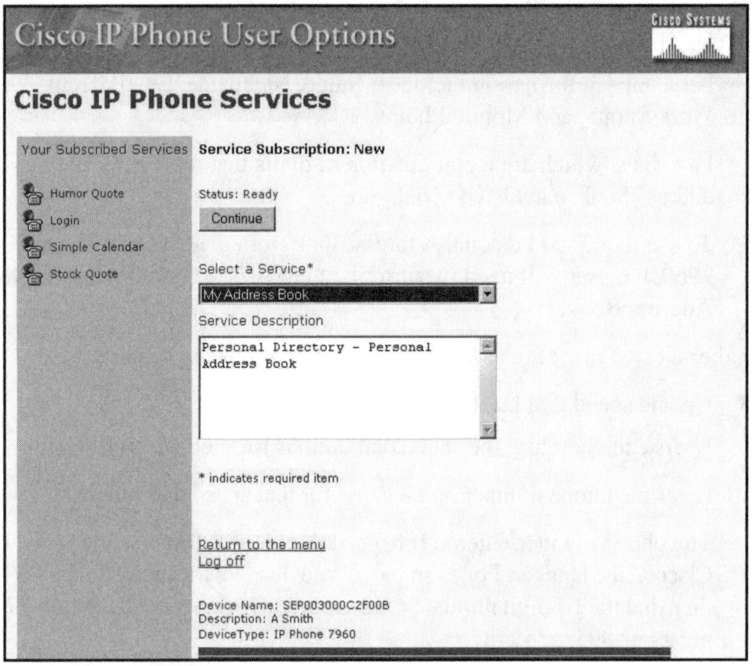

From the Cisco IP Phone Services page, users can

- Customize the name of the service as it appears on their services list; this is useful if users subscribe to the same service several times (like weather forecasts for various areas).
- Enter information to customize the service (if applicable).
- Subscribe to the service (subscriptions are made on a per-device basis).

Figure 5-16 shows a service that enables a user to customize it by adding service information.

Figure 5-16 *Completing a Service Subscription in the Cisco IP Phone User Options Web Page*

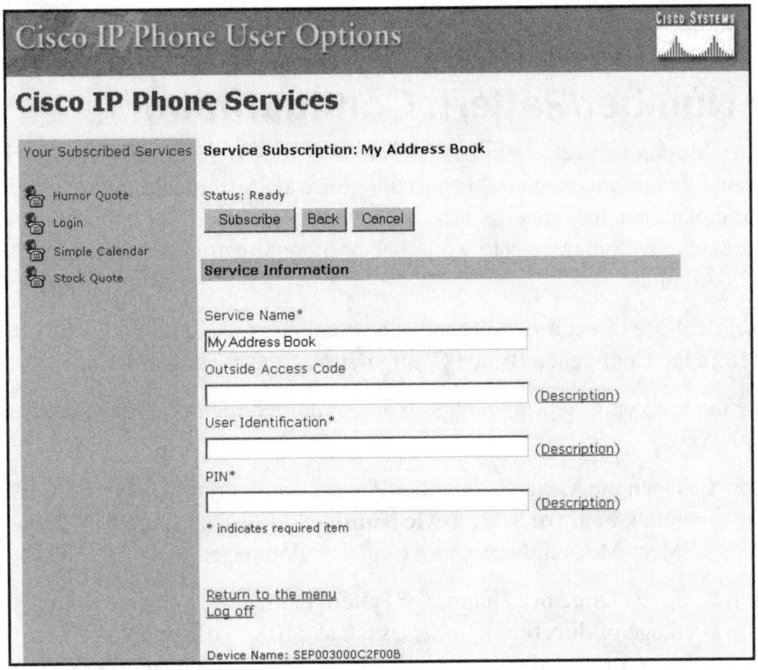

When the user clicks **Subscribe**, Cisco CallManager builds a custom URL and stores it in the database for this subscription. The service then appears on the user's phone in the list of services.

The Cisco IP Phone 7960 and 7940 model telephones have a button labeled **services**. When the user presses this button, the phone uses its HTTP client to load a specific URL that contains a menu of services to which the user has subscribed. To use a service, the user selects it from the listing. When a service is selected, the URL is requested via HTTP, and a server provides the content, which then updates the phone to display the requested service.

Typical services that might be supplied to a phone include weather information, stock quotes, and news quotes. Deployment of Cisco IP Phone services occurs using HTTP from standard web servers, such as the IIS. Phone services provide far greater value than the simple services that have been discussed so far. Companies and organizations have developed services that solve complex business problems (such as providing auto answer capabilities for a type of intercom system between two individuals), provide increased customer or user satisfaction, and even generate revenue. With a little ingenuity and the code provided by the Cisco XML SDK, you can turn the standard Cisco IP Phones in which you have already invested into robust business tools that do more than just provide voice communications.

Meet-Me Number/Pattern Configuration

Meet-Me conferences, described in Chapter 4, "Cisco CallManager Administration Service Menu," enable a user to establish a conference at a specified directory number. Conference participants can then dial the specified directory number to be connected to the conference. For Meet-Me conferences to work, one or more directory numbers must be configured in CCMAdmin.

The Meet-Me directory numbers can be configured using the link under the Feature menu or from the Conference Bridge Configuration page in CCMAdmin.

Use the following steps to configure one or more directory numbers for Meet-Me conference.

Step 1 Open the Meet-Me Number/Pattern Configuration page in CCMAdmin (click **Feature > Meet-Me Number/Pattern**). Figure 5-17 shows the Meet-Me Number/Pattern Configuration page.

Step 2 In the Directory Number or Pattern field, enter a directory number or a range of directory numbers, such as 510X. 510X would provide 10 directory numbers for use as Meet-Me conference numbers: 5100 through 5109.

Step 3 *(Optional)* Select the partition to which this Meet-Me number or pattern should belong.

Figure 5-17 *Meet-Me Number/Pattern Configuration*

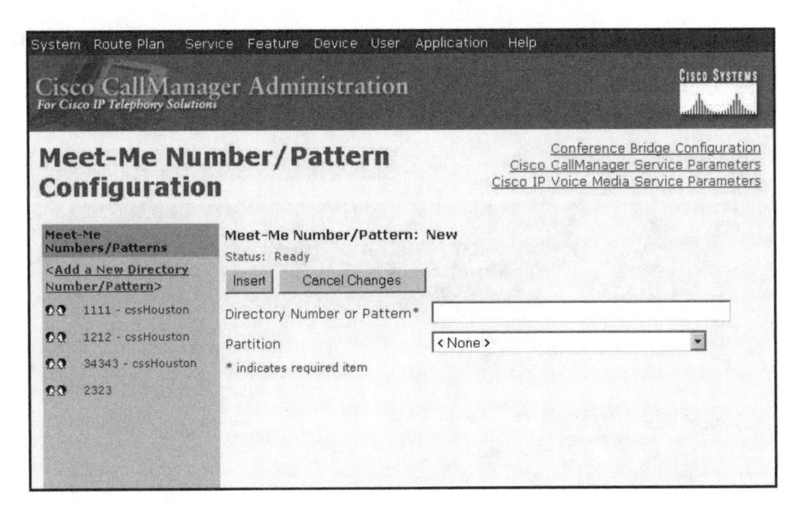

User Menu

Use the User menu in CCMAdmin to configure user information for use in the Cisco IP Telephony solution. The sections that follow discuss these menu items:

- Adding a new user
- Global directory

Adding a New User

The User area in CCMAdmin enables you to display and maintain information about Cisco CallManager users. Generally, completing user information is optional but recommended—the devices function whether or not you provide this information. However, information that you enter here is also accessed by Directory Services, Cisco WebAttendant, Cisco Personal Assistant, the IP Auto Attendant, and the Cisco IP Phone User Options Web page. If you want to provide any of these features to your users, you must complete the information in the User area for all users and their directory numbers and also for resources such as conference rooms or other areas with phones (this is useful for Cisco WebAttendant).

For users who have Web access to change their speed dial and forward numbers on the Cisco IP Phone User Options Web page, they must be associated with a device. When a user is associated with a device, the device's MAC address and directory number are provided in the User area.

The steps for entering user information are as follows:

Step 1 Open the User Information page in CCMAdmin (click **User > Add a New User**). Figure 5-18 shows the User Information page in CCMAdmin.

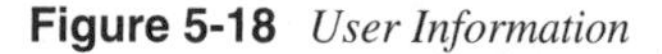

Figure 5-18 *User Information*

Step 2 Enter user information, completing the required information (denoted by an asterisk on the User Information page). Remember the User ID and the password.

Step 3 Click **Insert**.

Step 4 Associate the user to a device. In the User Information page for the user you just added, click the link in the left column for Device Association.

Step 5 Define a search parameter, if desired, and then click **Select Devices** from the Available Device List Filters section. Figure 5-19 shows the search query and the results.

Figure 5-19 *User Information*

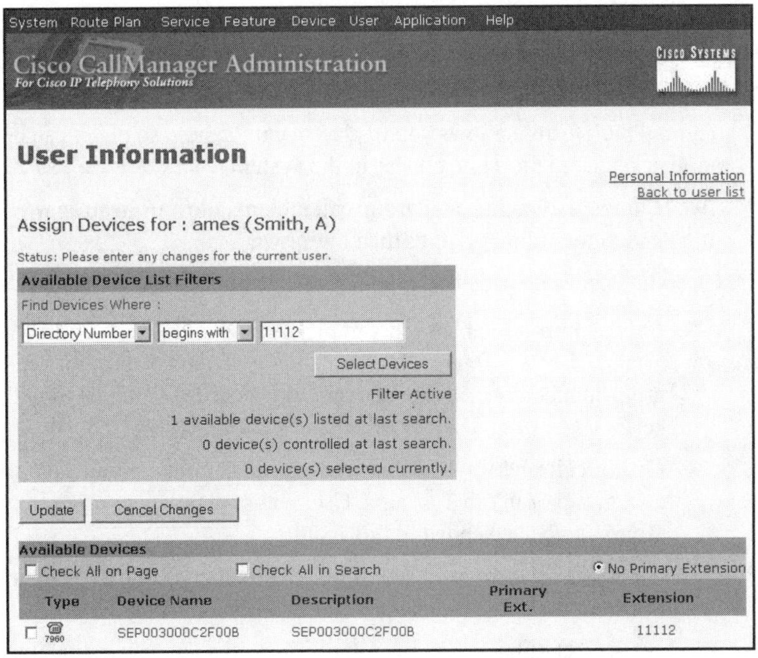

Step 6 Check the box next to the device you want to associate to the user and click **Update**.

Result: The user and device are now associated.

TIP You can add a large number of users at one time by using the Bulk Administration Tool as discussed in Chapter 7, "Understanding and Using the Bulk Administration Tool (BAT)."

Global Directory

The Global Directory for Cisco CallManager (Release 3.0 and later) contains every user within a Cisco CallManager directory. Cisco CallManager uses the LDAP to interface with a directory that contains user information. This is an embedded directory supplied with Cisco CallManager. Its primary purpose is to maintain the associations of devices with users. As of Cisco CallManager Release 3.0(10), Cisco CallManager can interface with third-party LDAP directories, such as Netscape Directory Server and Microsoft Active Directory.

You can access the Global Directory by performing a search for users.

Summary

This chapter covered the menu items provided in the Feature and User menus of CCMAdmin. Under the Feature menu, you learned about call park, call pickup groups, Cisco IP Phone services, and Meet-Me number/pattern configuration.

- The call park feature allows you to place a call on hold so that it can be retrieved from another phone in the same cluster in the system.
- Two features, call pickup and group call pickup, allow users to answer calls that come in on a directory number other than their own.

  Cisco IP Phones provide two types of call pickup:

  — **Call pickup**—Enables users to pick up incoming calls within their own group. Cisco CallManager automatically dials the appropriate call pickup group number when a user activates this feature by pressing the **PickUp** soft key.

  — **Group call pickup**—Enables users to pick up incoming calls within their own group or in other groups. Users must dial the appropriate call pickup group number when using this feature.
- Cisco IP Phone services are XML applications that enable the display of interactive content with text and graphics on Cisco IP Phone models 7960 and 7940.
- The Meet-Me Number/Pattern Configuration area lets you define one or more directory numbers for use as Meet-Me conferences.

The User area in CCMAdmin enables you to display and maintain information about Cisco CallManager users. Information that you enter here is also accessed by other Cisco IP Telephony components, including Directory Services, Cisco WebAttendant, IP Auto Attendant, Cisco Personal Assistant, and the Cisco IP Phone User Options Web page.

After completing this chapter, you should be able to perform the following tasks:

- Configure the call park feature.
- Identify and define the characteristics of each type of call pickup.
- Configure the call pickup feature.
- Configure Cisco IP Phones services and access those services from the Cisco IP Phones.
- Add users to the directory and access the Cisco IP Phone User Options Web page.

Post-Test

Use this section to test yourself on how well you learned the concepts discussed in this chapter. Use the following scale to decide whether to re-read, skim this chapter, or go on to the next chapter. If you answer one to three questions correctly, we recommend that you re-read the chapter. If you answer four to seven questions correctly, we recommend you skim the headers and read the sections related to the questions you missed. If you answer 8 to 10 questions correctly, go on to the next chapter. You can find the answers to the post-test for this chapter in Appendix B, "Answers to Chapter Pre-Test and Post-Test Questions."

1 What is the benefit of call park?

____

____

____

2 After configuring a call park directory number or range of directory numbers, how many calls can be parked at a given call park extension?

____

____

____

3 Describe the two types of call pickup.

____

____

____

4 Describe what Cisco IP Phone services provide.

____

____

____

5 Which two Cisco IP Phone models support phone services?

____

____

____

6 Name three of the Cisco AVVID IP Telephony components that access user information you configure in CCMAdmin?

____

____

____

7 Assuming you are using a Cisco IP Phone model 7940, list four of the seven items a user can select from the Cisco IP Phone User Options Web page.

____

____

____

8 What do you need to do in CCMAdmin to ensure that users can access the Cisco IP Phone User Options Web page?

____

____

____

9 Once phone services have been subscribed to a given phone, how does the user access those services?

____

____

____

10 Name the two Web pages through which a system administrator can subscribe to Cisco IP Phone services.

____

____

____

Upon completing this chapter you will be able to do the following tasks:

- Given a Cisco IP Phone model, identify the model and describe the features of the Cisco IP Phone model.
- Given a Cisco IP Phone and a Cisco CallManager, add that phone to the Cisco CallManager database using Cisco CallManager Administration.
- Given a list of gateways, identify which type(s) of protocol(s) each gateway supports.
- Given a distributed call processing model with limited WAN bandwidth, add a gatekeeper in Cisco CallManager Administration to manage the WAN bandwidth.
- Given a Cisco CallManager and Cisco Unity, configure Cisco CallManager in preparation for integration with Cisco Unity.
- Given a list of applications that use computer telephony integration (CTI), define each application and how it is used in a Cisco IP Telephony (CIPT) network.

Cisco IP Telephony Devices

Devices in a CIPT network enable communication. Whether the communication is between two Cisco IP Phones on a LAN or a Cisco IP Phone through a gateway to a relative's home phone connected to the PSTN, devices allow communication to happen. This chapter discusses the devices that can be added to a CIPT network and shows how to add and configure those devices. This chapter discusses the following topics:

- Abbreviations
- Device Configuration
- Phone Button Template
- Adding and Configuring a Cisco IP Phone
- Adding and Configuring a Gateway
- Adding and Configuring a Gatekeeper
- CTI
- Cisco Voice Mail
- Adding and Configuring Device Profiles

Pre-Test

Do you already know this information? The pre-test is designed to help you gauge your knowledge about this chapter. Of the 10 questions, if you answer one to three questions correctly, we recommend that you read this chapter. If you answer four to seven questions correctly, we recommend that you skim through this chapter, reading those sections that you need to know more about. If you answer 8 to 10 questions correctly, you probably understand this information well enough to skip this chapter. You can find the answer to these questions in Appendix B, "Answers to Chapter Pre-Test and Post-Test Questions."

1 Label the model number for the following Cisco IP Phones:

a.

b.

c.

d.

2 In a CIPT environment, what are the three supported gateway protocols?

____

____

____

3 Which gateway protocols are supported by the WS-6608-T/E1 and DT24+ gateways when using Cisco CallManager 3.1 and higher?

____

____

____

4 Which gateway protocols are supported by the Cisco 2600 and 3600 Series routers?

____

____

____

5 In a distributed call processing deployment, what device can be used to provide call admission control between clusters?

____

____

____

List the four Cisco provided CTI applications that use a CTI port or CTI route point.

6

____

7

____

8

____

9

____

10 What are the Cisco CallManager service parameters that need to be configured while integrating Cisco Unity?

____

____

____

Abbreviations

This section defines the abbreviations used in this chapter. For more information about terms and abbreviations used in this chapter refer to the *IP Telephony Network Glossary* at the following URL:

www.cisco.com/univercd/cc/td/doc/product/voice/evbugl4.htm

Table 6-1 provides the abbreviation and the complete term.

Table 6-1 *Abbreviations and Complete Terms Used in the Chapter*

Abbreviation	Complete Term
API	application programming interface
AS	Analog Access Station
AT	Analog Access Trunk
BAT	Bulk Administration Tool
CO	central office
CRA	Customer Response Application
CTI	computer telephony interface
DE	digital access trunk-Euro
DT	digital access trunk
FQCN	fully qualified calling number
GK	gatekeeper
HSRP	Hot Standby Router Protocol
IOS	Internetworking Operating System
MAC	Media Access Control
MCM	multimedia conference manager
MGCP	Media Gateway Control Protocol

continues

Table 6-1 *Abbreviations and Complete Terms Used in the Chapter (Continued)*

Abbreviation	Complete Term
MWI	message waiting indicator
PRI	Primary Rate Interface
RAS	Registration, Admission, and Status Protocol
TAPI	Telephony Application Programming Interface
VG200	Voice Gateway 200
VGC	Voice Gateway Chassis
VIC	Voice interface card
VWIC	Voice/WAN interface card

Device Configuration

This section introduces you to device configuration in Cisco CallManager Administration. To ensure success, you should connect CIPT devices to the network prior to adding them in Cisco CallManager Administration.

Figure 6-1 shows the Device menu in Cisco CallManager Administration.

Before you can use devices such as gateways and Cisco IP Phones in your IP telephony network, you must add them to the Cisco CallManager database. Database modifications are effected through Cisco CallManager Administration. Cisco CallManager enables you to configure the following devices in your telephony network:

- CTI route point
- Gatekeeper
- Gateway
- Phone
- Cisco voice mail port

NOTE The Device Profile option in the Device menu enables you to configure a device profile, which is used with extension mobility. Extension mobility is not covered in this book.

At any time, you can restart or reset a device by clicking the **Reset** button on the Device page or by clicking the **Reset** icon on the page. You can restart a device without shutting it down by clicking the **Restart** button. You can shut down a device and bring it back up again by clicking the **Reset** button. If you want to return to the previous window without resetting or restarting the device, click **Close**.

Figure 6-1 *Device Configuration*

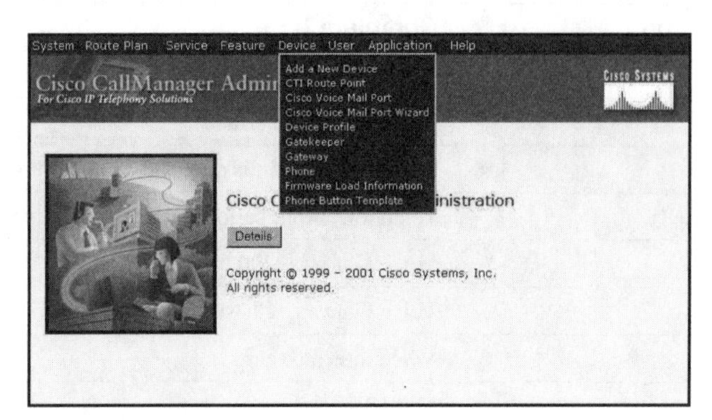

NOTE Restarting or resetting a gateway drops any calls in progress using that gateway. Other devices wait until calls are complete before restarting or resetting.

Phone Button Template

This section discusses phone button template configuration.

Figure 6-2 shows the Cisco IP Phone and the buttons on the phone affected when configuring phone button templates.

Creating and using templates provides a fast way to assign a common button configuration to a large number of phones.

Cisco CallManager includes several default phone button templates. When adding phones, you can assign one of these templates to the phones or create a new template. Figures 6-3 through 6-6 show the default phone button templates for the 12 SP+, 14-Button Expansion Module, 7910, and 7960 installed with Cisco CallManager. (Default phone button templates not shown are the 30 SP+ and 30 VIP.)

Figure 6-2 *Phone Button Templates*

Make sure all phones have at least one line assigned. Normally, this is button 1. You can assign additional lines to a phone, depending on the Cisco IP Phone model. Cisco IP Phones 7910, 7940, 7960, and expansion module 7914 can be configured with lines or speed dials. The Cisco IP Phone model 12 SP+ and 30 VIP buttons can be assigned with lines, speed dials, call park, call pickup, transfer, and other phone features.

Create phone button templates for all possible combinations for all phone models, before adding any phones to your system. All the possible combinations that are going to be used in your systems should be prepared. For example, a Cisco IP Phone 7960 can use the following phone button template combinations:

- One line, five speed dials
- Two lines, four speed dials (default)
- Three lines, three speed dials
- Four lines, two speed dial
- Five lines, one speed dials
- Six lines, zero speed dials

Figure 6-3 *Default 12 SP+ Phone Button Template*

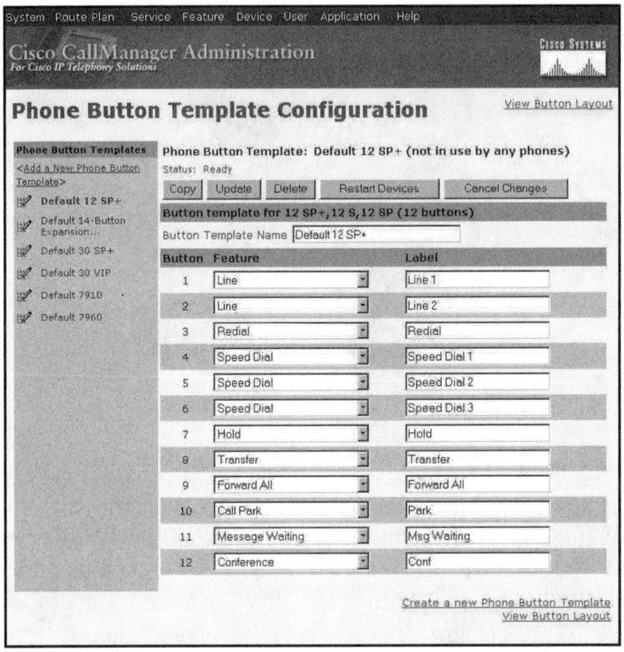

NOTE Other combinations are available for the Cisco IP Phone 7960, such as using just lines and no speed dials. For example, the Cisco IP Phone 7960 can have one line and no speed dials or it could have three lines and no speed dials. As long as the phone has one line assigned to it, it will be a valid phone button template.

When creating a phone button template, you should assign easily recognized names, for example if you have a Cisco IP Phone 7960 that has three lines and three speed dials, you could use "7960 3-3" for a name for the phone button template. Now if you have a special request or only a limited number of users that need three lines and three speed dials, you can quickly assign "7960 3-3" to the phone, rather than spend time creating a template for each phone and then assigning the template to the phone.

To create a template, copy an existing template and assign the template a unique name. You can make changes to the default templates included with Cisco CallManager or to custom templates you created. You can rename existing templates and modify them to create new ones; update custom templates to add or remove features (valid only for Cisco IP Phone 7910, Cisco IP Phone model 12 SP+ and 30 VIP), lines, or speed dials; and delete templates that are no longer used. When you update a template, the change affects all phones that use the template.

Figure 6-4 *Default 14-Button Expansion Module Phone Button Template*

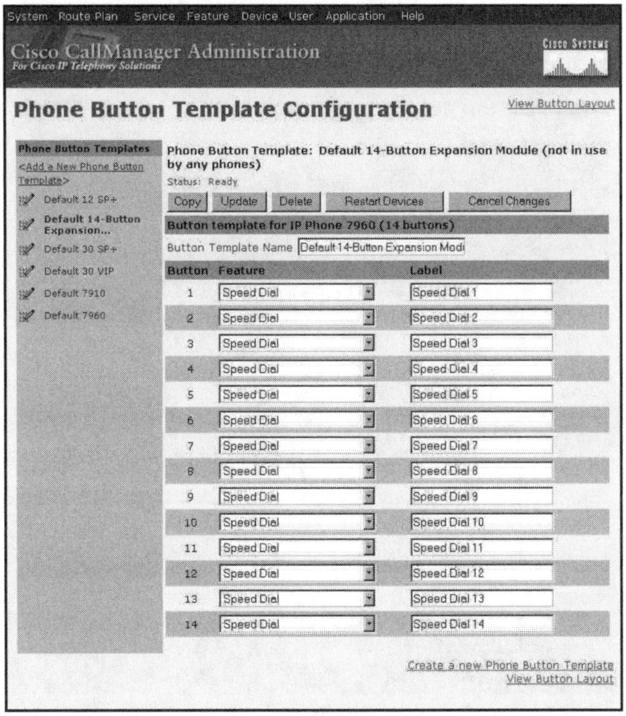

Renaming a template does not affect the phones that use that template. All Cisco IP Phones that use this template continue to use this template once it is renamed. You can use this feature to create a copy of an existing template that you can modify.

You can delete phone templates that are not currently assigned to any phone in your system. You cannot delete a template that is assigned to one or more devices. At this time, no features exist to easily query if a template is in use or not. You must reassign all Cisco IP Phones using the template you want to delete to a different phone button template before you can delete the template.

Figure 6-5 *Default 7910 Phone Button Template*

System Route Plan Service Feature Device User Application Help

Cisco CallManager Administration

For Cisco IP Telephony Solutions

Cisco Systems

Phone Button Template Configuration

View Button Layout

Phone Button Templates

<Add a New Phone Button Template>

Default 12 SP+

Default 14-Button Expansion...

Default 30 SP+

Default 30 VIP

Default 7910

Default 7960

Phone Button Template: Default 7910 (not in use by any phones)

Status: Ready

Copy | Update | Delete | Restart Devices | Cancel Changes

Button template for IP Phone 7910 (6 buttons)

Button Template Name: Default 7910

Fixed Button(s):	Line	Hold	Transfer	Settings
Button Arrangement:	1	2	3	
	4	5	6	

Button	Feature	Label
1	Message Waiting	Msg Waiting
2	Conference	Conference
3	Forward All	Forward All
4	Speed Dial	Speed Dial 1
5	Speed Dial	Speed Dial 2
6	Redial	Redial

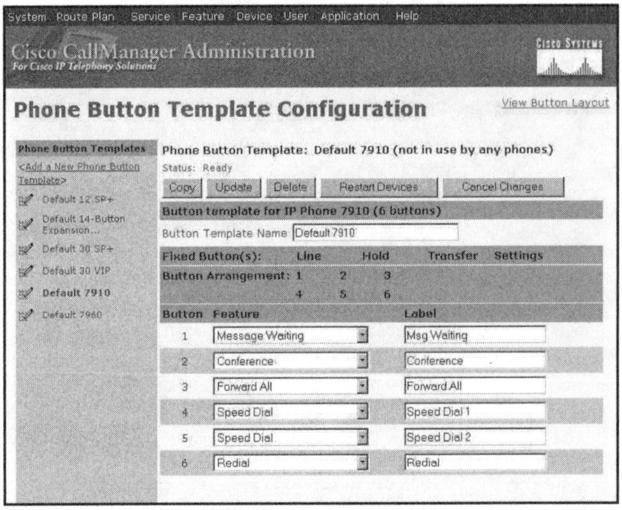

Figure 6-6 *Default 7960 Phone Button Template*

System Route Plan Service Feature Device User Application Help

Cisco CallManager Administration

For Cisco IP Telephony Solutions

Cisco Systems

Phone Button Template Configuration

View Button Layout

Phone Button Templates

<Add a New Phone Button Template>

Default 12 SP+

Default 14-Button Expansion...

Default 30 SP+

Default 30 VIP

Default 7910

Default 7960

Phone Button Template: Default 7960 (used by 2 phone(s))

Status: Ready

Copy | Update | Delete | Restart Devices | Cancel Changes

Button template for IP Phone 7960 (6 buttons)

Button Template Name: Default 7960

Button	Feature	Label
1	Line	Line 1
2	Line	Line 2
3	Speed Dial	Speed Dial 1
4	Speed Dial	Speed Dial 2
5	Speed Dial	Speed Dial 3
6	Speed Dial	Speed Dial 4

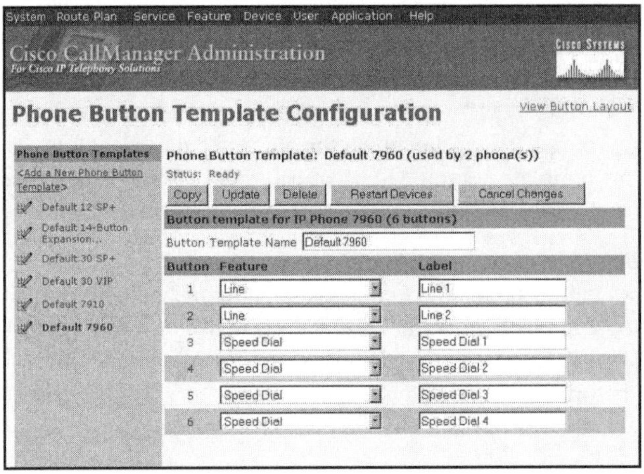

Figure 6-7 *Phone Button Template Configuration Page in CCMAdmin*

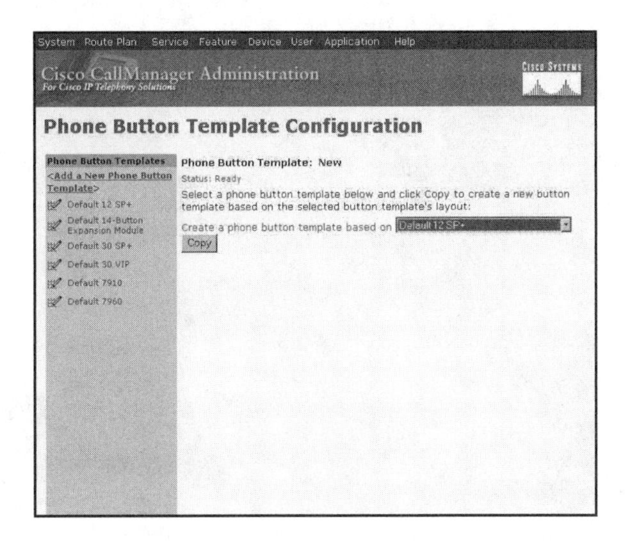

The steps for configuring phone button templates are as follows:

Step 1 Open the Phone Button Template Configuration page. In CCMAdmin go to **Device > Phone Button Template**.

Step 2 Create a new phone button template. Select a phone button template to configure by selecting an existing phone button template to copy from the **Create a phone button template based on** drop-down menu. When the type is selected, click on **Copy**.

Step 3 Enter a phone button template name and assign buttons to features. Enter a phone button template name. Then select the feature for that button number. Modify the label to match the features. Select **Insert**.

Step 4 Create more templates as needed. Repeat Steps 2 and 3.

Adding and Configuring a Cisco IP Phone

Adding a Cisco IP Phone to the network can be as easy with auto-registration and DHCP configured. This section discusses adding and configuring a Cisco IP Phone.

Figure 6-8 shows the different Cisco IP Phone models.

Because Cisco IP Phones are full-featured Ethernet telephones, they can plug directly into your IP network.

Figure 6-8 *Cisco IP Phones*

Cisco IP Phone 7910

Cisco IP Phone 7910+SW

Cisco IP Phone 7940

Cisco IP Phone Expansion Module 7914

Cisco IP Phone 7960

Cisco IP Conference Station 7935

You can use the Phone Configuration page in Cisco CallManager Administration to configure the following Cisco IP Phones and devices:

- Cisco IP Phone 79*xx* models
- H.323 clients
- CTI ports
- Cisco IP Phone models 12SP+ and 30 VIP
- Cisco VGC phone

You can add phones to the Cisco CallManager database automatically using auto-registration, manually using the Phone Configuration page or in groups with BAT. BAT is a Web-based application that enables you to perform batch add, update, and delete operations on large numbers of Cisco IP Phones (and other devices or users). Chapter 7, "Understanding and Using the Bulk Administration Tool (BAT)," discusses how to use BAT.

By enabling auto-registration, you can automatically add a Cisco IP Phone to the Cisco CallManager database when you connect the phone to your IP telephony network. During auto-registration, Cisco CallManager assigns the next available sequential directory number to the phone. In many cases, you might not want to use auto-registration. One such case is if you would want to assign specific directory numbers to phones or keep the prior numbers.

NOTE "Rogue phones," phones placed on the network without administrator knowledge, can be secured using auto-registration and calling search spaces (CSSs). Auto-registration is covered in Chapter 2, "Navigation and System Setup," and CSSs are covered in Chapter 3, "Cisco CallManager Administration Route Plan Menu."

NOTE If you do not use auto-registration, you must manually add phones to the Cisco CallManager database or use BAT.

Cisco IP Phone Features

Cisco CallManager Administration enables you to configure the following phone features on Cisco IP Phones:

- Call waiting
- Call forward
  - Call forward busy
  - Call forward no answer
  - Call forward all
- Call park
- Call pickup

NOTE Chapter 5, "Cisco CallManager Feature and User Menus," addresses the call park and call pickup features in more detail. This section discusses only call waiting and call forward.

Figure 6-9 shows an example of the configurable features of the Cisco IP Phones where Larry and Moe are having a conversation and Curly is calling Larry. If configured, Larry's phone can have call waiting to alert Larry a call is coming from Curly. If Larry does not pick up the call waiting, Larry's phone will use the call forward busy feature and forward the call to voice mail.

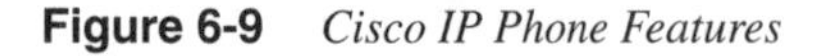

Figure 6-9 *Cisco IP Phone Features*

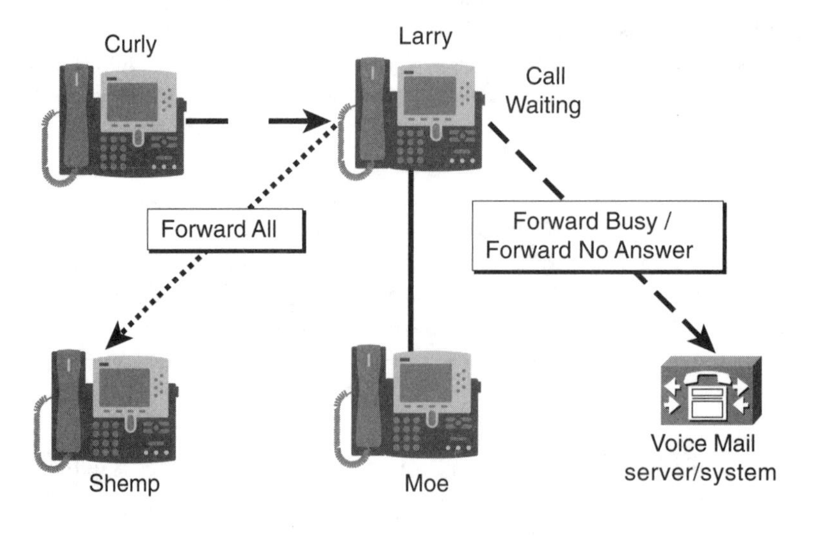

Call Waiting

Call waiting enables users to receive a second incoming call on the same line without disconnecting the first call. When the second call arrives, the user hears a brief call waiting indicator tone.

Configure call waiting on the Directory Number Configuration page in Cisco CallManager Administration as shown in Figure 6-10.

Call Forward

Call forward enables a user to configure a Cisco IP Phone so all calls destined for it are rerouted to an alternate directory number. Three types of call forward exist:

- **Call forward all**—Forwards all calls (user or administrator configurable).
- **Call forward busy**—Forwards calls only when the line is in use (administrator configurable).
- **Call forward no answer**—Forwards calls when the phone is not answered within the configurable setting in seconds. The default value is 12 seconds (administrator configurable).

Figure 6-10 *Call Waiting Setting*

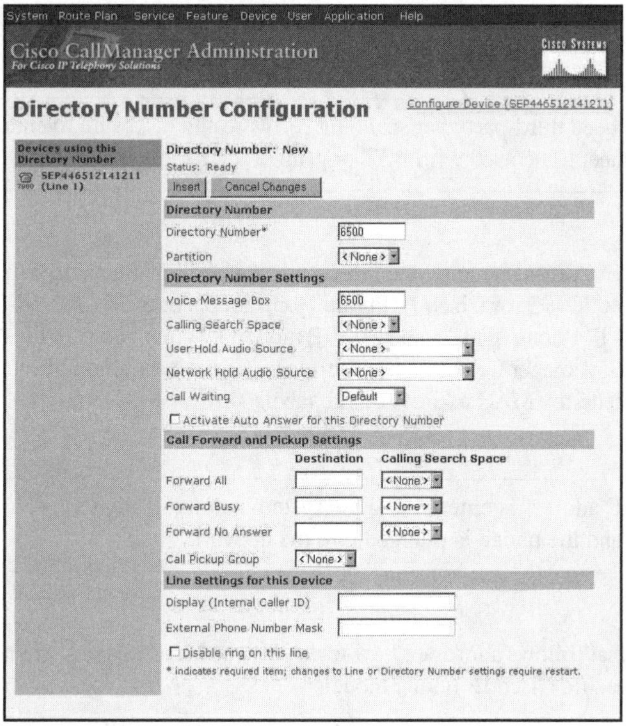

Call Forward All Example

In Figure 6-9 Larry could use the call forward all feature, where all calls to Larry are directed to Shemp's directory number. Larry can set this either from the phone or from the Cisco IP Phone User Options Web page.

Call Forward No Answer Example

In Figure 6-9 if Larry was not at his phone and Moe called, after a configured number of seconds (the default Cisco CallManager service parameter "Forward No Answer Timeout" is 12 seconds), Larry's phone would use the call forward no answer feature. This feature would forward the call to a configured directory number, usually voice mail.

Displaying the MAC Address

The MAC address is a required field for Cisco IP Phone configuration. The MAC address is a unique 12-character hexadecimal number that identifies a Cisco IP Phone or other hardware device.

If the phone is still in the box, on the outside of each phone box is the MAC address and a bar code that can be scanned. Scanning the MAC address can be very helpful when using BAT to input large numbers of phones before placing them at the final destination.

NOTE You will need third-party bar scanning software and hardware to enter MAC addresses into a speadsheet to be used with BAT or printed out for manual input.

The MAC address on the phone is located on a label on the back of the phone (for example, 000B6A409C405 for Cisco IP Phone 79*xx* models or SS-00-0B-64-09-C4-05 for Cisco IP Phone SP 12+ and 30 VIP models). When entering the MAC address in Cisco CallManager fields, do not use spaces or dashes and do not include the "SS" that may precede the MAC address on the label.

NOTE The MAC address is entered into the system automatically when auto-registration is enabled and the phone is plugged into the network.

The list that follows addresses how to display the MAC address when the phone is powered on for specific Cisco IP Phone models:

- **Cisco IP Phone 7960 and 7940 models**—Press **settings**, use the arrow buttons to highlight the **Network Configuration** menu item, and press the **Select** soft key.
- **Cisco IP Phone 7910**— Press **settings**; press **6** on the phone keypad to display the Network Config options. Use the down arrow button below the Volume label to quickly locate the MAC value. Press the **settings** button twice to exit the menu.
- **Cisco IP Phone Models 12 SP+ and 30 VIP**— Press ** on the keypad, and the MAC address appears in the LCD.

Directory Number Configuration

Using CCMAdmin, you can configure and modify directory numbers assigned to specific phones. To get to the Directory Number Configuration page, find a phone by going to **Device > Phone** and use the search criteria to list phones. Click on the phone you want to configure; the Phone Configuration page appears. From the left column click on a "Line" you want to configure, and the Directory Number Configuration page opens.

Use the Directory Number Configuration page of CCMAdmin to perform tasks such as adding or removing directory lines; configuring call forward, call pickup, and call waiting;

setting the display text that appears on the called party's phone when a call is placed from a line; and disabling ring on a line.

Directory number configuration affects the look and feel of the phone for the user. For instance, the display setting on the phone shows who is being called and who is calling.

The external phone number mask is displayed on the top line of the LCD and shows the fully qualified calling number (FQCN) or the user's full directory number (only the FQCN of line 1 is displayed on the phone). The number is the caller ID display that can be sent out through the PSTN.

Figure 6-11 shows how the directory number configuration settings are displayed on the phone in its idle and active stage.

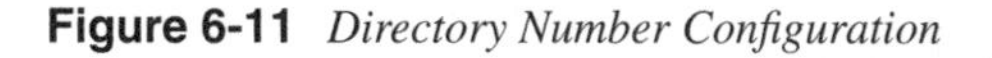

Figure 6-11 *Directory Number Configuration*

Shared Line Appearances

This section discusses shared line appearances in a Cisco CallManager cluster.

Figure 6-12 shows the display on the Directory Number Configuration page in Cisco CallManager Administration after adding a shared line in Cisco CallManager.

Figure 6-12 *Shared Line Appearance*

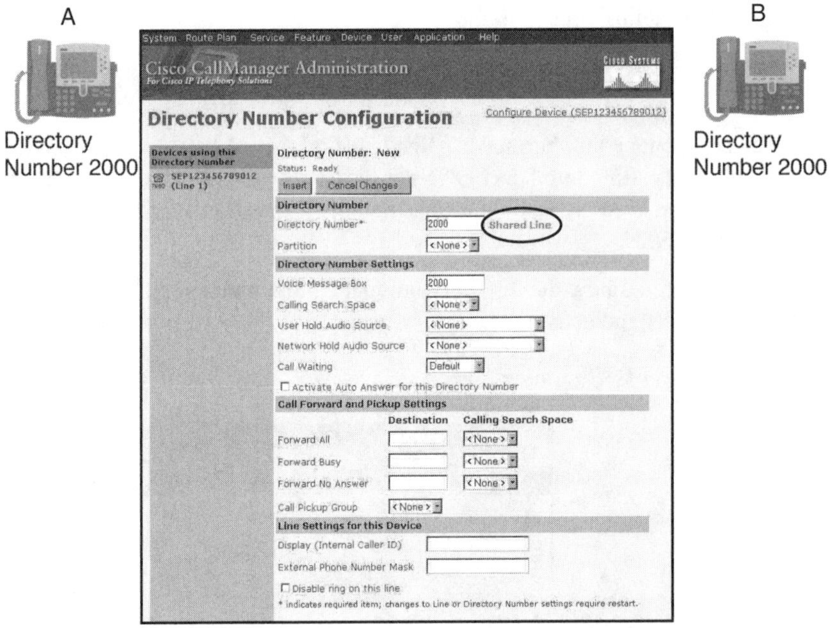

You can set up one or more lines with a shared line appearance. If a directory number in the same partition appears on more than one device it is considered a shared line appearance:

- Example 1: A directory number appears on line 1 of a manager's phone and also on line 2 of an assistant's phone.
- Example 2: A single incoming 800 number appears as line 2 on every sales representative's phone in an office.

The following notes and tips apply to using shared line appearances with Cisco CallManager:

- Create a shared line appearance by assigning the same directory number and partition to different lines on different devices.
- If other devices share a line, the words Shared Line appear in red next to the directory number in the Configure a Line Number page in CCMAdmin, as shown in Figure 6-12.
- If you change the calling search space, call waiting, or call forward and pickup settings on any device that uses the shared line, the changes apply to all devices that use that shared line.

- To stop sharing a line appearance on a device, perform the following actions:
  - If the shared line is not line 1, change the directory number or partition number for the line and update the device settings.
  - If the shared line is line 1, you must first delete the line from the device and then re-add the line with a new directory number or partition.
- In the case of a shared line appearance, **Delete** removes the directory number only on the current device. Other devices are not affected.
- Do not use shared line appearances on any phone that will be used with Cisco WebAttendant.
- Shared line appearances can also be made to an MGCP controlled analog gateway and to an H.323 "phone."
- If a shared line is in use, it is unavailable for others who share the line and the users with the shared line appearance who are not using that line will see the display, "Remotely in Use."

The sections that follow address the steps used to configure phones and users.

Steps for Manually Adding a Phone

Step 1 Open the Phone Configuration page. In CCMAdmin go to **Device > Add a New Device**. From the **Device type** drop-down menu select **phone** and then **Next**. From the **phone type** drop-down menu select 7960 and then **Next**.

Step 2 Enter Device Information. Enter the MAC address. Enter a description, usually a name or location of the phone. Select a device pool, CSS, media resource group list, user and network hold audio sources, and location for the phone to belong to.

Step 3 Enter phone button template and expansion module template information.

Step 4 View **Firmware**, **External Data Locations**, and **Product Specific Configuration**. You can use the default values that are systemwide or enter values that will only be used by this phone.

Step 5 Select **Insert** to complete phone configuration.

Step 6 Select **Line 1** from the left column. Enter a directory number and select a partition for this directory number.

Step 7 Provide directory number settings. Leave the directory number settings as is. By default the voice message box will be filled with the directory number. The CSS and hold audio sources will use the phone's configuration but can be added here if multiple lines on a phone will utilize different CSSs or hold audio sources.

NOTE If you do not configure any CSS for CFA, the CSS used will be what is configured for the line and the device.

Step 8 Configure **Call Forward and Pickup Settings**. Forward all can be left blank, but you may consider assigning a CSS to limit where the user can forward their calls. The Forward Busy, Forward No Answer, and Forward On Failure are usually assigned to the voice mail directory number. You can also assign this line to a call pickup group.

NOTE If the Call Forward All CSS is not used to limit call forwarding, an individual could call their office phone from their home and forward to international locations. So the user could be placing international calls on the company's dime, with only a local call being charged to their home phone.

Step 9 Configure line settings. Enter the user's name or location of device (copy what is entered in description) for **Display**. This information is seen on the called phone internally. Enter the fully qualified directory number for the external phone number mask.

Step 10 Finish directory number configuration. Select **Insert**.

Adding a User

Step 1 Open the User Information Configuration page. In CCMAdmin go to **User > Add a New User**.

Step 2 Enter user information. Enter first name, last name, user ID, user password, PIN, telephone number (directory number), manager user ID, and department number and choose to **Enable CTI Application Use**. Select **Insert**.

NOTE By checking the **Enable CTI Application Use** check box, you enable the user to use Cisco IP SoftPhone and have their directory information available in Auto Attendant. The password is used to log on to the Cisco IP Phone User Options Web page, and the PIN is used for extension mobility and accessing the address book.

Step 3 Go to the device association page. Select **Personal Information** from the top right of the page. Select **Device Association** from the left column.

Step 4 Select **Device Association** to assign phones to the user. Select the desired device utilizing the Available Device List Filters. When the search criteria has been entered, press the **Select Devices** button. From the resulting list, select the box next to the desired phone and select the **Primary Extension** if applicable. When finished, select **Update**.

Adding and Configuring a Gateway

Before Cisco CallManager can manage any IP telephony gateways on a network, each gateway must be added to the Cisco CallManager database. Certain gateways must also be configured on the command-line interface (CLI). Refer to the documentation for your particular gateway for more information about CLI.

The procedures, Web pages, and configuration settings for adding a gateway vary according to the model of gateway that you are adding.

There are a variety of configuration options and settings depending on which gateway you are adding to the system. Table 6-2 shows some of the different types of gateways.

Table 6-2 *Gateways for a Cisco IP Telephony Network*

Gateway	MGCP	H.323	Skinny Gateway Protocol
VG200	Yes, for FXS/FSO	Yes, with Cisco IOS Software 12.1(5)-XM1, the VG200 uses H.323 to support a wider range of digital and analog interfaces	No
DT-24+	Yes, with Cisco CallManager Release 3.1	No	Yes
827	No	Yes, for FXS	No
1750	No	Yes	No
3810 V3	Cisco IOS Software 12.1(3)T and Cisco CallManager Release 3.0(5)	Yes	No

continues

Table 6-2 *Gateways for a Cisco IP Telephony Network (Continued)*

Gateway	MGCP	H.323	Skinny Gateway Protocol
2600	Cisco IOS Software 12.1(3)T and Cisco CallManager Release 3.0(5) Analog Interfaces only no E&M T1 CAS – 12.1(5)XM & 12.2.1T Q.931 PRI Backhaul – 12.2.2T[1]	Yes	No
3600	Cisco IOS Software 12.1(3)T and Cisco CallManager Release 3.0(5) Analog Interfaces only no E&M T1 CAS – 12.1(5)XM & 12.2.1T Q.931 PRI Backhaul – 12.2.2T[1]	Yes	No
7200	Cisco IOS Software 12.2.(1)T[2]	Yes	No
7500	Undecided	Yes (Cisco IOS Software 12.1.5)	No
5300[3]	Yes (Cisco IOS Software 12.1(1)T	Yes	No
Catalyst 4000 WS-X4604-GWY Gateway Module	Cisco CallManager Release 3.1	Yes, for PSTN interfaces	Yes, for conferencing and MTP/transcoding services
Catalyst 6000 WS-X6608-x1 Gateway Module & FXS Module WS-X6624	Cisco CallManager Release 3.1 T1/E1 module supporting PRI and CAS FXS module	No	Yes, for FXS module and T1/E1 prior to Cisco CallManager Release 3.1

Table 6-2 *Gateways for a Cisco IP Telephony Network (Continued)*

Gateway	MGCP	H.323	Skinny Gateway Protocol
Catalyst 4224	Projected for Cisco CallManager Release 3.1	Yes	No

1 Cisco IOS Software Release 12.2.2T PRI Backhaul support for 26xx/36xx uses RUDP and is not compatible with Cisco CallManager. PRI backhaul with Cisco CallManager Release 3.1 as the Call Agent is scheduled for Cisco IOS Software Release 12.2.4T and uses TCP as the transport.

2 Not supported in Cisco CallManager.

Also note prior to any deployment consideration, it would be prudent to check the IOS Release Notes to confirm feature or interface support.

3 While the 5300 supports MGCP, it is as a Trunk Gateway module using SS7 signaling, which is not supported in Cisco CallManager.

Before adding a gateway in CCMAdmin, gather the important information. The MGCP and Skinny protocol gateways have all dial plan information in the Cisco CallManager database. The H.323 gateways have dial plan information configured internally; a Cisco CallManager route pattern directs the call to the H.323 gateway. Then the H.323 gateway takes the dial information and uses internal dial plan information to route the call.

A variety of gateway devices exist that enable the CIPT network to connect to various media and devices. The gateways use one of three protocols: H.323, MGCP, or Skinny gateway protocol. Be prepared with the proper information before configuring gateways.

In preparation for configuring gateways in Cisco CallManager, refer to Table 6-3.

Table 6-3 *Gateway Preparation*

Cisco IOS H.323 Gateway or Intercluster Trunk	Cisco IOS MGCP Gateway	Non-IOS MGCP Gateway	Skinny Gateway
H.225 or Intercluster Trunk	26XX 362X 364X 366X Catalyst 4000 Catalyst 4224 VG200	Catalyst 6000 E1 VoIP T1 VoIP	Catalyst 6000 24 port FXS Analog Interface Module AS-2, 4, 8 AT-2, 4, 8 DT-24+ DE-30+
IP Address	**Domain Name**	**MAC Address**	

This is required information for configuring a gateway. The information in Table 6-3 is specific to the gateway type you are configuring. For example, if you are configuring an H.323 gateway, you want to have the IP address of that gateway and enter it for the device name.

NOTE On each Gateway Configuration page, you will need to configure a number of settings that can be defined by accessing using the online help. When you are connecting to a CO, be sure to work with them for configuring clocking, PCM, protocol side and other settings.

The Description field on the Gateway Configuration page can be very helpful for troubleshooting and contacting the correct service provider if you think there is an issue with the line from the CO. For description information, you will want to include the circuit ID, physical location of the port, and phone number of the service provider. When calling in a trouble ticket to your provider, they will ask you for the circuit ID. Knowing where the port is physically located will help your service provider in troubleshooting any issues.

The sections that follow describe how to configure the four types of gateways listed in Table 6-2 and a gatekeeper in CCMAdmin.

Configuring IOS MGCP Gateways

Step 1 Open the Add a New Device page. In CCMAdmin go to **Device > Add a New Device**.

Step 2 **Select gateway from the list** of device types. Use the drop-down menu to select gateway. Select **Next**.

Step 3 Select a gateway type and device protocol. The following are IOS MGCP type gateways that do not require a device protocol:

— Cisco 26xx

— Cisco 362x

— Cisco 364x

— Cisco 366x

— Cisco Catalyst 4000 Access Gateway Module

— Cisco Catalyst 4224 Voice Gateway Switch

— Cisco VG200

Because of the variety of IOS MGCP gateways, we will discuss the 26xx so that we can compare configuring a 2600 as MGCP or H.323. After selecting the Cisco 26xx from the drop-down menu, press **Next**.

Step 4 Enter the Domain name and a description, and select a Cisco CallManager group. Enter the hostname of the device for the domain name.

NOTE If an IP domain name is configured on the router, it must be appended to the hostname.

Step 5 Select the **Installed Voice Interface Cards**. From the **Module in Slot 1** drop-down menu select the type of network module that is installed in the 2600. Select **Insert**.

Step 6 Select the sub-unit type. From the **Sub-Unit** drop-down menu select the type of voice interface card (VIC) or voice WAN interface cards (VWICs). Select **Update**.

Step 7 Select and configure an endpoint identifier. Configure the Gateway Information and Port Information. Select **Insert**.

NOTE Continue to follow these steps if you are configuring an analog port. If you are configuring a digital interface, after you select to configure an endpoint you select to use either T1-CAS or T1 PRI. After selecting the T1 type, you will open a Gateway Configuration page similar to a Catalyst 6000 T1/E1 Port Configuration page.

Step 8 Configure the directory number. Next to the endpoint identifier in the left-hand column, select **Add DN**. Configure directory number information for the endpoint and then **Insert**.

Step 9 Repeat Steps 6–8 to configure more endpoints.

Configuring Non-IOS MGCP Gateways

Step 1 Open the Add a New Device page. In CCMAdmin go to **Device > Add a New Device**.

Step 2 Use the drop-down menu to select a gateway from a list of device types. Select **Next**.

Step 3 Select a gateway type and device protocol. The following non-IOS MGCP gateways do require a device protocol:

— Cisco Catalyst 6000 24 port FXS Gateway

— Cisco Catalyst 6000 E1 VoIP Gateway

— Cisco Catalyst 6000 T1 VoIP Gateway

— DT-24+

— DE-30+

Depending on the gateway you select, the device protocol could be Analog Access, Digital Access PRI, or Digital Access T1. Select **Next** to continue.

Step 4 Configure the gateway. Enter appropriate settings for the gateway. For this type of gateway you will need to know the MAC address. For the description, a good rule to follow is to enter the following information:

— Circuit ID

— Physical location of port

— Contact number of service provider

Step 5 Complete the rest of the gateway settings and select **Insert**.

Configuring H.323 Gateways

Step 1 Open a Add a New Device page. In CCMAdmin go to **Device > Add a New Device**.

Step 2 Use the drop-down menu to select a gateway from the list of device types. Select **Next**.

Step 3 Select a gateway type and device protocol. Select **H.323** for the gateway type. For a remote Cisco CallManager, select **Inter-Cluster Trunk** for the device protocol. Use H.225 for non-Cisco CallManager devices. Select **Next**.

Step 4 Enter gateway configuration settings. Use an IP address for the device name and use the guidelines for the description. Configure the remaining gateway settings and then insert.

After configuring a gateway, you may need to assign that gateway to a route pattern or place the gateway into a route group. Refer to Chapter 4, "Cisco CallManager Administration Service Menu," for more information about route patterns, route groups, and route lists.

Cisco IOS MGCP gateways provide an opportunity to configure a directory number for the endpoint identifier. With the other gateways a route pattern must be associated to the gateway or a route list that includes the route group that the gateway is assigned.

A good reason for using a Cisco IOS MGCP gateway, rather than a H.323 gateway, is that the Cisco IOS MGCP gateway supports call preservation, which is discussed in Chapter 10, "Call Preservation."

Adding and Configuring a Gatekeeper

Within a distributed call processing environment, a gatekeeper is used to provide call admission control across the WAN. With the introduction of Cisco CallManager Release 3.0(5), a gatekeeper can simplify dial plans between Cisco CallManagers.

Figure 6-13 shows the deployment model that uses a gatekeeper.

A gatekeeper device (also known as a Cisco Multimedia Conference Manager or MCM) supports the H.225 Registration, Admission, and Status (RAS) message set used for call admission control, bandwidth allocation, and dial pattern resolution. Only one gatekeeper device per Cisco CallManager cluster is allowed. A second gatekeeper may be added for redundancy using HSRP.

Figure 6-13 *Gatekeeper Deployment Model*

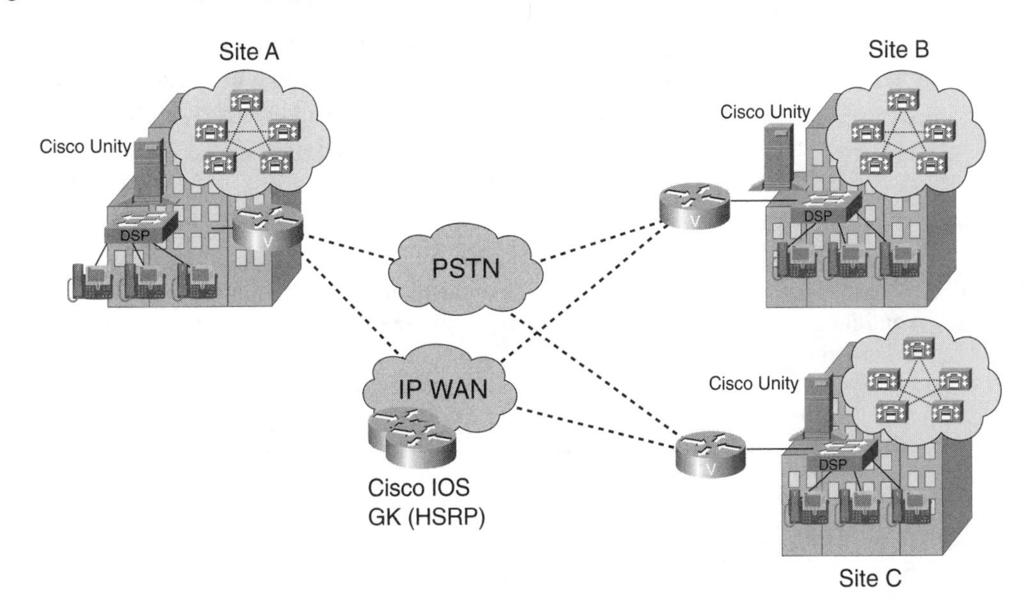

Two parts are required to configure the gatekeeper:

1. **Gatekeeper configuration**—This step is applicable when the network administrator configures a Cisco IOS MCM that acts as the gatekeeper. Recommended platforms include Cisco 2600, 3600, or 7200 routers with Cisco IOS Software Release 12.1(3)T or later.
2. **Cisco CallManager configuration**—Each Cisco CallManager or Cisco CallManager cluster must register with the gatekeeper as a single VoIP gateway.

NOTE The primary focus for using a gatekeeper device in a CIPT solution is for contolling bandwidth on small bandwidth WAN links in a distributed call processing deployment model.

A gatekeeper device in a distributed call processing deployment model can be configured one of two ways:

- **As a call admission control device**—As intercluster trunks are configured between clusters and before calls traverse the WAN, Cisco CallManager checks with the gatekeeper to determine whether there is enough bandwidth to make the call as shown in Figure 6-14 as "**Without** Anonymous Device Configured."
- **To control bandwidth and provide dial plan information**—If the gatekeeper device is configured as an anonymous device, it provides call admission control and dial plan information as shown in Figure 6-14 as "**With** Anonymous Device Configured."

Figure 6-14 shows a gatekeeper configured as both a gatekeeper and as an anonymous device in Cisco CallManager Administration.

NOTE When the gatekeeper is configured as an anonymous device and intercluster trunks are configured, the Cisco CallManager cluster will use the intercluster trunks for dial plan information, but will use the gatekeeper for call admission control. Cisco recommends that if the gatekeeper is configured as an anonymous device, then you should delete all intercluster trunks.

Configuring a Gatekeeper

Step 1 Open a Add a New Device page. In CCMAdmin go to **Device > Add a New Device**.

Step 2 Select gatekeeper from the list of device types. Select **Next**.

Figure 6-14 *Anonymous or Not?*

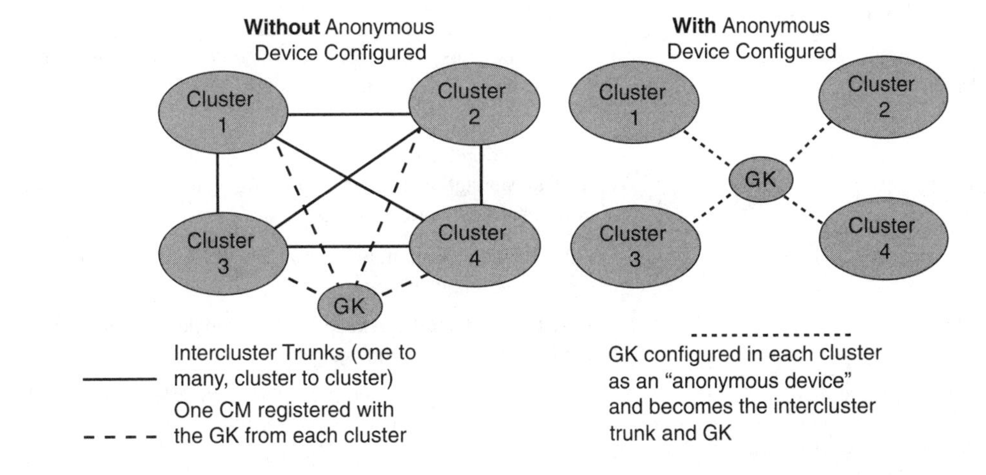

Step 3 Configure gatekeeper settings as shown in Figure 6-15 and detailed in Table 6-4. Enter the IP address of the gatekeeper as the **Gatekeeper Name**. (If you are using HSRP, use the HSRP IP address that is being shared.) For the **Description**, enter the physical location of the router.

Table 6-4 *Gatekeeper Configuration Settings*

Field	Description
Gatekeeper Name	Enter the IP address or DNS name of the gatekeeper. You can register only one gatekeeper per Cisco CallManager cluster.
Description	Enter a descriptive name for the gatekeeper.
Registration Request Time to Live	Do not change this value unless instructed to do so by a Cisco TAC engineer. Enter the time in seconds. The default value is 60 seconds. The Registration Request Time to Live field indicates the length of time that the gatekeeper considers a registration request (RRQ) valid. The system must send a KeepAlive RRQ to the gatekeeper before the RRQ Time to Live expires. Cisco CallManager sends an RRQ to the gatekeeper to register and subsequently to maintain a connection with the gatekeeper. The gatekeeper may confirm (RCF) or deny (RRJ) the request.

continues

Table 6-4 *Gatekeeper Configuration Settings (Continued)*

Field	Description
Registration Retry Timer	Do not change this value unless instructed to do so by a Cisco TAC engineer. Enter the time in seconds. The default value is 300 seconds. The Registration Retry Timer field indicates the length of time Cisco CallManager waits before retrying gatekeeper registration after a failed registration attempt.
Terminal Type	Use the Terminal Type field to designate the type for all devices controlled by this gatekeeper. Choose Gateway if all gatekeeper-controlled devices are gateways (including intercluster trunks). Choose Terminal if all gatekeeper-controlled devices are H.323 clients (for example, Microsoft NetMeeting devices). Make sure all gatekeeper-controlled devices are of the same type. Set this field to Gateway for normal gatekeeper call admission control.
Device Pool	Choose the appropriate device pool for the gatekeeper. A device pool specifies the collection of properties for the devices in that pool, such as Cisco CallManager group, date/time group, region, MRGL, and CSS for auto-registration.

Technology Prefix	Use this optional field to eliminate the need for entering the IP address of every Cisco CallManager when configuring the **gw-type-prefix** on the gatekeeper:
	
	• If you leave this field blank (the default setting), you must specify the IP address of each Cisco CallManager that can register with the gatekeeper when you enter the **gw-type-prefix** command on the gatekeeper.
	
	• When you use this field, make sure the value entered here exactly matches the *type-prefix* value specified with the **gw-type-prefix** command on the gatekeeper.
	
	For example, if you leave this field blank and you have two Cisco CallManagers with IP addresses of 10.1.1.2 and 11.1.1.3, enter the following **gw-type-prefix** command on the gatekeeper:
	
	`gw-type-prefix 1#* default-technology gw ip 10.1.1.`
	`gw ip 11.1.1.3`
	
	If you enter **1#*** in this field, enter the following **gw-type-prefix** command on the gatekeeper:
	
	`gw-type-prefix 1#* default-technology`

Table 6-4 *Gatekeeper Configuration Settings (Continued)*

Field	Description
Zone	Use this optional field to request a specific zone on the gatekeeper with which Cisco CallManager will register. The zone specifies the total bandwidth available for calls between this zone and another zone. • If you do not enter a value in this field, the **zone subnet** command on the gatekeeper determines the zone with which Cisco CallManager registers. Cisco recommends the default setting for most configurations. • If you want Cisco CallManager to register with a specific zone on the gatekeeper, enter the value in this field that exactly matches the zone name configured on the gatekeeper with the **zone** command. Specifying a zone name in this field eliminates the need for a **zone subnet** command for each Cisco CallManager registered with the gatekeeper. Refer to the command reference documentation for your gatekeeper for more information.
Allow Anonymous Calls	Check this box to enable Cisco CallManager to send calls to and receive calls from remote anonymous devices controlled by this gatekeeper. An anonymous device is one that you have not explicitly configured in the Cisco CallManager database. This setting eliminates the need for you to configure a separate H.323 device in the local Cisco CallManager cluster for each remote Cisco CallManager or H.323 gateway that it can call over the IP WAN. When you enable Allow Anonymous Calls, you must also fill in the configuration settings listed subsequently in this table. If you uncheck this box, you must configure a separate H.323 gateway for each remote device that the local Cisco CallManager can call over the IP WAN. The default setting disables Allow Anonymous Calls. When you enable Allow Anonymous Calls and fill in the remaining fields in this table, you essentially create a device (an intercluster trunk or gateway) named AnonymousDevice that can send calls to and receive calls from any remote Cisco CallManager controlled by the gatekeeper. The AnonymousDevice gets its device characteristics (such as Cisco CallManager group, region, and so on) from the gatekeeper device pool.

continues

Table 6-4 *Gatekeeper Configuration Settings (Continued)*

Field	Description
Device Protocol	Choose the appropriate protocol for the AnonymousDevice. Choose Intercluster Trunk if the AnonymousDevice is an intercluster trunk (Cisco CallManager), or choose H.225 if the AnonymousDevice is a gateway.
Calling Search Space	Choose the appropriate CSS for the AnonymousDevice. The CSS specifies the collection of route partitions searched to determine how to route a collected (originating) number.
Location	Choose the appropriate location for the AnonymousDevice. The location specifies the total bandwidth available for calls between this location and the central location or hub. A location setting of None specifies unlimited available bandwidth.
Caller ID DN	Enter the pattern, from 0 to 24 digits, that you want to use to format the caller ID on outbound calls from the AnonymousDevice. For example, in North America • 555XXXX = Variable Caller ID, where X represents an extension number. The CO appends the number with the area code if you do not specify it. • 5555000 = Fixed Caller ID. Use this form when you want the Corporate number to be sent instead of the exact extension from which the call is placed. The CO appends the number with the area code if you do not specify it.
Calling Party Selection	Choose the directory number sent on an outbound call on a gateway. The following options specify which directory number is sent: • Originator—Send the directory number of the calling device. • First Redirect Number—Send the directory number of the redirecting device. • Last Redirect Number—Send the directory number of the last device to redirect the call.
Presentation Bit	Choose whether the CO transmits or blocks caller ID. • Choose Allowed if you want the CO to send caller ID. • Choose Restricted if you do not want the CO to send caller ID.

Table 6-4 *Gatekeeper Configuration Settings (Continued)*

Field	Description
Display IE Delivery	Check this check box to enable delivery of the display information element (IE) in SETUP and CONNECT messages for the calling and called party name delivery service. The default setting leaves this check box unchecked.
Media Termination Point Required	Indicate whether a media termination point (MTP) is used to implement features that H.323v1 does not support, such as hold and transfer. You must use an MTP if you need a transcoder. Check the Media Termination Point Required check box if you want to use an MTP to implement features. Uncheck the MTP Required check box if you do not want to use an MTP to implement features. Use this check box only for H.323 clients and those H.323 devices that do not support the H.245 empty capabilities set (introduced in H.323v2).
Num Digits	Use this field only if you check the Sig Digits check box. Choose the number of significant digits, from 0 to 32, to collect for incoming calls to the AnonymousDevice. Cisco CallManager counts significant digits from the right (last digit) of the number called. This field processes incoming calls and indicates the number of digits starting from the last digit of the called number used to route calls coming into the H.323 device. *See* Prefix DN and Sig Digits.
Sig Digits	Check or uncheck this box depending on whether you want to collect significant digits. Choose significant digits to represent the number of final digits retained on inbound calls. A trunk with significant digits enabled truncates all but the final few digits of the address provided by an inbound call. If this box is unchecked, Cisco CallManager does not truncate the inbound number. If this box is checked, you also need to choose the number of significant digits to collect. (*See* Num Digits.)
Prefix DN	Enter the prefix digits that are appended to the called party number on incoming calls. Cisco CallManager adds prefix digits after first truncating the number in accordance with the Num Digits setting.

continues

Table 6-4 *Gatekeeper Configuration Settings (Continued)*

Field	Description
Run H225D On Every Node	This setting determines which Cisco CallManager in the cluster establishes the H.225 session. The default setting (checked) establishes the H.225 session on the Cisco CallManager where the calling device has registered. For most systems, the default setting works best. Unchecking this box establishes the H.225 session on the controlling Cisco CallManager in the same Cisco CallManager group and device pool as the H.225 gateway. Do not uncheck this box unless advised to do so by a Cisco TAC engineer.
Called party IE number type unknown	Choose the format for the type of number in called party directory numbers. Cisco CallManager sets the called directory number type. Cisco recommends that you do not change the default value unless you have advanced experience with dialing plans, such as NANP or the European dialing plan. You may need to change the default in Europe because Cisco CallManager does not recognize European national dialing patterns. You can also change this setting when connecting to PBXs using routing as a non-national type number. Choose one of the following options: • CallManager—Cisco CallManager sets the directory number type. • Unknown—The dialing plan is unknown. • National—Use when you are dialing within the dialing plan for your country. • International—Use when you are dialing outside the dialing plan for your country.
Calling party IE number type unknown	Choose the format for the type of number in calling party directory numbers. Cisco CallManager sets the calling directory number type. Cisco recommends that you do not change the default value unless you have advanced experience with dialing plans, such as NANP or the European dialing plan. You may need to change the default in Europe because Cisco CallManager does not recognize European national dialing patterns. You can also change this setting when connecting to PBXs using routing as a non-national type number.

Table 6-4 *Gatekeeper Configuration Settings (Continued)*

Field	Description
Calling party IE number type unknown (Cont.)	Choose one of the following options: • CallManager—Cisco CallManager sets the directory number type. • Unknown—The dialing plan is unknown. • National—Use when you are dialing within the dialing plan for your country. • International—Use when you are dialing outside the dialing plan for your country.
Called Numbering Plan	Choose the format for the numbering plan in called party directory numbers. Cisco CallManager sets the called directory number numbering plan. Cisco recommends that you do not change the default value unless you have advanced experience with dialing plans, such as NANP or the European dialing plan. You may need to change the default in Europe because Cisco CallManager does not recognize European national dialing patterns. You can also change this setting when connecting to PBXs using routing as a non-national type number. Choose one of the following options: • CallManager—Cisco CallManager sets the Numbering Plan in the directory number. • ISDN—Use when you are dialing outside the dialing plan for your country. • National Standard—Use when you are dialing within the dialing plan for your country. • Private—Use when you are dialing within a private network. • Unknown—The dialing plan is unknown.
Calling Numbering Plan	Choose the format for the numbering plan in calling party directory numbers. Cisco CallManager sets the calling directory number numbering plan. Cisco recommends that you do not change the default value unless you have advanced experience with dialing plans, such as NANP or the European dialing plan. You may need to change the default in Europe because Cisco CallManager does not recognize European national dialing patterns. You can also change this setting when connecting to PBXs using routing as a non-national type number.

continues

Table 6-4 *Gatekeeper Configuration Settings (Continued)*

Field	Description
Calling Numbering Plan (Cont.)	Choose one of the following options: • CallManager—Cisco CallManager sets the Numbering Plan in the directory number. • ISDN—Use when you are dialing outside the dialing plan for your country. • National Standard—Use when you are dialing within the dialing plan for your country. • Private—Use when you are dialing within a private network. • Unknown—The dialing plan is unknown.

Figure 6-15 *Gatekeeper Configuration Page*

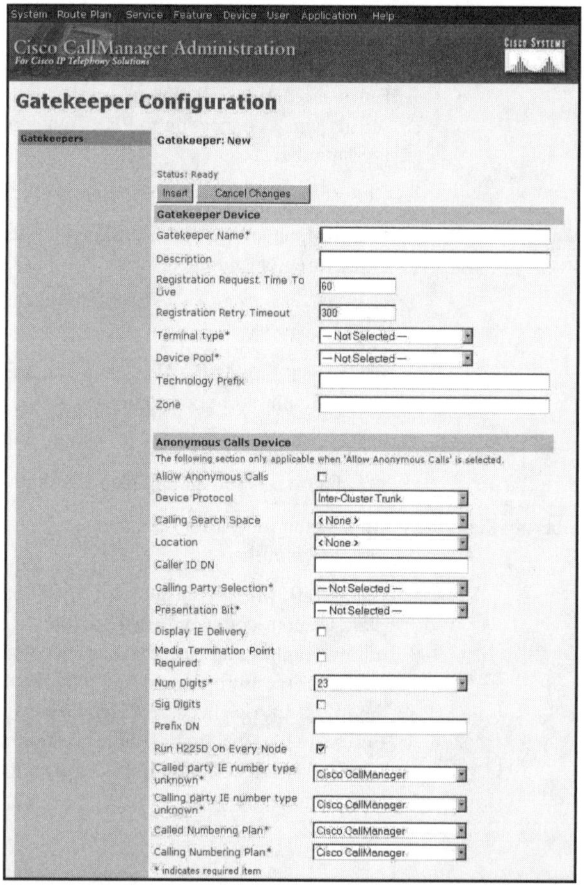

System Route Plan Service Feature Device User Application Help

Cisco CallManager Administration
For Cisco IP Telephony Solutions

Cisco Systems

Gatekeeper Configuration

Gatekeepers

Gatekeeper: New

Status: Ready

Insert Cancel Changes

Gatekeeper Device

Gatekeeper Name*

Description

Registration Request Time To Live 60

Registration Retry Timeout 300

Terminal type* — Not Selected —

Device Pool* — Not Selected —

Technology Prefix

Zone

Anonymous Calls Device

The following section only applicable when 'Allow Anonymous Calls' is selected.

Allow Anonymous Calls

Device Protocol Inter-Cluster Trunk

Calling Search Space < None >

Location < None >

Caller ID DN

Calling Party Selection* — Not Selected —

Presentation Bit* — Not Selected —

Display IE Delivery

Media Termination Point Required

Num Digits* 23

Sig Digits

Prefix DN

Run H225D On Every Node

Called party IE number type unknown* Cisco CallManager

Calling party IE number type unknown* Cisco CallManager

Called Numbering Plan* Cisco CallManager

Calling Numbering Plan* Cisco CallManager

* indicates required item

NOTE If you are using the gatekeeper only for call admission control with intercluster trunks, complete the Gatekeeper Device section only.

If you are using the gatekeeper for call admission and dial plan processing, complete the Gatekeeper Device section and the Anonymous Calls Device section.

Step 4 Complete the configuration and select **Insert**.

After adding a gatekeeper as an anonymous device, be sure to create a route pattern that points to the anonymous device. Refer to Chapter 4 for more information about configuring route patterns.

Computer Telephony Interface (CTI)

Computer telephony interface (CTI) is a software driver that enables telephony applications to connect to a server and other network applications and devices.

CTI applications enable you to perform such tasks as retrieving customer information from a database based on information provided by caller ID. CTI applications can also enable you to use information captured by an interactive voice response (IVR) system so that the call can be routed to the appropriate customer service representative or so that the information is provided to the individual receiving the call.

The following list contains descriptions of some of the available Cisco CTI applications that use CTI route points and CTI ports:

- **Cisco IP SoftPhone**—A desktop application that turns your computer into a full-feature telephone with the added advantages of call tracking, desktop collaboration, and one-click dialing from online directories. Use Cisco IP SoftPhone in tandem with a Cisco IP Phone to place, receive, and control calls from your desktop PC. All features function in both modes of operation.
- **Cisco IP AutoAttendant**—This application works with Cisco CallManager to receive calls on specified telephone extensions and to allow the caller to select an appropriate extension.
- **Cisco WebAttendant**—This application provides a Web-based interface for controlling a Cisco IP Phone to perform attendant console functions.
- **Personal Assistant**—This application is a virtual secretary or personal assistant that can selectively handle your incoming calls and help you make outgoing calls.

NOTE When you create a user in CCMAdmin, and the user is going to use a CTI application, be sure to check the **Enable CTI Application Use** check box on the Add a User page. If you do not check this check box, the CTI application does not work properly for that user.

Figure 6-16 shows how CTI is used in a CIPT network.

Figure 6-16 *Computer Telephony Interface*

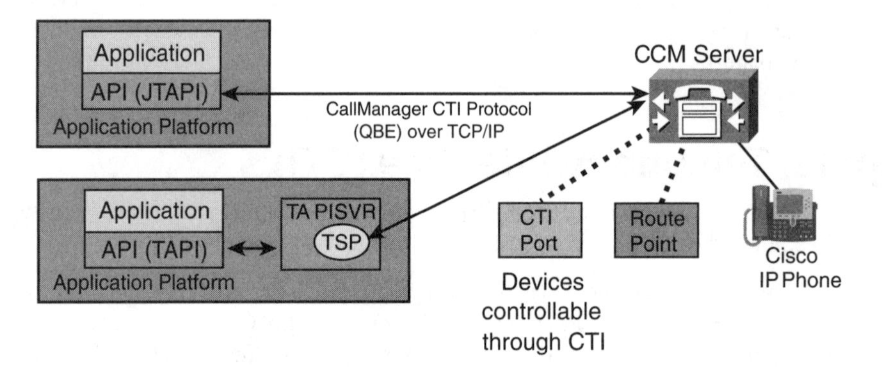

Applications that are identified as users can control CTI devices. When users have control of a device, they can control certain settings for that device, such as speed dial and call forwarding.

A CTI route point can receive multiple, simultaneous calls for application-controlled redirection. You can configure one or more lines on a CTI route point that users can call to access the application.

CTI route points must associate with device pools containing the list of eligible Cisco CallManagers for those devices.

For first-party call control, you must add a CTI port for each active voice line.

Cisco Voice Mail

The optional Cisco Unity software, available as part of Cisco IP Telephony solution, provides voice messaging capability for users when they are unavailable to answer calls.

The Cisco Voice Mail Port Wizard enables you to quickly add and delete ports associated with a Cisco voice mail server to the Cisco CallManager database.

Cisco Unity Service Parameters

This section discusses the service parameters in Cisco CallManager that need to be configured to integrate with Cisco Unity.

Figure 6-17 shows the main service parameters related to Cisco Unity.

Figure 6-17 *Cisco Unity Service Parameters*

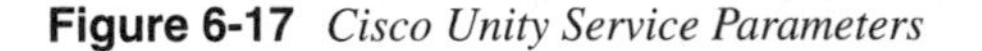

The following Cisco CallManager service parameters must be set up when configuring Cisco CallManager to work with Cisco Unity:

- **MessageWaitingOnDN and MessageWaitingOffDN**—Cisco Unity uses the MWI On and MWI Off directory numbers specified by these two service parameters to turn the message waiting indicator (MWI) on a user's phone on or off. The values for these parameters should match the CMMWIOffNumber value and the CMMWIOnNumber value in Cisco Unity.

NOTE For Cisco IP Phone 12 SP+ and 30 VIP models, the phone button template for the user's phone must have a button configured for Message Waiting for this feature to be available. This is also true for the Cisco IP Phone 7910, if you do not use the default 7910 phone button template.

- **VoiceMail-Voice mail pilot number (the number users dial to call in to the voice mail system)**—Setting this parameter enables you to configure a single button on users' phones for automatically dialing the voice mail pilot number (for example, the **messages** button on a Cisco IP Phone 79xx).
- **ForwardNoAnswerTimeout**—Specifies the number of seconds to wait before forwarding on a No Answer condition. The suggested and default value is 24 seconds.
- **ForwardMaximumHopCount**—Specifies the maximum number of attempts to extend a forwarded call. The default value is 12. This number should equal to the number of Cisco Unity ports plus one.

NOTE You must set the MWI On/Off service parameters for each Cisco CallManager in the cluster.

Adding and Configuring Device Profiles

Figure 6-18 shows the process used to authenticate a user that has a device profile configured to use extension mobility.

Figure 6-18 *Device Profile*

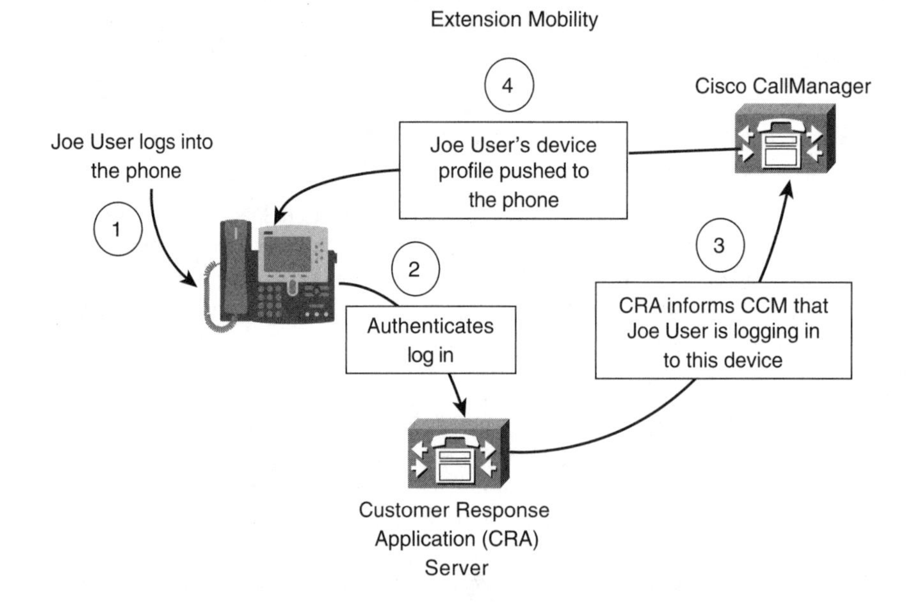

A device profile comprises the set of attributes (services and or features) associated with a particular device. Device profiles are used with the extension mobility feature that provides users the ability to log in and out of Cisco IP Phones and have their settings appear on the phone they are logged into. Device profiles include name, description, phone template, expansion modules, directory numbers, subscribed services, and speed dial information. Two kinds of device profiles exist: autogenerated and user. You can assign the user device profile to a user so that when the user logs into a device the user device profile you have assigned to that user loads onto that device as a default login device profile. Once a user device profile is loaded onto the phone, the phone picks up the attributes of that device profile.

You can also assign a user device profile to be the default logout device profile for a particular device. When a user logs out of a phone, for instance, the logout device profile loads onto the phone, giving that phone the attributes of the logout device profile. You can create, modify, or delete the user device profile in CCMAdmin.

The autogenerated device profile generates when you update the phone settings and choose a current setting to generate an autogenerated device profile. The autogenerated device profile associates with a specific phone to be the logout device profile. You can modify the autogenerated device profile but not delete it.

NOTE You may assign a default user device profile to a user for extension mobility purposes. If no profile is specified at the time of the login, Cisco CallManager uses the default profile.

Summary

Cisco IP Phones are full-featured telephones that plug directly into your IP network. You can use the Phone Configuration page in CCMAdmin to configure the following Cisco IP Phones and devices:

- Cisco IP Phone 79*xx* Family and 12SP+ and 30VIPs
- H.323 Clients
- CTI Ports

There are a variety of configuration options and settings depending on which gateway you are adding to the system. Table 6-5 reiterates the different types of gateways.

Table 6-5 *Gateways for a Cisco IP Telephony Network*

Gateway	MGCP	H.323	Skinny Gateway Protocol
VG200	Yes, for FXS/FSO	Yes, with Cisco IOS Software 12.1(5)-XM1, the VG200 uses H.323 to support a wider range of digital and analog interfaces	No
DT-24+	Yes, with Cisco CallManager Release 3.1	No	Yes
827	No	Yes, for FXS	No
1750	No	Yes	No
3810 V3	Cisco IOS Software 12.1(3)T and Cisco CallManager Release 3.0(5)	Yes	No
2600	Cisco IOS Software 12.1(3)T and Cisco CallManager Release 3.0(5) Analog Interfaces only no E&M T1 CAS – 12.1(5)XM & 12.2.1T Q.931 PRI Backhaul – 12.2.2T[1]	Yes	No
3600	Cisco IOS Software 12.1(3)T and Cisco CallManager Release 3.0(5) Analog Interfaces only no E&M T1 CAS – 12.1(5)XM & 12.2.1T Q.931 PRI Backhaul – 12.2.2T[1]	Yes	No
7200	Cisco IOS Software 12.2.(1)T[2]	Yes	No

Table 6-5 *Gateways for a Cisco IP Telephony Network (Continued)*

Gateway	MGCP	H.323	Skinny Gateway Protocol
7500	Undecided	Yes (Cisco IOS Software 12.1.5)	No
5300[3]	Yes (Cisco IOS Software 12.1(1)T	Yes	No
Catalyst 4000 WS-X4604-GWY Gateway Module	Cisco CallManager Release 3.1	Yes, for PSTN interfaces	Yes, for conferencing and MTP/transcoding services
Catalyst 6000 WS-X6608-x1 Gateway Module & FXS Module WS-X6624	Cisco CallManager Release 3.1. T1/E1 module supporting PRI and CAS FXS module	No	Yes, for FXS module and T1/E1 prior to Cisco CallManager Release 3.1
Catalyst 4224	Projected for Cisco CallManager release 3.1	Yes	No

1 Cisco IOS Software Release 12.2.2T PRI Backhaul support for 26xx/36xx uses RUDP and is not compatible with Cisco CallManager. PRI backhaul with Cisco CallManager Release 3.1 as the Call Agent is scheduled for Cisco IOS Software Release 12.2.4T and uses TCP as the transport.

2 Not supported in Cisco CallManager.

3 While the 5300 supports MGCP, it is as a Trunk Gateway module using SS7 signaling, which is not supported in Cisco CallManager.

A gatekeeper device (also known as an MCM) supports the H.225 RAS message set used for call admission control, bandwidth allocation, and dial pattern resolution. You can configure only one gatekeeper device per Cisco CallManager cluster.

CTI enables you to leverage computer-processing functions while making, receiving, and managing telephone calls. The following list contains descriptions of some of the available Cisco CTI applications:

- **Cisco IP SoftPhone**—A desktop application that turns your computer into a full-feature telephone with the added advantages of call tracking, desktop collaboration, and one-click dialing from online directories. You can also use Cisco IP SoftPhone in tandem with a Cisco IP Phone to place, receive, and control calls from your desktop PC. All features function in both modes of operation.

- **Cisco IP AutoAttendant**—This application works with Cisco CallManager to receive calls on specific telephone extensions and to allow the caller to select an appropriate extension.
- **Cisco WebAttendant**—This application provides a Web-based interface for controlling a Cisco IP Phone to perform attendant console functions.
- **Personal Assistant**—A virtual secretary or personal assistant that can selectively handle your incoming calls and help you make outgoing calls.

Post-Test

How well do you think you know this information? The post-test is designed to help you gauge your knowledge about this chapter. Of the 10 questions, if you answer one to three questions correctly, we recommend that you re-read this chapter. If you answer four to seven questions correctly, we recommend that you review those sections that you need to know more about. If you answer 8 to 10 questions correctly, you probably understand this information well enough to move on to the next chapter. You can find the answer to these questions in Appendix B, "Answers to Chapter Pre-Test and Post-Test Questions."

1 Label the model number for the following Cisco IP Phones:

a.

b.

c.

d.

2 In a CIPT environment, what are the three supported gateway protocols?

3 Which gateway protocols are supported by the WS-6608-T/E1 and DT24+ gateways when using Cisco CallManager 3.1 and higher?

4 Which gateway protocols are supported by the Cisco 2600 and 3600 Series routers?

5 In a distributed call processing deployment, what device can be used to provide call admission control between clusters?

____

____

____

List the four Cisco provided CTI applications that use a CTI port or CTI route point.

6

____

____

____

7

____

____

____

8

____

____

____

9

____

____

____

10 What are the Cisco CallManager service parameters that need to be configured while integrating Cisco Unity?

____

____

____

Upon completing this chapter you will be able to do the following tasks:

- Given a Cisco CallManager server and the Bulk Administration Tool (BAT) executable files, install the BAT application.
- Given a list of phones and BAT, create a phone template and a phone CSV file, and add those phones using BAT.
- Given BAT and a list of gateways or ports to add to Cisco CallManager, create templates and CSV files and add the gateways or ports using BAT.
- Given a list of users, associated phones, and BAT, create a CSV file of phones and users, and then add those users and phones using BAT to the Cisco CallManager database.

CHAPTER 7

Understanding and Using the Bulk Administration Tool (BAT)

Use the Bulk Administration Tool (BAT), a Web-based application, enables you to perform bulk transactions. These transactions include adding, updating, or deleting a large number of phones, users, CTI ports, Cisco VG200 gateways and ports, and ports on a Cisco Catalyst 6000 FXS analog interface module to the Cisco CallManager database. Where this was previously a manual operation, BAT helps you automate the process and achieve much faster add, update, and delete operations. BAT also provides the Tool for Auto-Registered Phone Support (TAPS), an optional component of BAT.

When used with TAPS, BAT further reduces the manual labor involved in administering a large system by enabling you to add phones with dummy MAC addresses instead of entering each MAC address in the CSV file. Using TAPS, you can correct the dummy MAC addresses in the Cisco CallManager database later simply by dialing into the TAPS directory number and following a few voice prompts. You must individually update each phone that was added using a dummy MAC address, but you can pass this task onto the phone's user by providing simple instructions on how to use TAPS.

This chapter discusses the following topics:

- Abbreviations
- BAT/TAPS Overview
- BAT/TAPS Installation
- BAT/TAPS Features
- Templates
- CSV Files
- Adding Phones and Users (Example Transaction)
- Updating, Deleting, or Querying in BAT

Pre-Test

Do you already know this information? The pre-test is designed to help you gauge your knowledge about this chapter. Of the 10 questions, if you answer one to three questions correctly, we recommend that you read through this chapter. If you answer four to seven questions correctly, we recommend that you skim through this chapter, reading those sections that you need to know more about. If you answer 8 to 10 questions correctly, you probably understand this information well enough to skip this chapter. You can find the answers to the pre-test for this chapter in Appendix B, "Answers to Chapter Pre-Test and Post-Test Questions."

1 On what server should BAT be installed?

a. Primary

b. Publisher

c. Subscriber

d. Any of the above

2 Where are the BAT Excel template files located?

____

____

____

3 Describe the purpose of TAPS.

____

____

____

4 Describe the steps you must complete to add phones and users to the Cisco CallManager database using BAT.

a.

b.

c.

d.

5 When creating the CSV file for users or while inserting the users into the Cisco CallManager database using BAT, are password and PIN values required?

____

____

____

6 For values that appear on both the BAT template and the CSV file, which takes precedence?

____

____

____

7 Using BAT, can you add, update, or delete Cisco Catalyst 6000 Analog Interface Modules?

____

____

____

8 True or False? Auto-registration does not need to be enabled to use TAPS.

____

____

____

9 How does BAT save time over the usual method of adding, updating, or deleting devices or users in CCMAdmin?

____

____

____

10 During BAT installation you have the option of also installing TAPS. Other than BAT and Cisco CallManager, name the required component TAPS needs to function.

____

____

____

Abbreviations

This section defines the abbreviations used in this chapter. For more information about terms and abbreviations used in this chapter refer to the *IP Telephony Network Glossary* at the following URL:

www.cisco.com/univercd/cc/td/doc/product/voice/evbugl4.htm

Table 7-1 provides the abbreviation and the complete term.

Table 7-1 *Abbreviations and Complete Terms Used in the Chapter*

Abbreviation	Complete Term
BAT	Bulk Administration Tool
CCMAdmin	Cisco CallManager Administration
CRA	Customer Response Application
CSV	comma-separated value
CTI	Cisco Telephony Interface
IIS	Microsoft Internet Information Service
MAC	Media Access Control
TAPS	Tool for Auto-Registered Phone Support

BAT/TAPS Overview

BAT Release 4.2(1) is compatible with Cisco CallManager Release 3.1(1). You must install BAT on the same server as the Publisher database for Cisco CallManager. The BAT application, along with the TAPS application, uses approximately 27 MB of disk space for the applications and the online documentation.

Only Cisco CallManager system administrators require access to BAT; however, end users can use TAPS when instructed to do so by the system administrator.

BAT, a Web-based application, requires Internet Explorer 4.01 Service Pack 2 or later or Netscape 4.5 or later with the exception of Netscape 6.0. CCMAdmin provided the model for the look and feel of BAT. You can access BAT from CCMAdmin and vice versa using the Application menu.

CAUTION Use BAT only during off-peak hours. Otherwise, bulk transactions could affect the Cisco CallManager performance, and call processing may be adversely affected.

BAT/TAPS Installation

BAT/TAPS must be installed on the same server as the Publisher database for Cisco CallManager. During BAT/TAPS installation or reinstallation, the setup program halts the following services:

- IIS Admin
- World Wide Web publishing
- FTP publishing

These services automatically restart once the installation is complete.

TAPS requires a two-part installation:

1. You must install the first part on the same server as the Publisher database for Cisco CallManager; you can do this during BAT installation.
2. You must install the second part on the Cisco CRA server. You must have purchased Cisco CRA to use TAPS.

NOTE Template migration is provided if you are upgrading to BAT Release 4.2(1) from BAT Releases 4.0(1) or 4.1(1).

BAT does not support backward template migration; if you have installed BAT Release 4.2(1), no template migration will occur if you reinstall BAT Release 4.0(1) or 4.1(1).

If you are currently running BAT Release 3.0(3) and upgrade to BAT Release 4.2(1), no template migration will occur.

The BAT/TAPS installation process includes BAT Excel template files that are located in the C:\CiscoWebs\BAT\ExcelTemplate folder. Because you should not have Microsoft Excel installed on the Publisher database server, copy and paste these templates where you have Microsoft Excel installed and use the templates from that location.

BAT/TAPS Features

Using BAT, you can add, update, or delete large numbers of devices or users (or combinations of devices and users) in the Cisco CallManager database. You effect these add, update, or delete operations using a BAT template (in all cases except for users) and a CSV file. The BAT template sets up the basic values for the transaction, while the CSV file enables you to specify individual criteria for each user or device (or combination of the two). Values specified in the CSV file override values specified in the BAT template.

For example, say you want to add 200 phones and users. Of the 200, five belong to managers. For 195 phones, you want the Forwarding settings to be the general voice mail number. However, for the five manager phones, you want the Forward settings to roll to an assistant's phone rather than to voice mail. To do this, you can add all 200 phones in one transaction, specifying the general Forward Busy and Forward No Answer destinations in the line details for the BAT phone template. Then in the CSV file for those five phones, you specify the assistant's directory number in the Forward Busy and Forward No Answer fields. For the 195 phones that do not have an entry in the Forward Busy and Forward No Answer fields in the CSV file, the Forward settings will be the same as those you specified in the BAT template. For the other five phones, the Forward settings in the BAT template will be overridden by the settings you specified in the CSV file.

Use BAT Release 4.2(1) to perform the following operations to the Cisco CallManager database:

- Add, update, or delete phones or CTI ports.
- Add phones and users in one transaction or CTI ports and users in one transaction.
- Add users.
- Add or delete Cisco VG200 gateways and ports; you can work with FXS, FXO, T1-PRI, T1-CAS, or E1-PRI trunk interfaces. For FXS trunks, you can create a Gateway Directory Number template for directory numbers on POTS ports.
- Add, update, or delete ports on Cisco Catalyst 6000 analog interface modules. You can effect changes for ports only on existing Catalyst 6000 modules. BAT Release 4.2(1) cannot add or delete the modules themselves. You can create a Gateway Directory Number template for directory numbers on POTS ports.

Using TAPS, you can update dummy MAC addresses in the Cisco CallManager database. TAPS also enables you to:

- Protect directory numbers you specify from being used by TAPS (this is called *Secure TAPS*).
- Limit TAPS to updating only phones that were added to the Cisco CallManager database using BAT (this is called *Configure TAPS*).

NOTE To use TAPS, auto-registration must be enabled in Cisco CallManager and you must have configured TAPS on your Cisco CRA server.

BAT Templates

BAT provides Microsoft Excel templates that you use to perform bulk transactions. The templates used in BAT are different than the phone button templates used in CCMAdmin.

Each template provides fields based on the type of template—phones, gateways, or ports. To give you an idea of how BAT works, this book describes the process for adding phones and users in one transaction. Refer to the BAT documentation for detailed instructions for using each of the different templates:

www.cisco.com/univercd/cc/td/doc/product/voice/sw_ap_to/admin/bulk_adm/4_2_1/index.htm

CSV Files

CSV files are used in conjunction with BAT templates (in all cases except for users) when adding, updating, or deleting users or devices or combinations of users and devices. You provide information in the CSV file for each of the users or devices that you want to add or update. Information provided in CSV files, such as call forwarding designation, overrides information provided in BAT templates for the same fields. See the earlier section, "BAT/TAPS Features," for an example.

You can create a CSV file using two methods:

- Use the Microsoft Excel template called BAT.xlt.
- Create the CSV using a sample text file.

Cisco recommends you use the BAT.xlt template because the data is validated automatically when you export to CSV format.

Understanding the BAT.xlt File

The BAT.xlt file simplifies the creation of CSV files. It provides validation and error checking automatically to help reduce configuration errors. The BAT.xlt file provides several tabs (along the bottom edge of the file) that enable you to create CSV files for the various devices and user combinations in BAT. Figure 7-1 shows an example of a CSV file in BAT.xlt.

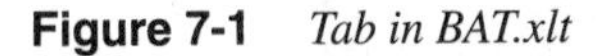

Figure 7-1 *Tab in BAT.xlt*

To use the BAT.xlt file to create a CSV file, first click the tab for the type of device with which you want to work. For example, to add phones and users all at once, click the tab marked **Phones-Users**. Each tab provides the field name (whether it is a required or optional field) and the maximum number of characters allowed.

The CSV file works in combination with the BAT template. For example, on the Phone tab in the BAT.xlt file, you can leave Location, Forward Busy Destination, or Call Pickup Group for any record on the CSV file blank. The values from the BAT phone template will be used for these fields. However, if you specify values in the CSV file for those fields, those values override the values for these fields that were set in the BAT phone template.

Creating a Text-Based CSV Text File

If you do not use the BAT.xlt file for data input when adding or updating devices or users, you must create the CSV file using lines of ASCII text with values separated by commas. See the BAT documentation, *Bulk Administration Tool User Guide*, for detailed instructions. The documentation is available from the Help menu in BAT or on CCO at the link supplied earlier in this chapter.

Adding Phones and Users (Example Transaction)

The BAT phone template and CSV file work together in bulk transactions. Based on the type of phone you want to add in a batch, you can create a template that has features that are common to all the phones in that batch, such as model, device pool, and so on. The system stores these templates, so that they are reusable for future bulk transactions. For example, you can configure a template for the Cisco IP Phone 7960 with two lines and another Cisco IP Phone 7960 template with four lines configured. Then when you need to add a large number of phones with the same configuration, you can reuse the existing template.

The CSV file stores the details for each individual phone, such as its MAC address, description, and so on. Because you customize CSV files for each bulk transaction, you are less likely to reuse them than BAT templates.

You create a phone template by specifying values in the phone template fields. You can also specify line attributes, Cisco IP Phone services, and speed dials (if applicable). The phone settings for the BAT phone template require similar values to those you enter when adding a phone in CCMAdmin; however, you must use the BAT phone template when performing bulk operations in BAT.

NOTE Make sure phone settings such as device pool, location, calling search space, and button template have already been configured in CCMAdmin prior to creating the template. You cannot create new settings in BAT.

Use the following steps to create the phone template. You can then add lines, services, and speed dials.

Step 1 Start BAT and click **Configure > Template > Phone**. Figure 7-2 shows the Phone Template Configuration Web page.

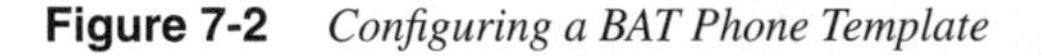

Figure 7-2 *Configuring a BAT Phone Template*

Step 2 In the Device Information area, enter the settings for the phone model for which you are creating the template.

Step 3 Click **Insert** to create the BAT phone template. If an insert was not successful, you can click the **View Log** button to view details about the insert attempt in the log file.

Step 4 Once the status indicates the insert completed, scroll down the page to the Line Details area to add line attributes (if applicable). The button template in use for this BAT template determines the number of lines you can add or update.

Step 5 Click **Add Line**. A popup window appears. Figure 7-3 shows the Line Details popup window.

Figure 7-3 *Line Details Window*

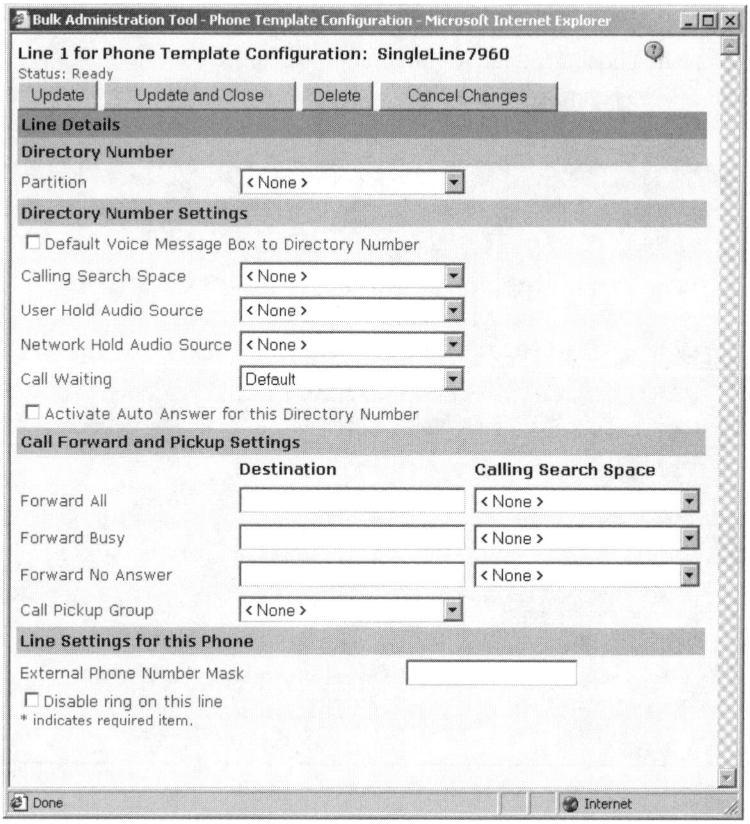

Step 6 Enter or choose the appropriate values for the line settings. Remember that all phones in this batch for this line will use the settings that you choose here. All fields are optional.

Step 7 Click **Update and Close**. BAT inserts the line settings to the database, and the popup window closes.

Step 8 Repeat Steps 5 through 7 to add settings for any additional lines.

Step 9 If you want to add Cisco IP Phone services to the template (Cisco IP Phone models 7960 and 7940 only), complete the remaining steps. If you do not want to specify services but do want to specify speed dials, skip to Step 17.

Step 10 You can subscribe Cisco IP Phone services to the phones. Only Cisco IP Phone models 7960 and 7940 include this feature. Click **Update Services** in the upper right corner of the window.

A popup window appears. In this window, you can subscribe to Cisco IP Phone services. Figure 7-4 shows the Line Details popup window.

Figure 7-4 *Cisco IP Phone Services Window*

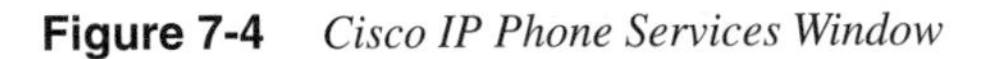

Step 11 In the **Select a Service** box, choose a service to which you want all phones to be subscribed. The Service Description box displays details about the service you selected.

Step 12 Click **Continue**.

Step 13 In the Service Name field, you can modify the name of the service, if desired.

Step 14 Click **Subscribe** to subscribe all phones to this service.

Step 15 Repeat Steps 10 through 14 to add more services.

Step 16 Close the popup window.

Step 17 You can designate speed dials for the phones if the Phone Button Template has provided speed dial buttons. If you want to add speed dials to the template, complete the following steps. Otherwise, skip Steps 18 through 22.

Step 18 Click **Update Speed Dial Buttons** in the upper right corner of the window.

A popup window appears. In this window, you can designate speed dial buttons for base Cisco IP Phones and expansion modules. The number of speed dial buttons available for this template depends on the Phone Button Template in use for this BAT template. Expansion module sections do not display for phone models other than 7960. Figure 7-5 shows the Speed Dials popup window.

Figure 7-5 *Speed Dials Window*

Step 19 In the Speed Dial Settings for Base Phone area, enter the number in the **Speed Dial Number** fields, including any access or long distance codes.

Step 20 In the **Speed Dial Label** fields, enter a corresponding label for each speed dial number you entered.

Step 21 Repeat Steps 19 through 21 to set speed dials for any expansion modules, if applicable (Cisco IP Phone 7960 templates only).

Step 22 Click **Update and Close**.

BAT inserts the speed dial buttons to the database, and the popup window closes.

You have successfully created a BAT phone template and provided line details, services, and speed dials where applicable. Now you must specify phone and user details in the CSV file you will create for this bulk transaction.

Creating a CSV File for Phones and Users

The CSV file for phones contains information about each phone as a record. Make sure all phones in a CSV file are the same model and have the same number of configured lines. You can associate the phones to an existing user. To associate more than one phone to an existing user, you need to write the required information in separate records. For example, to associate two new Cisco IP Phone 7960s to an existing user, you need to write two records in the CSV file—one for each Cisco IP Phone 7960 but each with the same user ID.

The CSV file contains duplicates of some of the values from the BAT template. Values in the CSV file override any values set in the BAT phone template. For example, you can set speed dial buttons and labels in the BAT phone template, as well as in the CSV file. This override feature allows for special configuration in some cases. For example, if you want most of the phones in the bulk-add transaction to be redirected to voice mail, you can set the Call Forward Busy (CFB) and Call Forward No Answer (CFNA) fields to the voice mail number. But for a few phones in the bulk-add transaction, you want the calls to be redirected to a secretary instead of voice mail; for only those phones, you can specify the secretary's directory number in the CFB and CFNA fields in the CSV file. This enables most of the phones to use the CFB and CFNA values from the BAT phone template, but select phones use the secretary's directory number as specified in the CSV file instead.

The CSV file for phones can contain multiple directory numbers depending on whether the BAT phone template in question supports multiple lines.

NOTE The number of directory numbers entered in the CSV file must equal the number of lines configured in the phone template, or an error will result.

You can create a CSV file either by using the Microsoft Excel template called BAT.xlt or by using a sample text file. Because Cisco recommends you use the BAT.xlt template, that is how it will be explained in this chapter.

The BAT.xlt file provides data file templates with macros, support for multiple phone lines, and error checking, and exports the values into CSV files for phones, users, CTI ports, phone/user combinations, CTI port/user combinations, Cisco VG200 gateways, and FXS ports on Cisco Catalyst 6000 analog interface modules.

The information you provide in the CSV file, in combination with the information provided in the BAT template for phones, is used to add the Cisco IP Phones to the Cisco CallManager database and associate them with the specified users.

The list that follows details the procedure for creating a CSV file for phones and users.

Step 1 The BAT.xlt file resides on the Publisher database server; however, you should not have Microsoft Excel running on the Publisher database server. You must copy the file from the Publisher database server to the local machine on which you plan to work.

Using a floppy disk or a mapped network drive, open the path C:\CiscoWebs\BAT\ExcelTemplate on the Publisher database server, and copy the file BAT.xlt to a local machine where Microsoft Excel is installed.

Step 2 Double-click BAT.xlt.

Step 3 When prompted, click **Enable Macros**.

Step 4 Click the tab for the type of CSV file you want to add, in this case **Phones-Users**. Figure 7-6 shows the Phones-Users tab in the BAT.xlt file.

Step 5 Scroll to the right side of the template until you see the **Number of Phone Lines** box. In that box, enter the number of lines equal to the number of directory numbers.

NOTE The number of lines you specify here must match the number of lines configured in the BAT template, or an error will result when you attempt to insert a BAT phone template and CSV with mismatched number of lines.

Figure 7-6 *Creating the Phones-Users CSV file in BAT.xlt*

First Name (String[50] MANDATORY)	Last Name (String[50] MANDATORY)	User Id (String[30] MANDATORY)	Password (String[20] OPTIONAL)	Manager (String[30] OPTIONAL)	Department Number (String[50] OPTIONAL)
John	Smith	jsmith		paino	
Pierre	Renault	prenault		tmulder	
Vishwa	Patel	vipatel		rohitsri	
Jonathan	Miller	jonathan		vincem	
Sally	Jones	sallyj		paino	
Chris	Peterson	cpeterso		paino	
Kim	Lee	kimxlee		vincem	
Manuel	Rivas	mrivas		rohitsri	
Jack	Hu	jackhu		rohitsri	
Anton	Smit	asmit		beasley	
Albert	Morris	amorris		tmulder	
Cindy	Perez	cperez		vincem	
Susie	Paino	paino		beasley	
Lola	Kendrick	kendrick		akanksha	
Lyle	Strand	lstrand		akanksha	
Sandy	Schneider	sschneid		alsmith	
Hans	Andersen	handerse		paino	
Johan	Lang	jlang		alsmith	
Vince	Messersmith	vincem		alsmith	
Joe	Sanderson	sanderso		akanksha	

Step 6 In the **Number of Speed Dials** box, enter the number of speed dial buttons that are configured on the BAT phone template.

NOTE The number of speed dials you specify here cannot exceed the number of speed dials configured in the BAT template or an error will result when you attempt to insert the BAT phone template and CSV file.

Step 7 Complete all mandatory fields and any relevant optional fields. Each column heading specifies the length of the field and whether it is required or optional. If you have multiple devices, several fields will appear multiple times—once for each device.

NOTE The system treats blank rows in the spreadsheet as "End of File" and discards subsequent records.

Step 8 In each row, provide the information documented in Table 7-2.

Table 7-2 *CSV Fields*

Field	Description
First Name	The first name of the user to whom this phone will be issued, up to 50 characters.
Last Name	The last name of the user to whom this phone will be issued, up to 50 characters.
User ID	The user ID for the user to whom this phone will be issued.
Password	The password the user needs to access the Cisco IP Phone User Options Web page. Although considered optional in the CSV file, you must provide a password. You can specify the password either on the CSV file or during user insertion in BAT. If you want to apply individual passwords for each user, specify the password in the CSV file. If you want to use a default password that can be used by all users, do not specify the password in the CSV and instead provide this information when you insert the users in BAT.
Manager	The manager's user ID for the user to whom this phone will be issued.
Department	The department number for the user to whom this phone will be issued.
PIN	The PIN to be used for extension mobility. Although considered optional in the CSV file, you must provide a PIN. You can specify the PIN either on the CSV file or during user insertion in BAT. If you want to apply individual PINs for each user, specify the PIN in the CSV file. If you want to use a default PIN that can be used by all users, do not specify the PIN in the CSV and instead provide this information when you insert the users in BAT.
User Device Profile	The user device profile for this user. A user device profile specifies basic device information, such as the phone button template, and is used in connection with the extension mobility feature.
MAC Address	The MAC address. This is optional if you plan to use dummy MAC addresses.
Description	Description of the phone, such as the MAC address preceded with "SEP," or something more descriptive like "Conference Room A" or "John Smith" if the phone is going to be placed in a conference room or given to a specific user.

Table 7-2 *CSV Fields (Continued)*

Field	Description
Location	If you provided a location in BAT phone template, you can leave this field blank to use the value from the BAT phone template. A location indicates the remote location accessed using restricted bandwidth connections.
Directory Number	The directory number for the phone.
Voice Message Box	The directory number used to identify the voice mail box. This will be the same as the directory number, unless the voice mail box number and directory number for a user are different.
Display	The text that you want to appear on the called party's phone display, such as the user's name (John Smith) or the phone's location (Conference Room 1).
Forward Busy Destination	The directory number to which calls should be forwarded when the phone is busy. To use the value provided in the BAT phone template, leave this field blank.
Forward No Answer Destination	A directory number assigned to phones that allow the user to answer a call that comes in on a directory number other than his or her own. To use the value provided in the BAT phone template, leave this field blank.
Call Pickup Group	The number that can be dialed to answer calls to this directory number. To use the value provided in the BAT phone template, leave this field blank.
Speed Dial	The complete number, including access or long distance codes, that you want users to be able to dial at the press of the speed dial button.
Speed Dial Label	A description of the speed dial number; for example, "555-1234," "Security," or "Cafeteria."

Step 9 (*Optional*) To use the dummy MAC address option, check the **Create Dummy MAC Address** box. You must enter the MAC address or use the dummy MAC address option. If you choose the dummy MAC address option, you can update the phones later with the correct MAC address by manually entering this information into CCMAdmin for each phone or by using the TAPS tool.

Step 10 Click **Export to BAT Format** to transfer the data from the BAT Excel spreadsheet into a CSV file. The system saves the file to C:\XLSDataFiles (or to your choice of another existing folder) as

```
tabname#timestamp.txt
```

where *tabname* represents the type of CSV file you created (such as phones, phones-users) and *timestamp* represents the precise date and time the file was created.

Step 11 To be accessed by BAT, the CSV file must reside on the Publisher database server. However, you normally would not have Microsoft Excel running on the Publisher database server. So this step assumes that you have saved the CSV file to the local machine (not the Publisher database server). In that case, you must copy the file to Publisher database server.

Using a floppy disk or a mapped network drive, copy the CSV file from C:\XLSDataFiles\ to C:\BATFiles\PhonesUsers folder on the server running the Publisher database for Cisco CallManager.

Step 12 For information on how to read the exported CSV file, click the link to **View Sample File** in the Insert Phones/Users window in BAT (**Configure > Phones/Users**).

Adding Phones/User Combination to Cisco CallManager

Once you have created the BAT phone template and the Phones-Users CSV file, you are ready to insert the phones into the Cisco CallManager database using BAT as documented in the following series of steps.

Step 1 Click **Configure > Phones/Users**. Figure 7-7 shows the **Insert Phones/Users** window in BAT.

Step 2 In the **File Name** field, choose the CSV file that you created for this type of bulk transaction.

Step 3 In the **Phone Template Name** field, choose the BAT template that you created for this type of bulk transaction.

NOTE If you want to insert phones that require different phone templates, you must create separate CSV files. The Line Details link shows how many lines are configured for the selected template.

Step 4 If you did not enter individual MAC addresses in the CSV file, you must check the **Create Dummy MAC Address** check box. You can update the phones or devices later with the correct MAC address using TAPS or by manually entering this information into CCMAdmin for each phone.

Figure 7-7 *Inserting Phones/Users to the Cisco CallManager Database*

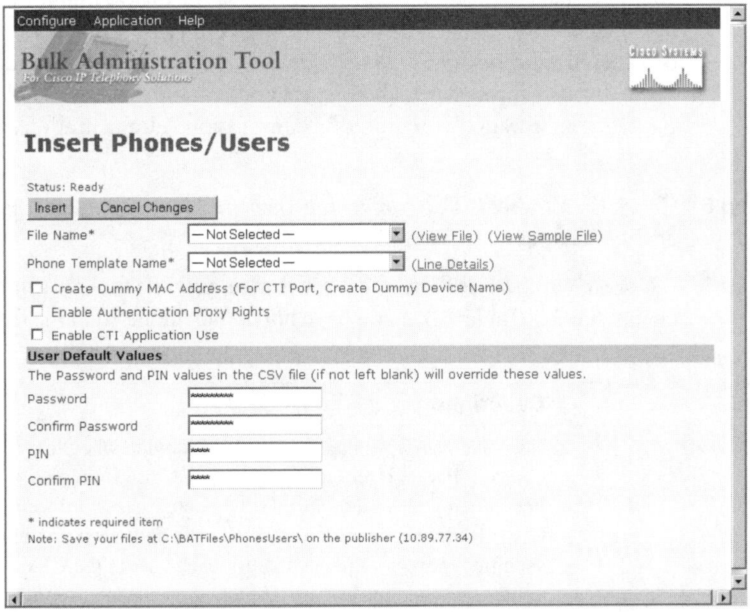

Configure Application Help

Bulk Administration Tool
For Cisco IP Telephony Solutions

Insert Phones/Users

Status: Ready

Insert | Cancel Changes

File Name* — Not Selected — (View File) (View Sample File)

Phone Template Name* — Not Selected — (Line Details)

☐ Create Dummy MAC Address (For CTI Port, Create Dummy Device Name)

☐ Enable Authentication Proxy Rights

☐ Enable CTI Application Use

User Default Values

The Password and PIN values in the CSV file (if not left blank) will override these values.

Password

Confirm Password

PIN

Confirm PIN

* indicates required item

Note: Save your files at C:\BATFiles\PhonesUsers\ on the publisher (10.89.77.34)

If you are adding CTI ports, the dummy MAC address option provides a unique device name for each CTI port in the form of fake MAC addresses. If you did not provide device names in the CSV file, then you must check the **Create Dummy MAC Address** box.

This field automatically generates fake MAC addresses in the following format:

XXXXXXXXXXXX

where **X** is any 12-character, hexadecimal (0–9 and A–F) number.

— Choose this option only when auto-registration is enabled.

— Choose this option if you do not know the MAC address of the phone that will be assigned to the user. Once the phone is plugged in, a MAC address registers for that device.

— Do not choose this option if you supplied MAC addresses or device names in the CSV file.

When phones are given to users, remember to update the phone records with the valid MAC address either by using TAPS or by manually updating in CCMAdmin.

Step 5 Check the **Enable Authentication Proxy Rights** check box if you want all users added in this transaction to be able to log on to a phone on behalf of someone else. Users with authentication proxy rights enabled are considered *super users* or *admin users* who act as the single point of authentication through which all users connect for extension mobility. Further configuration is required in Application Administration on the Cisco CRA server.

Step 6 Check the **Enable CTI Application Use** check box to enable the use of applications such as Cisco IP SoftPhone.

Step 7 (*Optional*) In the **User Default Values** area, provide the information documented in Table 7-3 if you have not already done so in the CSV file:

Table 7-3 *Insert Phones/Users Fields*

Field	Description
Password	Enter the password that users should provide when logging on to the Cisco IP Phone User Options Web page. You should only specify a value here when you want to specify the default password for access to the Cisco IP Phone User Options Web page and when you have not already specified individual passwords for each user in the CSV file. Password values specified in the CSV file take precedence over any values you enter here.
Confirm Password	Reenter the password.
PIN	Enter the PIN that users should provide when logging in to a Cisco IP Phone 7960 or 7940 for extension mobility. You should only specify a value here when you want to specify the default PIN for extension mobility and when you have not already specified individual PINs for each user in the CSV file. PIN values specified in the CSV file take precedence over any values you enter here.
Confirm PIN	Reenter the PIN.

Step 8 Click **Insert**. A message appears advising you of approximately how long it will take to insert the records to the Cisco CallManager database. You can cancel the transaction if you feel it may cause performance degradation.

Step 9 Click **OK** to insert the phones or click Cancel to cancel the transaction. If you clicked OK, a Transaction Status window displays. You can click the **Show Latest Status** button to see the transaction in progress.

NOTE If any line information for a phone record fails, BAT does not insert that phone record.

Step 10 When the transaction completes, you can click **View Latest Log File** to see a log file indicating the number of records added and the number of records failed, including an error code. For more information on log files, see the BAT documentation, *Bulk Administration Tool User Guide*. The documentation is available from the Help menu in BAT or on CCO at the link supplied earlier in this chapter.

Updating, Deleting, or Querying in BAT

You can bulk-update, bulk-delete, or search for a list of like devices in BAT. This is useful when you want to change one of the settings for many like phones, such as device pool or calling search space. Creating a query requires defining a filter. You can also create complex queries by clicking either the AND or the OR button.

To perform an update or a delete operation, you must first query the records you want to affect. BAT enables you to query phone records, lines, VG200 gateways, and ports on Catalyst 6000 FXS analog interface modules. Once you have specified a query to run, BAT returns the results of the query, showing all phones, lines, gateways, or ports that matched the specified details.

You can run a query in BAT simply to see a list of the devices that share the criteria you specified. This is useful when you are trying to locate a phone that was added in BAT using the dummy MAC address option. Device names for those phones are always prefixed by BAT so it is simple to search for and retrieve a list of all phones with dummy MAC addresses. If you want to delete VG200 gateways, you would run a query to find the gateways you want to delete.

The following instructions show you how to update a group of phone records. Specifying a query is fairly simple business. This example should give you enough information about the process so that you can run a query for any other device (gateway, port, phone, or line) for the purpose of updating, deleting, or simply viewing a list of related devices.

Step 1 Start BAT.

Step 2 Click **Configure > Phones** and click the link to **Update Phones** in the upper, right corner of the window. Figure 7-8 shows the Update Phones window.

Step 3 To locate the records you want to update, define the filter.

CAUTION If no filter is defined, BAT applies the changes to all phone records.

Figure 7-8 *Running a Query for All Phones Whose Device Names Begin with 'BAT'*

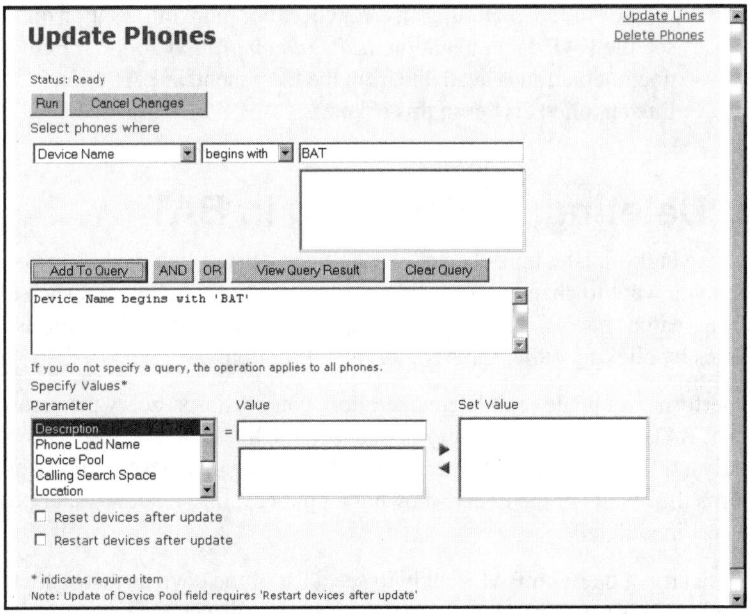

Step 4 In the first drop-down list box, choose the field to query such as Model, Device Name, and so on.

Step 5 In the second drop-down list box, choose the search criteria such as begins with, contains, is empty, and so on.

Step 6 In the search field/list box, either choose or enter the value that you want to locate, such as a specific phone model.

Step 7 Click **Add To Query** to add the defined filter to the query. Click **AND** or **OR** to add multiple filters and repeat Steps 2 through 5 to further define your query. If you make a mistake, click the **Clear Query** button to remove the query; then return to Step 3 and start over.

Step 8 Click **View Query Result** to display the records that are going to be affected. Specify the setting you want to update for all the records you have defined in your query.

Step 9 In the **Parameter** list box, choose a setting from the list box.

Step 10 In the **Value** field, enter the new value or choose a value from the list box.

Step 11 Click the arrow pointing toward the **Set Value** box to add the specified parameter and value to the Set Value box. Values in the **Set Value** box will be applied to the records you have defined in your query. You can remove values by choosing the value you want to remove from the **Set Value** box and clicking the arrow facing the **Value** field. You can choose multiple parameters to update. Repeat Steps 7 through 10 to add more parameters.

Step 12 *(Optional)* Check the **Reset devices after update** box to reset (power-cycle) the phones as soon as the update completes (if you are updating the device pool) or check the **Restart devices after update** box if you want to reset phones without power-cycling (for update of fields other than device pool).

NOTE If you want to reset or restart the devices at a later time, do not check either check box. When you click **Run**, the records update with the specified parameters; however, no changes take effect until the devices are reset or restarted.

Step 13 Click **Run** to apply the updates to the records.

Summary

BAT reduces the amount of time you have to spend administering your Cisco IP Telephony solution by allowing you to perform bulk add, update, or delete operations. You can reuse templates to further increase efficiency.

With BAT Release 4.2(1), you can work with phones, users, CTI ports, Cisco VG200 gateways and ports, and ports on Cisco Catalyst 6000 24-Port FXS Analog Interface Modules.

If you have a Cisco CRA server installed and configured, you can also use TAPS, which can be installed at the same time as BAT. TAPS enables you to add phones to the Cisco CallManager database with dummy MAC addresses. You, or the end user, can later use TAPS to update the MAC address in the Cisco CallManager database. Auto-registration in Cisco CallManager must be enabled to use TAPS.

Post-Test

How well do you think you know this information? The post-test is designed to help you gauge your knowledge about this chapter. Of the 10 questions, if you answer one to three questions correctly, we recommend that you re-read this chapter. If you answer four to seven questions correctly, we recommend that you review those sections that you need to know more about. If you answer eight to 10 questions correctly, you probably understand this information well enough to move on to the next chapter. You can find the answers to the post-test for this chapter in Appendix B, "Answers to Chapter Pre-Test and Post-Test Questions."

1 On what server should BAT be installed?

a. Primary

b. Publisher

c. Subscriber

d. Any of the above

2 Where are the BAT Excel template files located?

____

____

____

3 Describe the purpose of TAPS.

____

____

____

4 Describe the steps you must complete to add phones and users to the Cisco CallManager database using BAT.

a.

b.

c.

d.

5 When creating the CSV file for users or while inserting the users into the Cisco CallManager database using BAT, are password and PIN values required?

____

____

____

6 For values that appear on both the BAT template and the CSV file, which takes precedence?

____

____

____

7 Using BAT, can you add, update, or delete Cisco Catalyst 6000 Analog Interface Modules?

____

____

____

8 True or False? Auto-registration does not need to be enabled to use TAPS.

____

____

____

9 How does BAT save time over the usual method of adding, updating, or deleting devices or users in CCMAdmin?

____

____

____

10 During BAT installation you have the option of also installing TAPS. Other than BAT and Cisco CallManager, name the required component TAPS needs to function.

____

____

____

Upon completing this chapter you will be able to perform the following tasks:

- Given a Cisco Media Convergence Server (MCS), identify the MCS model and the hardware components of that model.
- Given a MCS-7800 series server and the correct installation CD-ROMs, install a Cisco CallManager server to provide IP telephony in a brand new or existing cluster.
- Given a running Cisco CallManager server, backup and restore the Cisco CallManager database to and from a location you have specified.
- Given a Cisco CallManager cluster deployment model, describe the process used to upgrade a Cisco CallManager cluster.

CHAPTER 8

Installation, Backups, and Upgrades

This chapter describes how to install and configure Cisco CallManager Release 3.1 on a Cisco Media Convergence Server or a customer-provided Compaq DL380 and DL320 server. Two IBM servers, IBM xSeries 330 and 340, are also approved hardware platforms. After you install Cisco CallManager on your server, the server becomes a Cisco IP Telephony Server. Cisco CallManager, along with the Cisco IP Telephony Applications Server, enables the conversion of conventional, proprietary, circuit-switched telecommunication systems to multiservice, open LAN systems. This chapter discusses the following topics:

- Abbreviations
- Supported Hardware Platforms
- Prerequisite Operations Before Installing Cisco CallManager
- CD-ROMs
- Installation Configuration Information
- Post Installation
- Upgrading

Pre-Test

Do you already know this information? The pre-test is designed to help you gauge your knowledge about this chapter. Of the 10 questions, if you answer one to three questions correctly, we recommend that you read this chapter. If you answer four to seven questions correctly, we recommend that you skim through this chapter, reading those sections that you need to know more about. If you answer 8 to 10 questions correctly, you probably understand this information well enough to skip this chapter. You can find the answers to the pre-test for this chapter in Appendix B, "Answers to Chapter Pre-Test and Post-Test Questions."

1 When you are installing a cluster of Cisco CallManagers, what must be the first server installed?

a. Subscriber

b. Primary

c. Call Processor

d. Publisher

2 List at least five pieces of information you should have prior to starting an installation.

____

____

____

3 List at least two recommended post-installation tasks.

____

____

____

4 If you choose a network directory as the backup destination, how should you configure the directory in Windows 2000?

____

____

____

5 List at least five unnecessary services to be set to manual and stopped on all servers in a cluster, and list the two additional services to be set to manual and disabled on Subscriber servers.

____

____

____

6 When you are upgrading a cluster, which server in the cluster must you upgrade first?

____

____

____

This question is worth four points. Describe the four steps used to upgrade a Cisco CallManager cluster of three servers supporting 2500 users.

7

____

____

____

8

____

____

____

9

____

____

____

10

____

____

____

Abbreviations

This section defines the abbreviations used in this chapter. For more information about terms and abbreviations used in this chapter refer to the *IP Telephony Network Glossary* at the following URL:

www.cisco.com/univercd/cc/td/doc/product/voice/evbugl4.htm

Table 8-1 provides the abbreviation and the complete term for those abbreviations used frequently in this chapter.

Table 8-1 *Abbreviation with Definition*

Abbreviation	Complete Term
ATA	advanced technology attachment
DAT	digital audio tape

continues

Table 8-1 *Abbreviation with Definition (Continued)*

Abbreviation	Complete Term
DNS	Domain Name System
MCS	Media Convergence Server
NIC	network interface card
PCI	peripheral component interconnect
RAID	redundant arrays of independent disks
RAM	random-access memory
SCSI	small computer system interface
TCP	Transmission Control Protocol
USB	universal serial bus

Supported Hardware Platforms

Figure 8-1 shows the positioning of two Media Convergence Servers.

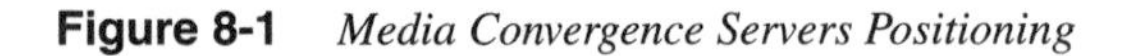

Figure 8-1 *Media Convergence Servers Positioning*

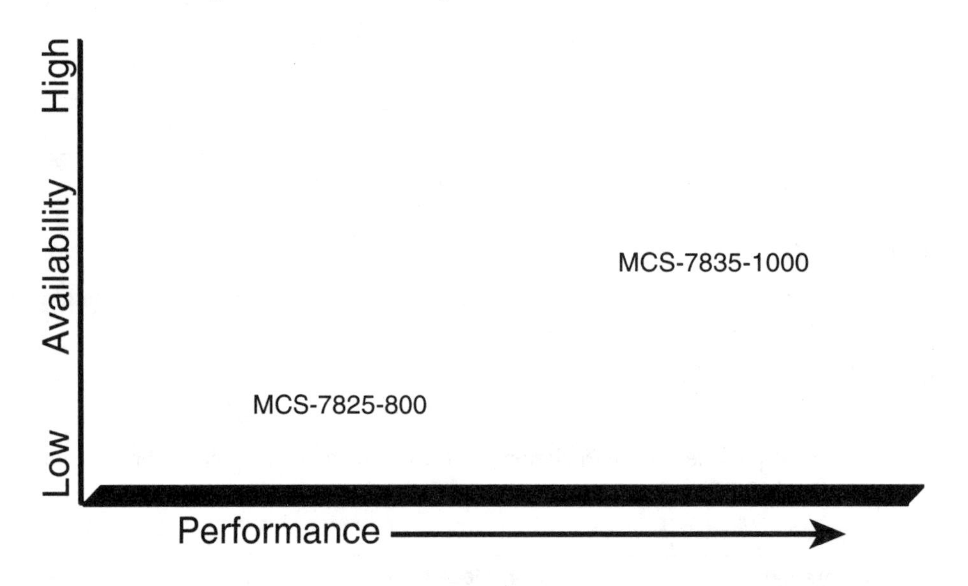

Cisco requires that any Cisco software purchased for installation on a customer-provided server be installed on a server meeting approved Cisco configuration standards.

Cisco CallManager installed on a Cisco supported server provides a network business communications system for high-quality telephony over IP networks.

You do not receive a monitor with any Cisco Media Convergence Servers. The MCS-7820 and MCS-7822 include a keyboard and mouse, but the MCS-7830, MCS-7825, and MCS-7835 do not. During initial startup and configuration of a Cisco supported server and Cisco CallManager, you will need to use a monitor, keyboard, and mouse. It is conceivable to remove the monitor and mouse after initial installation.

NOTE At the time of writing only the MCS-7835-1000 and the MCS-7825-800 are orderable from Cisco. Check with your local Cisco Systems, Inc., sales representative for the current orderable hardware platforms.

Table 8-2 provides a detailed checklist of components on the various MCS platforms.

Table 8-2 *Cisco Media Convergence Server Hardware Specifications*

Component	MCS-7835[1]	MCS-7835-1000[2]	MCS-7830	MCS-7825-800[3]	MCS-7822	MCS-7820
Intel Pentium III processor	733 MHz	1 GHz	500 MHz	800 MHz	550 MHz	500 MHz
Registered ECC SDRAM	1 GB	1 GB	512 MB[4]	512 MB	512 MB	512 MB[4]
10/100BaseTX Ethernet controller	X	X	X	X	X	X
Integrated dual-channel wide Ultra SCSI-3 controller	X	X	X			
Integrated Ultra ATA/100 controller module (ATA models)				X		
Integrated wide Ultra2 SCSI adapter					X	X
Dual 18.2-GB Ultra3 SCSI hot-plug drives		X				

continues

Table 8-2 *Cisco Media Convergence Server Hardware Specifications (Continued)*

Component	MCS-7835[1]	MCS-7835-1000[2]	MCS-7830	MCS-7825-800[3]	MCS-7822	MCS-7820
Dual 18.2-GB Ultra2 SCSI hot-plug drives	X					
Dual 9.1-GB Ultra2 SCSI hot-plug drives			X			
Single 9.1-GB Ultra2 SCSI non-hot-plug drive					X	X
Single 20-GB Ultra ATA/100 7200 rpm non-hot plug "1 drive"				X		
1.44-MB floppy disk	X	X	X	X	X	X
Preinstalled high-speed IDE CD-ROM drive	X	X	X		X	X
Removable CD-ROM/diskette drive assembly				X		
Hot-plug redundant 275-watt power supply	X	X	X			
180-watt PFC Power Supply				X		
200-watt power factor corrected, CE mark-compliant power supply					X	X
Integrated video card	X	X	X	X	X	X
RAID controller	X	X	X			

Table 8-2 *Cisco Media Convergence Server Hardware Specifications (Continued)*

Component	MCS-7835[1]	MCS-7835-1000[2]	MCS-7830	MCS-7825-800[3]	MCS-7822	MCS-7820
12/24-GB internal DAT drive	**optional**	**optional**				

[1] Unless otherwise specified in this document, all further references to the MCS-7835 apply to the MCS-7835, which contains a 733-MHz processor the MCS-7835-1000, which contains a 1-GHz processor and the customer-provided DL380.

[2] The same Cisco-approved hardware configuration standards apply for both the MCS-7835-1000 and the customer-provided Compaq DL380.

[3] All references to the MCS-7825 in this document apply to the MCS-7825-800, which contains a 800-MHz processor, and the customer-provided Compaq DL320. The same Cisco-approved hardware configuration standards apply for both the MCS-7825-800 and the customer-provided Compaq DL320.

[4] The MCS-7830 and MCS-7820 require memory upgrades to meet the 512-MB RAM minimum specification. The MCS-7830 is optionally upgradeable to 1 GB.

Table 8-3 shows the requirements needed for the IBM supported hardware platforms.

Table 8-3 *IBM xSeries Platforms Hardware Specifications*

Component	IBM xSeries 330	IBM xSeries 340
Intel Pentium III processor	800 MHz[1]	1000 MHz[1]
Registered ECC SDRAM	512 MB[2]	1 GB
10/100BaseTX protocol control information unshielded (PCI UTP) controller(s)	X	X
Integrated wide Ultra160 SCSI controller(s)	X	X
Dual 18.2-GB Ultra160 SCSI hot-plug drives		X
ServeRaid-4L Ultra160 SCSI		X
Single 18.2-GB Ultra160 SCSI non-hot-plug drive	X	
1.44-MB floppy disk	X	X
Preinstalled high-speed IDE CD-ROM drive	X	X
Removable CD-ROM/diskette drive assembly	X	X
Hot-plug redundant 270-watt power supply		X
200-watt PFC, CE mark-compliant power supply	X	
xSeries Cable Chain Tech Kit	X	

continues

Table 8-3 *IBM xSeries Platforms Hardware Specifications (Continued)*

Component	IBM xSeries 330	IBM xSeries 340
Integrated video card	X	X
20/40-GB internal DAT drive		optional

1 The IBM xSeries 330 and xSeries 340 require the minimum processing speed stated in the table.

2 Both the IBM xSeries 330 and xSeries 340 require memory upgrades to meet the minimum RAM specifications stated in the table.

For more information on IBM xSeries server hardware specifications or to navigate to the IBM website, go to www.cisco.com/go/swonly/.

NOTE The MCS platforms are the focus of this chapter. You can use guidelines for the MCS-7835 when using a Compaq DL380 or an IBM xSeries 340. You can use guidelines for the MCS-7825 when using a Compaq DL320 or an IBM xSeries 330.

MCS-7835, Compaq DL380, and IBM xSeries 340

These high-availability server platforms for Cisco AVVID (Architecture for Voice, Video and Integrated Data) are an integral part of a complete, scalable architecture for a new generation of high-quality IP voice solutions that run on enterprise data networks. At only 3U high, the servers pack tremendous power in a low-profile chassis that minimizes rack space. All three servers can run a variety of Cisco AVVID applications, such as Cisco CallManager, Cisco Unity, Customer Response Application, and the Cisco IP Interactive Voice Response (IP IVR) solution.

Key Benefits and Features: Performance

All three servers, MCS-7835-1000, Compaq DL380, and IBM xSeries 340, host a 1-GHz Intel Pentium III processor and ship with 1 GB of 133-MHz registered SDRAM, extending the high performance that you need to roll out current and future Cisco AVVID applications. In addition, hardware RAID support for dual 18.2-GB Ultra3 SCSI hot-plug hard drives (10,000 rpm) improves overall system performance.

Key Benefits and Features: High Availability

Availability, or the percentage of time that a system is available to provide service, was assumed in old world networks. In new world networks built by Cisco Systems, availability is a key requirement. The three servers come with redundant hot-plug power supply and two redundant 18.2-GB SCSI hot-plug hard drives running RAID-1 disk mirroring to ensure maximum availability. If a hard drive or power supply fails, it can be replaced without powering down the server and without affecting service. In the case of the SCSI drive, as soon as the replacement drive is inserted, the integrated RAID controller copies the image from the primary drive to the new drive without any user intervention.

Key Benefits and Features: Scalability

Whether you start your Cisco IP Telephony network with 10 telephones or 10,000, these three servers seamlessly enable you to grow the network at your pace. As a Cisco CallManager server, each of these servers can handle up to 2500 Cisco IP Phones (the total number of Cisco IP Phones is dependent on N+1 redundancy configuration, discussed in Chapter 2, "Navigation and System Setup"). Remote sites can be interconnected with a standards-based H.323 interface using an H.323 gatekeeper.

Key Benefits and Features: Flexibility

The three servers can run Cisco CallManager software, Cisco IP IVR software, or Cisco Unity. These servers are also designed to run future Cisco application packages for the Cisco AVVID solution. The MCS-7835-1000 and Compaq DL380 have an optional internal 12/24-GB DAT drive to back up your critical data and rack mount hardware that enables server installation in a variety of industry-standard racks.

Key Benefits and Features: System Backup and Restore

Every server includes custom backup and restore functionality that is configured when you run the automatic installation software. Simply specify a file location on another server on your IP network or use the optional internal DAT drive. The MCS-7835-1000 does the rest. All relevant data files are stored nightly at 2:00 a.m. (or another time you choose) on the DAT or to another networked file server. In case of a failure, you can run the restore routine, specify the file location of the backup, and be back online quickly.

Figure 8-2 shows the front view and calls out the highlighted features of the MCS-7835-1000.

Figure 8-2 *MCS-7835-1000 Front View*

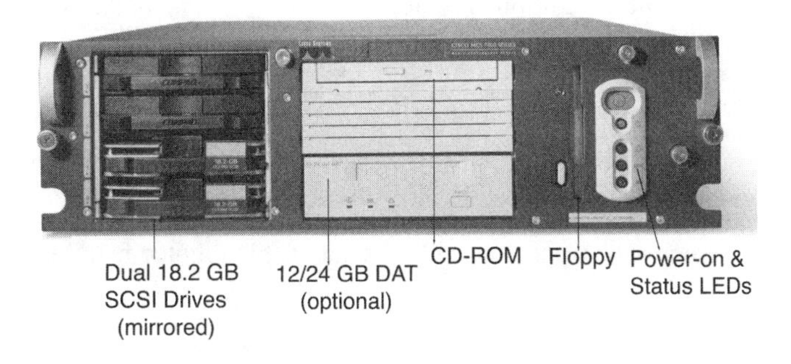

Figure 8-3 shows the rear view and calls out the highlighted features of the MCS-7835-1000.

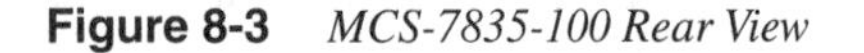

Figure 8-3 *MCS-7835-100 Rear View*

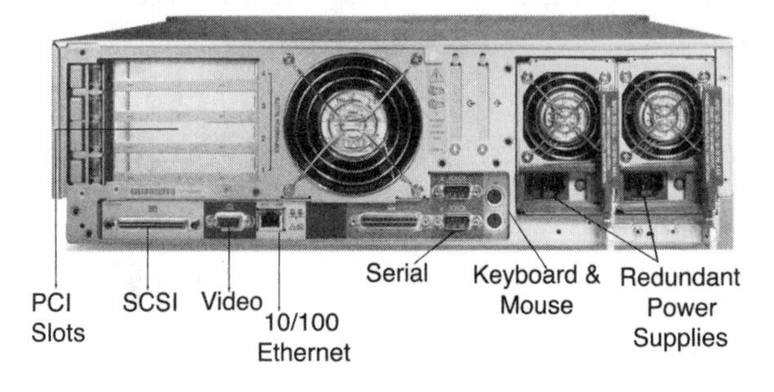

MCS-7825-800, Compaq DL320, and IBM xSeries 330

The Cisco MCS-7825-800, Compaq DL320, and IBM xSeries 330 are powerful platforms for Cisco AVVID. At only one rack unit (1U) high, these servers are the most space-efficient member of the Cisco supported servers. You can configure the MCS-7825-800 to ship with Cisco CallManager or Cisco IP Interactive Voice Response (IP IVR), either of which can be loaded via a fast-running installation script to make the deployment of IP telephony simple and cost effective.

Key Benefits and Features: Performance

These server platforms are designed for today's IP telephony applications. The processor is an Intel Pentium III with a clock speed of 800 MHz. Memory is a robust 512 MB of error-correcting code (ECC) SDRAM.

Key Benefits and Features: Availability

These three servers, MCS-7825, DL320, and xSeries 330 support high-availability for enterprise IP telephony. Same as the servers mentioned earlier, these servers can participate in clustered, multiple call processing.

Key Benefits and Features: Scalability

To help achieve the ultimate scalability that your organization requires, the compact size of these servers allow you to deploy up to 42 servers in a single 19-inch rack.

Figure 8-4 shows the front view and calls out the highlighted features of the MCS-7825-800.

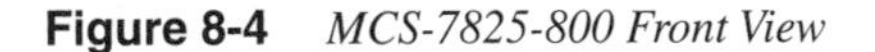

Figure 8-4 *MCS-7825-800 Front View*

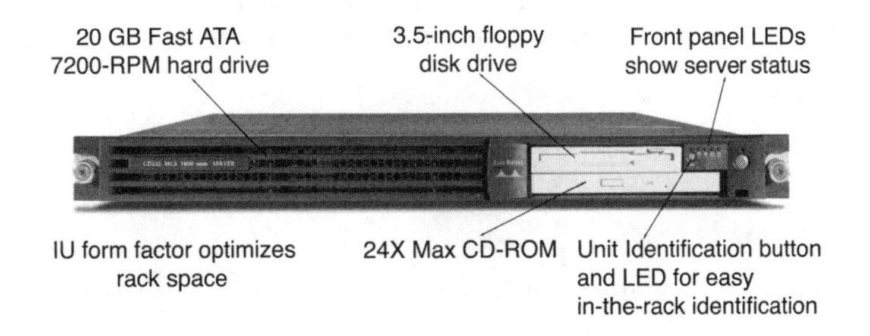

Figure 8-5 shows the rear view and calls out the highlighted features of the MCS-7825-800.

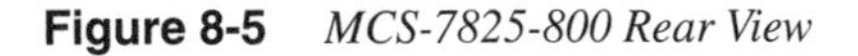

Figure 8-5 *MCS-7825-800 Rear View*

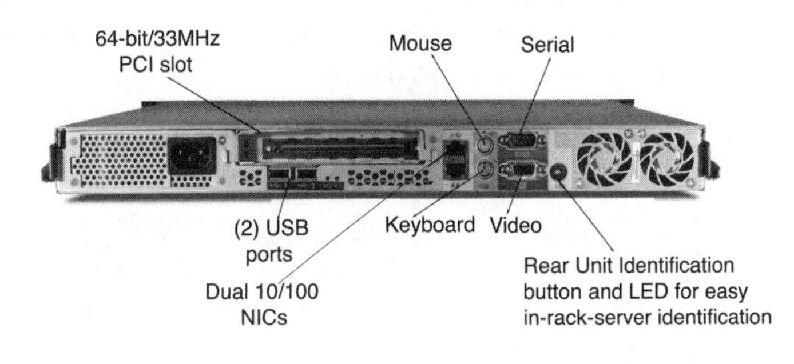

Prerequisite Operations Before Installing Cisco CallManager

Before installing Cisco CallManager for a distributed call processing system, plan the system configuration. A cluster comprises a set of Cisco CallManagers that share the same database.

In a Cisco CallManager distributed system, one server maintains the master (or Publisher) database and all others in the cluster maintain Subscriber databases. Subscriber databases are duplicates of the master database. The duplicate databases on the Subscribers are only used if the Subscriber cannot access the Publisher's master database. During normal operation, all Cisco CallManager servers in the cluster read data from and write data to the Publisher database. Periodically, Cisco CallManager automatically updates the Subscriber copies of the database from the Publisher database.

At a minimum, determine how many Cisco CallManager servers the cluster will contain, which server will house the Publisher database, and where backup tasks will be performed.

Additional planning should include a strategy for distributing the devices (such as phones or gateways) among the Cisco CallManagers in the cluster to achieve the type of distribution you want. For more information on planning a Cisco CallManager distribution system, refer to online help in Cisco CallManager or one of the guides:

- *Cisco CallManager System Guide:*

  www.cisco.com/univercd/cc/td/doc/product/voice/c_callmg/3_1/sys_ad/adm_sys/ccmsys/index.htm

- *Cisco IP Telephony Network Design Guide*:

  www.cisco.com/univercd/cc/td/doc/product/voice/ip_tele/network/

Make sure you connect your server to the network before you begin the installation. Windows 2000 will not install if you do not connect the server to the network.

NOTE The MCS-7825 contains two NICs, but Cisco CallManager supports only one NIC. When you connect the server to the network, use the lower NIC connector because the upper connector is disabled during the installation.

CAUTION Installing or using Netscape Navigator on the Cisco CallManager server can cause severe performance problems. Cisco strongly recommends against installing Netscape Navigator or any other application software on the Cisco CallManager server.

NOTE You must have the proper NetBIOS name resolution when installing a Subscriber. Accomplish this by using WINS, LMHOSTS, or having the Subscriber on the same subnet as the Publisher.

Cisco CallManager CD-ROMs

The Cisco Media Convergence Server ships with a blank hard drive. When you install Cisco CallManager Release 3.1, you use three of the following six CD-ROMs, depending on your server type:

- Hardware Detection CD-ROM (CD #1). Required for all servers.
- Service Pack CD-ROM (CD #2). Used, but not necessarily required, for upgrades via CD-ROM.
- Operating System Installation and Recovery CD-ROM for IBM servers (CD #3).

- Operating System Installation and Recovery CD-ROM for MCS-7825 and MCS-7835 (CD #4).
- Operating System Installation and Recovery CD-ROM for MCS-7820, MCS-7822, and MCS-7830 (CD #5).
- Cisco CallManager 3.1 Installation and Recovery CD-ROM (not numbered). Required for upgrades, backups, restorations, and installations on all servers.

Installation Configuration Information

During a Cisco CallManager installation, you will need to be prepared to enter configuration information required to complete the installation. This section discusses the installation configuration information you will need for a successful Cisco CallManager server installation.

Table 8-4 shows the configuration information required for installing software on your server. Complete all fields unless otherwise noted. Gather this information for each Cisco CallManager server you are installing in the cluster. Make copies of this table and record your entries for each server in a separate table. Have the completed lists with you when you begin the installation.

For a printable copy of this table go to page 10, table 3 of the following document:

> www.cisco.com/univercd/cc/td/doc/product/voice/c_callmg/3_1/install/cm311ins.htm#xtocid18583

Cisco Product Key

Cisco supplies you with a Cisco Product Key when you purchase a Cisco IP Telephony product. The product key, based on a file encryption system, enables installation of only the components you have purchased, and it prevents other supplied software from being installed for general use. The product key comprises alphabetical letters only; it contains no numbers or special characters. The product key provided in this document allows you to install Cisco CallManager Release 3.0 and 3.1 only. To install another Cisco IP Telephony product, you must purchase the product and obtain the appropriate product key.

Table 8-4 *Configuration Data for Cisco Media Convergence Server*

Configuration	Data Entry
Cisco Product Key	BTOO VQES CCJU IEBI
User Name	
Name of your organization	
Computer name	

continues

Table 8-4 *Configuration Data for Cisco Media Convergence Server (Continued)*

Configuration	Data Entry
Workgroup	
NT domain (optional)	
DNS Domain suffix	
Current time zone, date, and time	
DHCP parameters	Cisco recommends that you program a fixed IP address in TCP/IP properties for the server instead of using DHCP.
TCP/IP properties (required if DHCP is not used) • IP address • Subnet mask • Default gateway	
DNS servers (optional) • Primary • Secondary WINS servers (optional) • Primary • Secondary	
Database server (specify one) • Publisher • Subscriber If you are configuring a Subscriber server, supply the user name and password of the Publisher database server: — User name of Publisher — Password of Publisher	
Backup (specify one or both) • Server • Target	
New system administrator password	

User and Organization Name

Registering the software product you are installing requires user and organization name.

Computer Name

Assign a unique network name of 15 characters or less for this server. It may contain alphanumeric characters, hyphens (-), and underscores (_), and it must begin with an alphabetical character. You should follow your local naming conventions, if applicable.

CAUTION Due to a restriction in Microsoft SQL Server 7.0, you cannot change the Windows 2000 computer name after installation. If the computer name is changed, you must reinstall the server. Make sure the name you assign is the permanent server name.

Workgroup

This entry records the name of the workgroup of which this computer is a member. A workgroup comprises a collection of computers that have the same workgroup name. Ensure this entry of 15 characters or less follows the same naming conventions as the computer name.

Domain Suffix

Always enter the Domain Name System (DNS) domain suffix in the format *mydomain.com* or *mycompany.mydomain.com*. If you are not using DNS, use a fictitious domain suffix, such as acme.com.

TCP/IP Properties

Assign an IP address, subnet mask, and default gateway to the Cisco CallManager server. Because the IP addresses you assign are permanent properties, you should not change them after installation.

Cisco recommends choosing static IP information, which ensures that the Cisco CallManager server obtains a fixed IP address. With this selection, Cisco IP Phones can register with Cisco CallManager when you plug the phones into the network.

If you choose to use DHCP, Cisco Technical Assistance Center (TAC) insists that you reserve an IP address for each Cisco CallManager server in the DHCP server scope. This action prevents the release or reassignment of IP addresses. If you do not reserve IP addresses through the DHCP server scope, the DHCP server may assign a different address to the Cisco CallManager server if the server is disconnected from and then reconnected to the network. To return the Cisco CallManager server to its original IP address, you have to reprogram the IP addresses of the other devices on the network. For information on DHCP option settings, refer to the *Cisco CallManager Administration Guide* at

www.cisco.com/univercd/cc/td/doc/product/voice/c_callmg/3_1/sys_ad/adm_sys/ccmcfg/index.htm.

Domain Name System (DNS)

Identify a primary DNS server for this optional field.

NOTE Before you begin installing multiple servers in a cluster, you must have a name resolution method in place, such as DNS, WINS, or local naming using a configured lmhosts file. If you use DNS, make sure the DNS server contains a mapping of the IP address and hostname of the server you are installing before you begin the installation. If you use local name resolution, ensure the lmhosts file is updated on the existing servers in the cluster before beginning the installation on the new Subscriber server; you must then add the same information to the lmhosts file on the new server during installation, as instructed in the procedure.

If NetBIOS name resolution is not possible, the installed servers will not replicate all the database information.

By default, the phones attempt to connect to the Cisco CallManager using DNS. Therefore, if you use DNS, make sure the DNS contains a mapping of the IP address and the fully qualified domain name of the Cisco CallManager server. If you do not use DNS, use the server IP address instead of a server name for the phones to register with the Cisco CallManager server. If you choose not to use DNS, install Cisco CallManager normally and then refer to the "Server Configuration" section in the *Cisco CallManager Administration Guide* or online help in the Cisco CallManager Administration for information on changing the server name.

Directory Manager Password

This password allows access DC Directory, Cisco CallManager's LDAP directory. This is where user information is stored and accessed for the Directory's URL and other applications.

Database Server

Determine whether this server will be configured as a Publisher database server or a Subscriber database server. This decision designates a permanent selection. If you want to reassign the database server type at a later date, you must reinstall the Cisco CallManager server.

CAUTION If you are configuring a Subscriber database server, make sure the Publisher database server for that cluster is installed, connected to the network, and configured properly to work as a distributed system. When configuring a Subscriber database server, ensure the server you are installing can connect to the Publisher database server during the installation. This connection facilitates copying the master database from the Publisher server to the local drive on the Subscriber server. You must supply the name of the Publisher database server and a user name and password with administrator access rights on that server. If the Publisher server cannot be authenticated during the installation for any reason, the installation will not continue.

Backup Server or Target

Determine whether this Cisco CallManager server will be configured as a backup server or a backup target.

The backup server actually performs the backup operation. It stores the backup data in the local directory, local tape drive, or network destination that you specify. If you select a network area as the backup server, the directory must be shared in Windows 2000.

A backup target contains the data to be backed up. You can select more than one target but it is recommended to have only one server. If a server is configured as a backup server, Cisco CallManager will automatically add it to the backup target list.

New Password for the System Administrator

Cisco CallManager Release 3.0 and later supports password protection. At the end of the installation, a prompt asks you to supply a new password for the system administrator.

You must enter the same Administrator password for the Publisher and all Subscribers in the cluster so that Cisco CallManager database replication occurs.

Configuration Process

When using DNS or WINS, select the **Use the following DNS and WINS server addresses** button; then enter the IP addresses of the primary and secondary DNS servers and primary and secondary WINS servers. Click **Next**.

If you are not using DNS, leave the DNS and WINS fields empty. Make sure that the **Use the following DNS and WINS server addresses** button is selected; then, click **Next**.

If you did not enter DNS or WINS server information in the previous window and if you are installing multiple servers in a cluster, you must configure local name resolution by updating the lmhosts file. The updated lmhosts file must contain a mapping of the IP address and hostname of each server in the cluster. Perform the following steps to configure the lmhosts file:

Step 1 On the LMHost window, select the **Check if you want to edit LMHosts file** check box.

Step 2 Enter the IP address and server name. For example,

172.16.0.10 dallascm1

Step 3 Click **Add Server**.

Step 4 Click **Next** to continue.

Figure 8-6 shows the Cisco CallManager components that can be selected to install.

Figure 8-6 *Cisco CallManager Components*

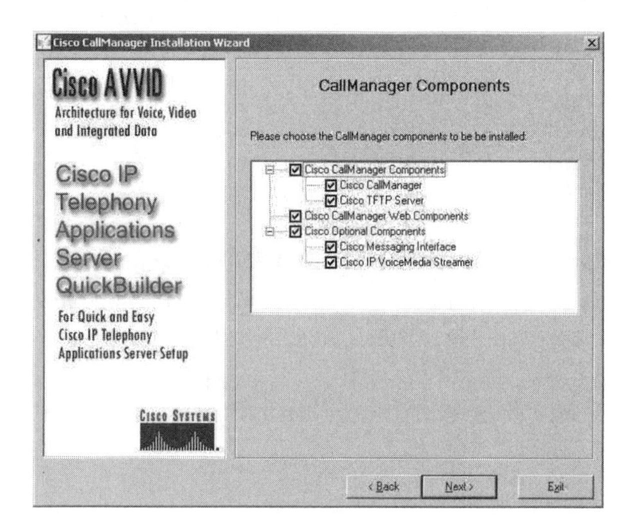

Cisco IP Telephony Applications Server QuickBuilder then automatically installs the following software applications:

- Microsoft Windows 2000 Server
- Microsoft SQL Server 7.0 Standard Edition, Service Pack 2
- DC Directory 2.4
- Cisco CallManager Release 3.1

With the appropriate CD-ROM in place, the server automatically begins installing the operating system, essential software, and Cisco CallManager, prompting you to insert the Cisco CallManager 3.1 Installation and Recovery CD-ROM when necessary. When you choose to install the Cisco CallManager service, you automatically install the following services/features:

- Cisco CallManager Serviceability
- Cisco Telephony Call Dispatcher
- Cisco CTI Manager
- Cisco RIS Data Collector
- The user login/logout phone service associated with Extension Mobility

During the installation process, you receive prompts telling you to enter important configuration information about the server, such as the server name and IP address. You can complete the initial power up more efficiently if you gather all the necessary configuration information before beginning the installation process.

The Cisco IP Voice Media Streaming Application contains the MTP, MOH, and conference bridge services.

If you are installing TFTP only, you automatically install the database option when you choose Cisco TFTP. TFTP support requires database installation.

If you are configuring a Subscriber database server, make sure the server you are installing can connect to the Publisher database server before the installation can continue. This connection is necessary because the Subscriber server attempts to connect to the Publisher server so that the Publisher database can be copied from that server to the local drive on the Subscriber server. When the server finishes building Windows 2000, it will ask you to insert the Cisco CallManager CD. To make sure a good connection exists between the servers, issue a **ping** command from the Subscriber server to the Publisher server before you try to authenticate to it. If you are using DNS, use a fully qualified domain name (for example, *hostname.cisco.com*) with the **ping** command. If the **ping** command is not successful, you must exit the installation program, fix the problem, and begin the installation process again.

NOTE You cannot rely on DNS; you must have a way to resolve NetBIOS names using either WINS, lmhosts, or the two machines must be on the same subnet.

Figure 8-7 shows an example of the Cisco CallManager Database Distribution dialog box.

Figure 8-7 *Publisher or Subscriber*

The Publisher database serves as the master database for all servers in the cluster. All servers except the Publisher database server maintain Subscriber databases, which are copies of the Publisher database.

You must decide whether you are upgrading the Publisher database server or Subscriber database server. If you are upgrading the Publisher database server, choose **I am upgrading/ installing the CallManager Publisher**. Click **Next** to continue.

If a Publisher database already exists and you are configuring a Subscriber database server, choose **I am upgrading/installing the CallManager Subscriber**. By default, the Publisher database server associated with the Subscriber appears in the hostname Publisher field. You may choose another Publisher by clicking **Browse**. After you choose the appropriate Publisher, click **Next**.

If you chose **I am upgrading/installing the CallManager Subscriber**, an authentication window opens and prompts you to enter user names and passwords for a Windows 2000 user with administrative privileges and a SQL Server user with administrative privileges on the Publisher database server. Enter the account information in the appropriate fields as prompted and then click **Next**. The system connects to the Publisher database server.

Figure 8-8 shows the dialog box where you select the server to be a backup or target server.

The Cisco IP Telephony Applications Server Backup Utility Setup loads automatically. You must specify whether this server will act as a backup target or the backup server during the backup and restore operation.

Figure 8-8 *Backup Server or Target*

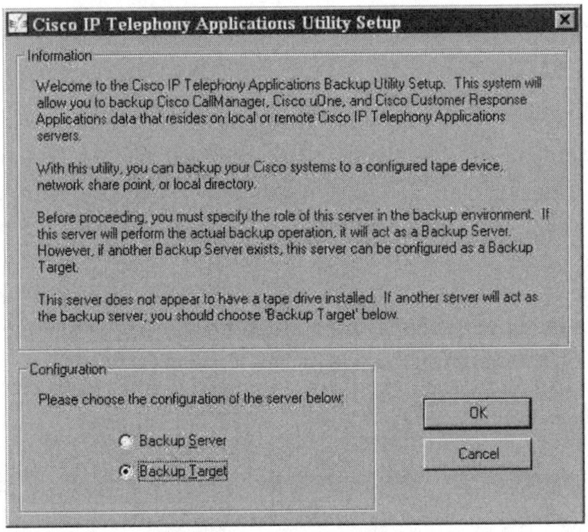

The backup server actually performs the backup operation. It stores the backup data in the directory or tape drive destination that you specify. If a server is configured as a backup server, Cisco CallManager automatically adds it to the backup target list.

A backup target acts as a server that contains data to be backed up. You can choose more than one target, but you can choose only one server.

Choose either **Backup Server** or **Backup Target** and then click **OK**.

CAUTION Cisco strongly recommends that you use the Cisco IP Telephony Applications Server Backup Utility to perform backups and that you do not use third-party backup software. However, if you do not want to use the Cisco IP Telephony Applications Server Backup Utility, complete the installation as instructed, and then stop the service called "stiBack for Cisco MCS." To stop the service, choose **Start > Run**, enter **services.msc /s**, choose the service in the main window, and click **Stop Service**.

NOTE If you want to use a third-party backup utility, you can use the Cisco Backup utility to put the backup on a network drive and then use the third-party software on the server where the network drive resides.

If you chose **Backup Target** in the previous step, a message appears indicating that the setup is complete. You will need to configure this server as a CallManager target on the backup server.

If you chose **Backup Server** in the previous step, complete the backup configuration settings from the Backup Utility Configuration window.

On the Backup Utility Configuration dialog box, add the name of any additional Cisco CallManager servers you want to back up by clicking **Add**, entering the server name, and then clicking **OK**. Click **Delete** to remove servers from the target list. If you want to add a remote server, you must install it and connect it to the network before you add it to the target list. A prompt asks you to enter a user name and password with administrator access rights on the remote server and then click **Verify**. Figure 8-9 shows the Backup Utility Configuration dialog box.

Figure 8-9 *Backup Configuration Settings*

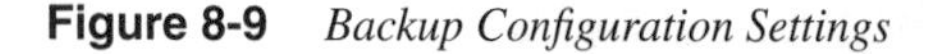

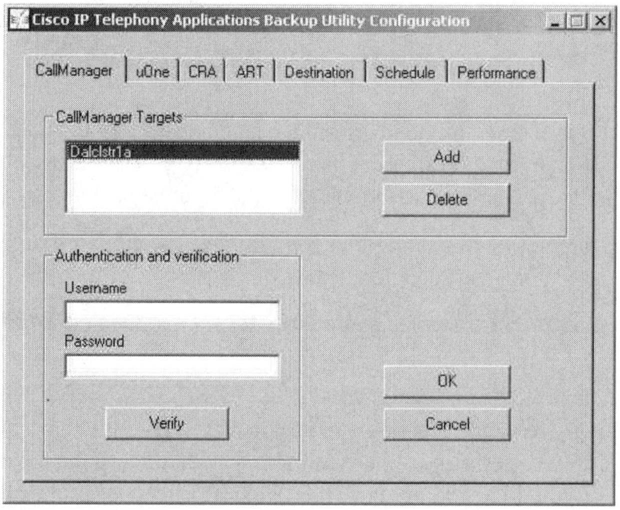

The Backup Utility attempts to connect to the remote server. If the remote server is not found, the authentication fails. The server name remains in the target list but may not be accessible.

Click the **Destination** tab. Choose a destination location where the backup data will be stored as shown in Figure 8-10. You must choose a destination for backups that are to be performed; no default destination setting exists.

If you choose a network directory as the backup destination, the directory must be shared in Windows 2000. To share a directory, log in on that server, right-click the directory folder icon you want to share, click **Sharing**, click **Share this folder**, and then click **OK**.

Figure 8-10 *Destination for Backup*

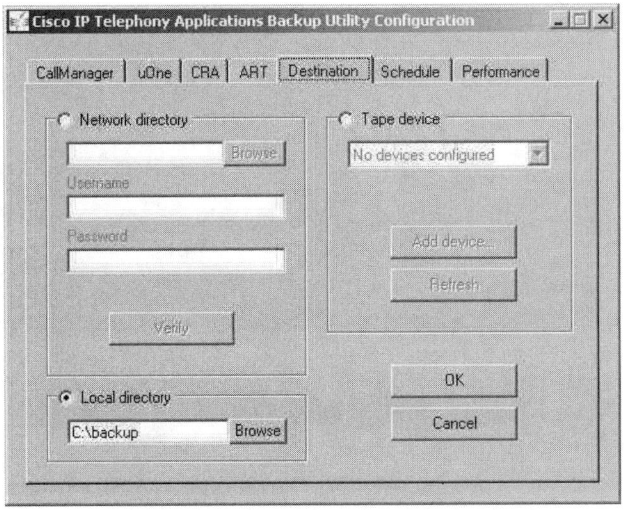

Choose **Network directory**, **Local directory**, or **Tape device** to specify where backup files will be backed up. To enter a network or local directory, you can enter the path and directory name or choose it from a browse box by clicking **Browse**. For network directories, you must supply a user name and password with administrator access rights on that server. In the tape device box, only tape drives configured in Windows NT or Windows 2000 appear. To add new devices, click **Add device.**

NOTE You may click the tape device box if you have a MCS-7835. Only the MCS-7835 has a tape drive available.

NOTE You must choose **Network directory** or **Tape device** to use the Cisco IP Telephony Applications Server Restore Utility on your server.

Click the **Schedule** tab. Figure 8-11 shows an example of the Schedule tab in the Cisco MCS Backup Utility Configuration dialog box.

Select the **Schedule** tab in Cisco MCS Backup Utility Configuration dialog box.

Choose the days and times you want the backup to run. The default is 2:00 a.m. Tuesday through Saturday. You can also choose the length of the system log in days. Click **OK**.

Figure 8-11 *Schedule Tab for Backup*

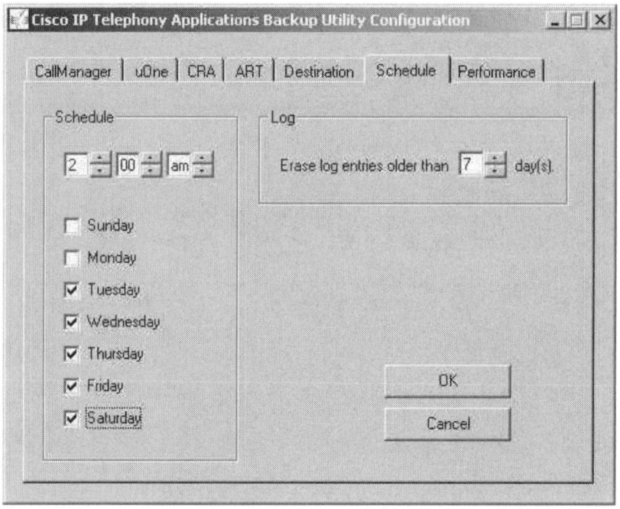

NOTE You should schedule Cisco CallManager backups to occur during off-peak hours because CPU utilization is high during the backup process and performance degradation may occur.

Click **OK** to save your settings and exit the Backup Utility Configuration.

Completing the Installation

When entering passwords for the local administrator and SA (SQL Server system administrator) accounts, do not use the apostrophe ('). Enter the same administrator password for the Publisher and all Subscribers in the cluster so that Cisco CallManager database replication occurs.

A prompt asks you to enter a new password for the local administrator account. Enter the new password in the **New Password** field and then enter it again in the **Retype Password** field. Click **OK**.

A prompt asks you to enter a new password for the SA (SQL Server system administrator) account. Enter a new password of at least five characters in the **New Password** field and then enter it again in the **Retype Password** field. Click **OK**. Table 8-5 summarizes password details for CCMAdmin and SQLSvc administrator accounts.

Table 8-5 *CCMAdmin and SQLSvc Administrator Account Passwords*

Application	Password Functionality
CCMAdmin	Provides access to all services on the server. To view all services on all servers in the cluster, passwords need to be identical.
SQLSvc	All passwords need to be identical. During upgrades, password resets to original default.

Cisco CallManager Post-Installation Procedures

After installing Cisco CallManager on your server, you must set some configuration parameters for Cisco CallManager and perform other post-installation tasks before you can begin using it. Perform these tasks for each server you install and complete them before or after the other servers in the cluster are installed.

This section explains how to perform the following operations:

- Change passwords for Cisco CallManager Administration and SQLSvc accounts
- Configure DNS
- Configure the database
- Activate Cisco CallManager services

After you have performed these tasks and your Cisco CallManager is operational, you should back up your Cisco CallManager data.

Changing Passwords

At installation, these user accounts receive a default password. To eliminate a security risk, change the passwords after installation.

CAUTION Make sure you set the same password for all the CCMAdmin accounts and the same password (different than the CCMAdmin accounts) for all the SQLSvc accounts for each server in the cluster.

Cisco CallManager Administration

The CCMAdmin user provides a common administrator login account for all Cisco CallManagers in the cluster. For example, if you are logged in as CCMAdmin, you can view all the services in all the servers in the cluster in one Control Center window. To have access to all servers in the cluster simultaneously, you must ensure the password for CCMAdmin is identical on every server in the cluster.

Perform the following steps to change the CCMAdmin password:

Step 1 Choose **Start** > **Programs** > **Administrative Tools** > **Computer Management**.

Step 2 Choose the **Users** folder located in the following path: System Tools/Local Users & Groups. The users display in the main window.

Step 3 Right-click on **CCMAdmin** and choose **Set Password**.

Step 4 Enter the new password in the **New Password** and **Confirm Password** fields and click **OK**.

SQLSvc

Only the Microsoft SQL Server service can use the SQLSvc account. The server uses it as the core account for server-to-server interaction by the Cisco CallManager system. For the service to function properly, ensure the password for the SQLSvc account is identical on all servers in the cluster.

The default SQLSvc password that is set up during installation provides an encrypted password that is unique to the cluster. It is identical on all servers in the cluster. If you change the password, make sure the passwords are identical on all servers in the cluster.

Perform the following steps to change the SQLSvc password:

Step 1 Choose **Start** > **Programs** > **Administrative Tools** > **Computer Management**.

Step 2 Choose the Users folder located in the following path: System Tools/Local Users & Groups. The users display in the main window.

Step 3 Right-click **SQLSvc** and choose **Set Password**.

Step 4 Enter the new password in the **New Password** and **Confirm Password** fields and click **OK**.

Step 5 A window confirms that the password was changed. Click **OK**.

Step 6 Click on the **Services** icon located in Services and Applications. The services display in the main window.

Step 7 Right-click **MSSQLServer** and choose **Properties**. Click the **Log On** tab.

Step 8 Enter the same password as in Step 4 in the **Password** and **Confirm Password** fields and click OK.

Step 9 Right-click **SQLServerAgent** and choose **Properties**.

Step 10 Click the **Log On** tab.

Step 11 Enter the same password as in Step 4 in the **Password** and **Confirm Password** fields and click **OK**.

Step 12 Right-click **stiBack for Cisco MCS** and choose **Properties**.

Step 13 Click the **Log On** tab.

Step 14 Enter the same password as in Step 4 in the **Password** and **Confirm Password** fields and click **OK**.

Step 15 Close the Computer Management window.

Step 16 Choose **Start > Programs > Administrative Tools > Component Services**.

Step 17 Choose the COM+Applications folder located in the following path: Component Services/Computers/My Computer. The applications appear in the main window.

Step 18 Right-click **DBL** and choose **Properties**.

Step 19 Click the **Identity** tab.

Step 20 Enter the same password as in Step 4 in the **Password** and **Confirm Password** fields and click **OK**.

Step 21 Right-click **DBL** and choose **Shut Down**.

Step 22 Right-click **DBL** and choose **Start**.

Step 23 Close the Computer Management window.

Name Resolution Versus IP Addresses

Although DNS or WINS should be used during the installation of multiple Cisco CallManagers, Cisco strongly recommends against using a DNS name resolution server in a Cisco CallManager production environment unless you are confident about the reliability and availability of your DNS server and network structure.

Cisco recommends that you not force the Cisco IP Phones to rely on DNS to resolve IP addresses as it adds an additional point of failure unnecessarily. Figure 8-12 shows the layers of the OSI reference model needed to go through when using DNS versus the IP address.

If you use a DNS server, your call processing system depends on the accessibility of that server. For example, if the DNS server goes down or the connection between it and the Cisco CallManager is broken, the phones must rely on cached server name information. If the phones have been reset and have lost their cached information, they may be unable to resolve the DNS name of the Cisco CallManager server and call processing may be interrupted.

Figure 8-12 *IP Address Versus DNS*

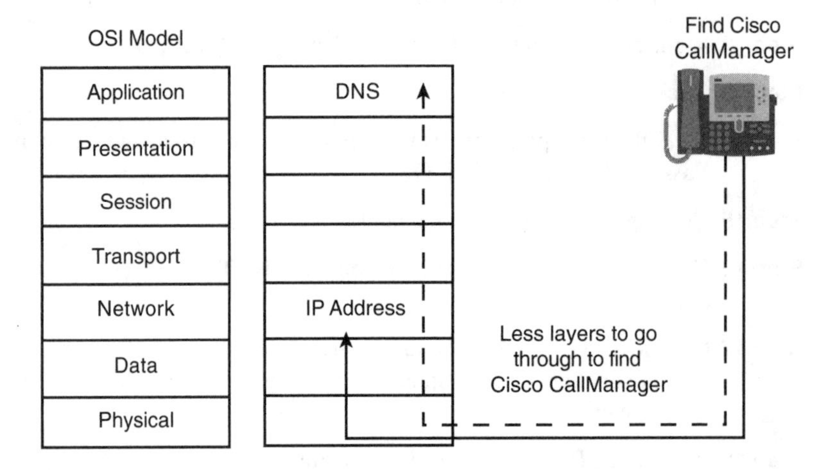

To prevent this situation, Cisco recommends that you disable DNS after installation and change the Cisco CallManager server name to an IP address in the Cisco CallManager Administration using the procedure that follows and as in Chapter 2.

CAUTION If you want to use DNS on your Cisco CallManager system, make sure that the Reverse DNS Lookup function is configured and enabled. This action is necessary whether you use the IP address or the DNS name for your Cisco CallManager server name.

To change the Cisco CallManager server name to the IP address, perform the following steps:

Step 1 Choose **Start** > **Programs** > **Cisco CallManager 3.1** > **CallManager Administration** and log in with administrator privileges.

Step 2 Click **System** > **Server**.

Step 3 From the list of Cisco CallManagers on the left, choose the server you are configuring.

Step 4 Edit the contents of the **DNS or IP Address** field so that it contains the server IP address. Click **Update.**

Re-Creating Subscriber Connections

The one-way replication of the database in a cluster from the Publisher to the Subscriber can best be described as follows:

- On Subscriber:
  - — Check status of Subscriber database
  - — Delete subscriptions
  - — Re-create subscriptions
- On Publisher:
  - — Reinitialize subscriptions
  - — Start the replication snapshot agent

If the connections between the Publisher database and the Subscribers within a cluster are broken for any reason, you cannot replicate the master database to the Subscribers. If you suspect a problem with Subscriber connections, you should verify the status of the subscriptions or the jobs. On the Publisher server, perform the following steps to verify the status of the Subscribers and jobs:

Step 1 Open SQL Server Enterprise Manager by choosing **Start** > **Programs** > **Microsoft SQL Server 7.0** > **Enterprise Manager**.

Step 2 To see the status of subscriptions, choose the **Pull Subscriptions** folder located in the following path:

Microsoft SQL Servers/SQL Server Group/<*this server's hostname*>/Databases/<*publication name*>

To see the status of jobs, choose the **Jobs** folder located in the following path:

Microsoft SQL Servers/SQL Server Group/<*this server's hostname*>/Management/SQL Server Agent.

The Expired Subscription Cleanup service could display a red X under normal operation. However, if a red X appears next to a Subscriber name or a job name other than Expired Subscription Cleanup, assume the Subscriber connection is broken. You must reinitialize it.

NOTE The event log on the Subscriber server lists SQL Server Agent errors. To view the event log, choose **Start > Programs > Administration tools > Event Viewer** on the Subscriber server.

If you determine that one or more subscription connections are broken, you must perform the following actions to reinitialize the connection:

- On each Subscriber server, delete and then re-create the subscriptions.
- On the Publisher server, reinitialize the subscriptions and start the replication snapshot agent.

On each Subscriber server, perform the following steps to delete and re-create the subscriptions:

Step 1 Open SQL Server Enterprise Manager by choosing **Start** > **Programs** > **Microsoft SQL Server 7.0** > **Enterprise Manager**.

Step 2 In the following path, choose the **Pull Subscriptions** folder:

Microsoft SQL Servers/SQL Server Group/*<this server's hostname>*/ Databases/*<the Publisher database name>*

> **NOTE** The default database is typically CCM0300. After an upgrade a new database is built. For example, if you upgrade Cisco CallManager, the most current database is CCM0301. If you upgrade again, the most current database will be CCM0302.

Step 3 In the main window, right-click the subscription name and choose **Delete**. Click **Yes** to confirm.

Step 4 Right-click the **Pull Subscription** folder and choose **New Pull Subscription**.

Step 5 The Welcome to the Pull Subscription Wizard window opens. Click **Next**.

Step 6 If your Publisher server does not appear in the publication list, click **Register Server**. Enter the name of the Publisher server in the **Server** field and choose **Use SQL Server authentication**. Enter the SQL Server system administrator username (SA) and the password in the appropriate fields and then click **OK**. When your server appears in the list, go to Step 7.

Step 7 Double-click the Publisher server and choose the publication name that matches the database you are configuring. Click **Next**.

Step 8 In the Specify Synchronization Agent Login window, choose **By impersonating the SQL Server Agent account** and then click **Next**.

Step 9 In the Specify Immediate-Updating Subscriptions window, choose **Yes**, make this an immediate-updating subscription, and click **Next**.

Step 10 Enter the password for the SA (SQL Server system administrator) user in the **Password** and **Confirm password** fields. (If you are not prompted for a password go to Step 11.)

Step 11 In the Initialize Subscription window, choose **Yes, Initialize the schema at the Subscriber** and click **Next**.

Step 12 In the Set Distribution Agent Schedule window, choose **Continuously** and click **Next**.

Step 13 In the Start Required Services window, click **Next**.

Step 14 Click **Finish**.

On the Publisher server, perform the following steps to reinitialize the subscriptions and start the replication snapshot agent:

Step 1 Open SQL Server Enterprise Manager by choosing **Start** > **Programs** > **Microsoft SQL Server 7.0** > **Enterprise Manager**.

Step 2 In the following path, choose the name of the Publisher database that you are configuring:

Microsoft SQL Servers/SQL Server Group/<*this server's hostname*>/Databases/<*the Publisher database name*>Publications

Step 3 In the main window, right-click the subscription name and choose **Reinitialize all Subscriptions**. Click **Yes** to confirm.

Step 4 In the following path, choose the **Snapshot Agents** folder:

Microsoft SQL Servers/SQL Server Group/<*this server's hostname*>/Replication Monitor/Agents

Step 5 Right-click the publication name that matches the database name you are configuring; then click **Start**.

To restore a server from a backup copy of the database, go to **Start** > **Programs** > **Cisco IP Telephony Applications Backup** and select **Restore Utility**.

In the Cisco IP Telephony Applications Restore Utility follow the steps in the dialog boxes, which are outlined as follows:

Step 1 Select the location of the backup as illustrated in Figure 8-13.

Step 2 Choose a target server to restore. Authenticate the user name and password (see Figure 8-14).

Step 3 Verify to continue.

Figure 8-13 *Restore from Backup—Step 1*

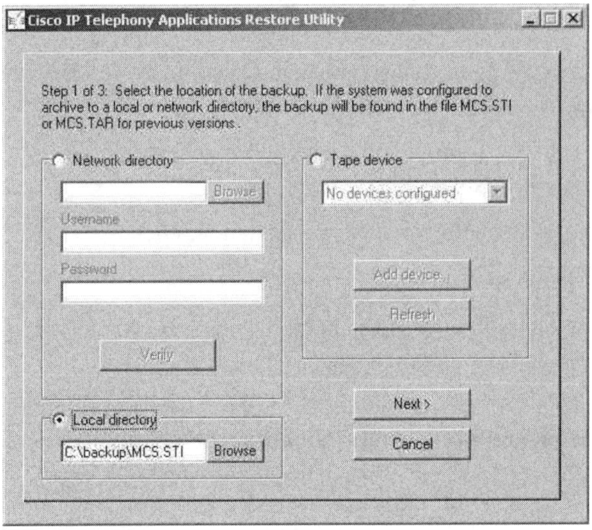

Figure 8-14 *Restore from Backup—Step 2*

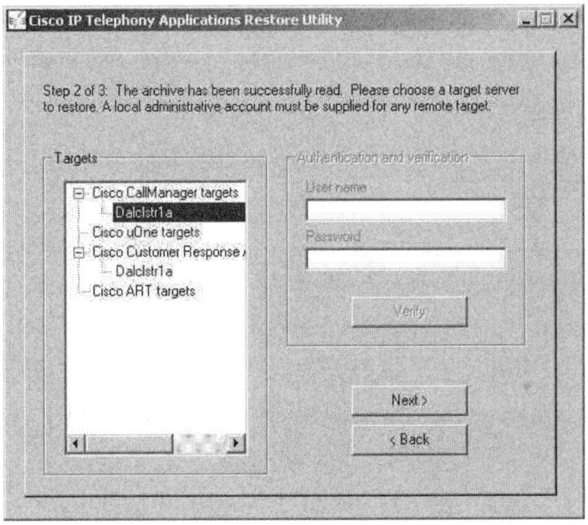

Stopping Unneccesary Services

One of the fundamental principles of securing a server is that each service running will expose potential security vulnerabilities. Because securing every service is difficult and time-consuming, it is a logical task to disable all services that are not mandatory, even if those services aren't immediately known to have a security hole.

Unless otherwise needed on the system, all the following services should be stopped and set to Manual Start status:

- DHCP Client
- FTP Publishing Service
- Alerter Service
- Computer Browser
- Distributed File System

In addition to the services in the preceding list, you should stop and set the following services to Manual Start on the Subscribers:

- IIS Admin Service
- World Wide Web Publishing Service

Both the FTP Publishing Service and World Wide Web Publishing Service depend on the IIS Admin Service. So when the IIS is stopped, FTP Publishing Service and World Wide Web Publishing Service are also stopped. You still need to go to the FTP Publishing Service and World Wide Web Publishing Service and set the services to **Manual**.

Frequently Accessed File Folders

There are file folders built during a Cisco CallManager installation that are accessed frequently. Adding audio files for MOH, accessing device configuration files, finding Web pages, and accessing trace files are the folders most commonly accessed in a Cisco CallManager server.

Figure 8-15 shows the DropMOHAudioSourceFilesHere folder. The path to this folder is C:\Cisco. This is where the audio files are placed so that the audio translator can access those files and convert them to be used for MOH.

Figure 8-16 shows the TFTPPath. The path to the TFTPPath is C:\Program Files\Cisco. The TFTPPath folder is where device configuration files are located. After the audio translator converts the audio files, the audio files can be found in the MOH folder in the TFTPPath.

Figure 8-17 shows CiscoWebs, which is located in the root of the C:\ drive. CiscoWebs contains all the Web files from Administration to the Admin Serviceability Tool (AST), also known as Real-Time Monitoring.

Figure 8-18 shows the Trace folder. Trace is located in C:\Program Files\Cisco. Trace is the default location for trace files used for troubleshooting or collecting to send to TAC for advanced troubleshooting.

Figure 8-15 *DropMOHAudioSourceFilesHere Folder*

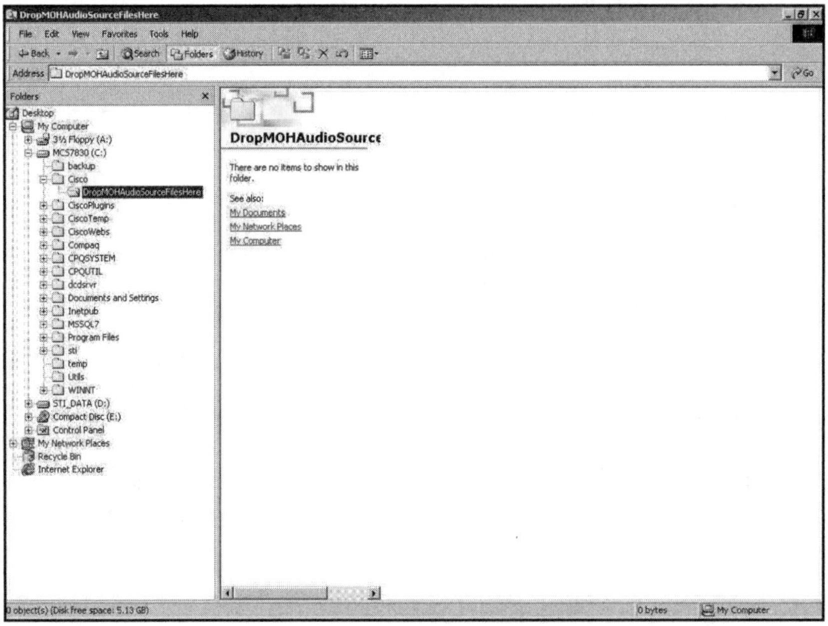

DropMOHAudioSourceFilesHere
File Edit View Favorites Tools Help
Back Search Folders History
Address DropMOHAudioSourceFilesHere Go
Folders
Desktop
My Computer
3½ Floppy (A:)
MCS7830 (C:)
backup
Cisco
DropMOHAudioSourceFilesHere
CiscoPlugins
CiscoTemp
CiscoWebs
Compaq
CPQSYSTEM
CPQUTIL
dcdsrvr
Documents and Settings
Inetpub
MSSQL7
Program Files
sti
temp
Utils
WINNT
STI_DATA (D:)
Compact Disc (E:)
Control Panel
My Network Places
Recycle Bin
Internet Explorer
DropMOHAudioSource
There are no items to show in this folder.
See also:
My Documents
My Network Places
My Computer
0 object(s) (Disk free space: 5.13 GB) 0 bytes My Computer

Figure 8-16 *TFTPPath Folder*

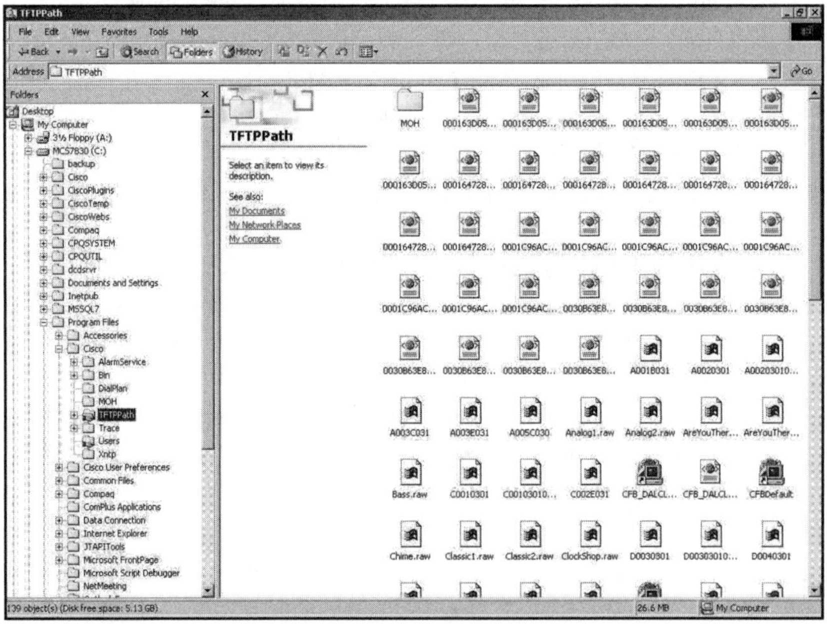

TFTPPath
File Edit View Favorites Tools Help
Back Search Folders History
Address TFTPPath Go
Folders
Desktop
My Computer
3½ Floppy (A:)
MCS7830 (C:)
backup
Cisco
CiscoPlugins
CiscoTemp
CiscoWebs
Compaq
CPQSYSTEM
CPQUTIL
dcdsrvr
Documents and Settings
Inetpub
MSSQL7
Program Files
Accessories
Cisco
AlarmService
Bin
DialPlan
MOH
TFTPPath
Trace
Users
Xntp
Cisco User Preferences
Common Files
Compaq
ComPlus Applications
Data Connection
Internet Explorer
JTAPITools
Microsoft FrontPage
Microsoft Script Debugger
NetMeeting
TFTPPath
Select an item to view its description.
See also:
My Documents
My Network Places
My Computer
MOH 000163D05... 000163D05... 000163D05... 000163D05... 000163D05... 000163D05...
000163D05... 000164728... 000164728... 000164728... 000164728... 000164728... 000164728...
000164728... 000164728... 0001C96AC... 0001C96AC... 0001C96AC... 0001C96AC... 0001C96AC...
0001C96AC... 0001C96AC... 0001C96AC... 0030B63E8... 0030B63E8... 0030B63E8... 0030B63E8...
0030B63E8... 0030B63E8... 0030B63E8... 0030B63E8... A001B031 A0020301 A00203010...
A003C031 A003E031 A005C030 Analog1.raw Analog2.raw AreYouTher... AreYouTher...
Bass.raw C0010301 C00103010... C002E031 CFB_DALCL... CFB_DALCL... CFBDefault
Chime.raw Classic1.raw Classic2.raw ClockShop.raw D0030301 D00303010:... D0040301
139 object(s) (Disk free space: 5.13 GB) 26.6 MB My Computer

Figure 8-17 *CiscoWebs Folder*

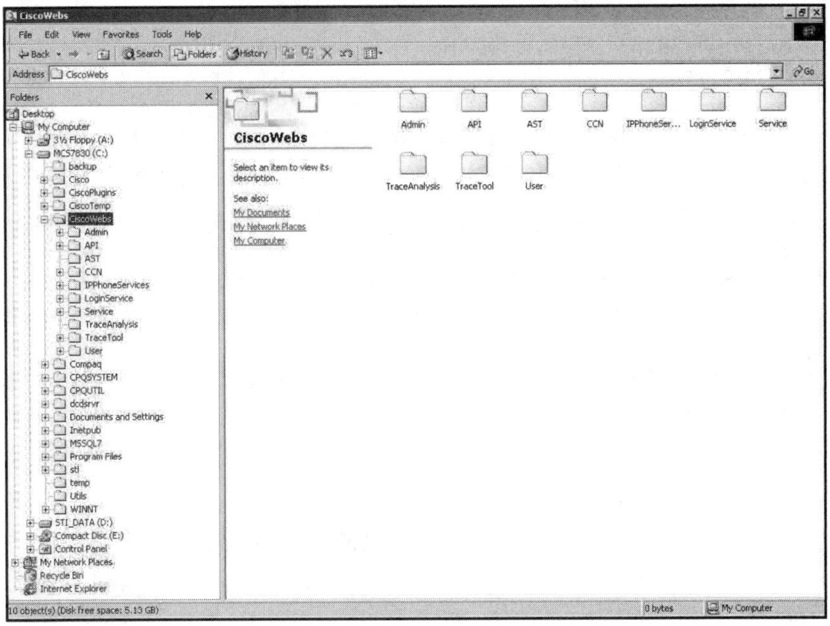

Figure 8-18 *Trace Folder*

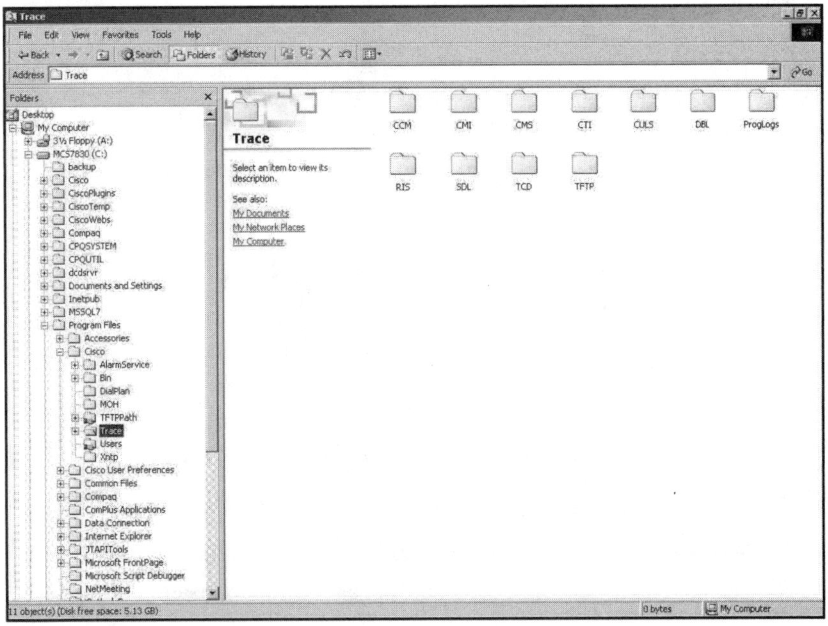

Upgrading Cisco CallManager

Because the Cisco CallManager cluster is based around a Publisher/Subscriber database relationship, upgrading the Publisher first is important. When an upgrade occurs, a new database is created on the Publisher and data from the old database is copied to it. This enables any new changes to the database schema to be easily handled. When Subscribers are added, they subscribe to the new database on the Publisher. That is why the Publisher must be upgraded first; a Subscriber cannot subscribe to a database that does not yet exist.

In a cluster, it is not necessary to bring down all the Cisco CallManagers in a cluster to perform an upgrade. In a large campus, it can be taxing on a DHCP server to be receiving IP address requests from thousands of phones for a couple of hours. It may be undesirable for all phone services to be down for an extended upgrade time. While upgrades should be performed after hours, keeping a part of the cluster running during an upgrade is possible in many cases. This can be done utilizing the Cisco CallManager groups. Essentially while one server is being upgraded, the backup supports all of the phones. The following case studies show three standard deployment models and processes used to upgrade the cluster.

Case Study 1: 2500 Users Using 3 Cisco CallManager Servers

The cluster in Figure 8-19 can support up to 2500 users in a highly redundant deployment; however, some implementations may support 2500 users with fewer servers.

Figure 8-19 *Case Study 1: 2500 Users, 3 Cisco CallManager Servers*

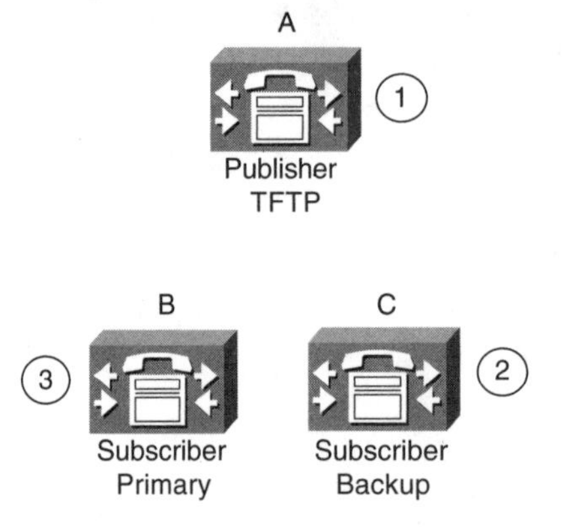

The following details the function of each server in the cluster:

- Server A is a dedicated Publisher and TFTP server.
- Server B is the primary Cisco CallManager for all registered devices.
- Server C is the backup Cisco CallManager for all registered devices.

The cluster configuration has only one Cisco CallManager group, which includes servers B and C.

The following is the process used to upgrade a Cisco CallManager cluster of three servers supporting 2500 users:

Step 1 Upgrade the Publisher (Server A). After the upgrade, reboot the server.

Step 2 Upgrade the backup Cisco CallManager (Server C). The backup is upgraded next because no devices are registered to it and service will not be interrupted. When the upgrade is complete, reboot the server.

Step 3 Upgrade the primary Cisco CallManager (Server B) by stopping the Cisco CallManager service to force devices to fail to the backup Cisco CallManager (Server C). A slight interruption in service occurs while the devices register and receive firmware updates. When the upgrade is complete, reboot the server.

Step 4 The final step of the upgrade process is to reboot all of the servers in the cluster, which synchronizes the DBConnection values in the registry. Start by rebooting the Publisher (Server A). Once the reboot is complete, reboot the primary Cisco CallManager (Server B). When the server comes back online, wait 5 to 10 minutes to enable the devices to begin the fail-back process. Finally, reboot the backup Cisco CallManager (Server C). The cluster upgrade is now complete.

Case Study 2: 2500 to 5000 Users Using 4 Cisco CallManager Servers

The cluster in Figure 8-20 can support up to 5000 users.

Figure 8-20 *Case Study 2: 5000 Users, 4 Cisco CallManager Servers*

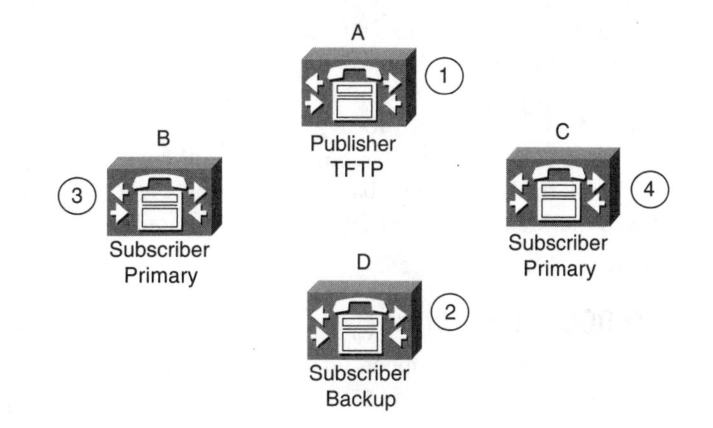

The following details the function of each server in the cluster:

- Server A is a dedicated Publisher and TFTP server.
- Server B is the primary Cisco CallManager for IP phones 1 through 2500.
- Server C is the primary Cisco CallManager for IP phones 2501 through 5000.
- Server D is the backup Cisco CallManager for all registered devices.

The cluster configuration has two Cisco CallManager groups, one includes servers B and D and the other is for servers C and D.

The following is the process used to upgrade a Cisco CallManager cluster of four servers supporting 5000 users:

Step 1 Upgrade the Publisher (Server A). After the upgrade, reboot the server.

Step 2 Upgrade the backup Cisco CallManager (Server D). The backup is upgraded next because no devices are registered to it and service will not be interrupted. When the upgrade is complete, reboot the server.

Step 3 Upgrade the primary Cisco CallManager (Server B) by stopping the Cisco CallManager service to force devices to fail to the backup Cisco CallManager (Server D). A slight interruption in service occurs while the devices register and receive firmware updates. When the upgrade is complete, reboot the server. When Cisco CallManager Server B is rebooted wait 5 to 10 minutes for the devices to fail-back.

Step 4 Upgrade the primary Cisco CallManager (Server C) by stopping the Cisco CallManager service to force devices to fail to the backup Cisco CallManager (Server D). A slight interruption in service occurs while the devices register and receive firmware updates. When the upgrade is complete, reboot the server. When Cisco CallManager server C is rebooted wait 5 to 10 minutes for the devices to fail-back.

Step 5 The final step of the upgrade process is to reboot all of the servers in the cluster, which synchronizes the DBConnection values in the registry. Start by rebooting the Publisher (Server A). Once the reboot is complete, reboot the primary Cisco CallManager (Server B). When the server comes back online, wait 5 to 10 minutes to allow the devices to begin the fail-back process. Next, reboot Cisco CallManager server C and wait until the server comes back online. Finally, reboot backup Cisco CallManager (Server D). The cluster upgrade is now complete.

Case Study 3: 10,000 Users Using 8 Cisco CallManager Servers

The cluster in Figure 8-21 can support up to 10,000 users.

The following details the function of each server in the cluster:

- Server A is a dedicated Publisher
- Server F is a dedicated TFTP server.
- Server B is the primary Cisco CallManager for Cisco IP Phones 1 through 2500.
- Server C is the primary Cisco CallManager for Cisco IP Phones 2501 through 5000.
- Server D is the backup Cisco CallManager for Cisco IP Phones 1 through 5000.
- Server E is the primary Cisco CallManager for Cisco IP Phones 5001 through 7500.
- Server G is the primary Cisco CallManager for Cisco IP Phones 7501 through 10,000.
- Server H is the backup Cisco CallManager for Cisco IP Phones 5001 through 10,000.

Figure 8-21 *Case Study 3: 10,000 Users, 8 Cisco CallManager Servers*

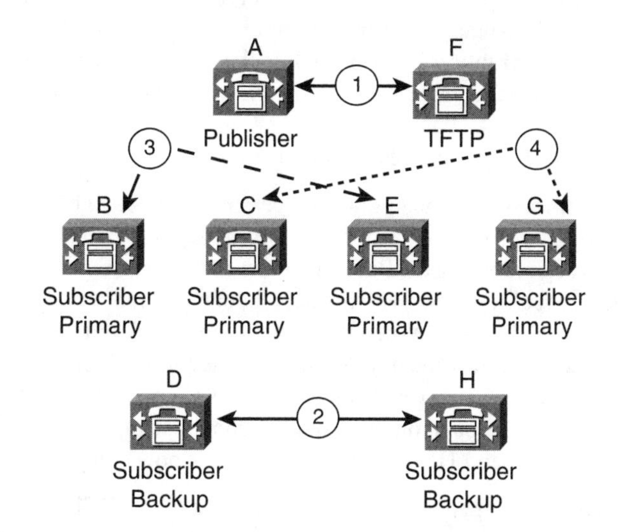

The cluster configuration has four Cisco CallManager groups, configured in the following manner:

- BD
- CD
- EH
- GH

The following is the process used to upgrade a Cisco CallManager cluster of eight servers supporting 10,000 users:

Step 1 Upgrade the Publisher (Server A). After the upgrade, reboot the server. Upgrade the TFTP (Server F), and after the upgrade, reboot the server.

Step 2 Upgrade the backup Cisco CallManagers (Servers D and H). The backup servers are upgraded next because no devices are registered to them and service will not be interrupted. When the upgrade is complete, reboot the servers.

Step 3 Upgrade the primary Cisco CallManagers (Servers B and E) by stopping the Cisco CallManager service to force devices to fail to the backup Cisco CallManager (Servers D and H, respectively). A slight interruption in service occurs while the devices register and receive firmware updates. When the upgrade is complete, reboot the servers. When Cisco CallManager Servers B and E are rebooted wait 5 to 10 minutes for the devices to fail-back.

Step 4 Upgrade the primary Cisco CallManagers (Servers C and G), by stopping the Cisco CallManager service to force devices to fail to the backup Cisco CallManager (Servers D and H, respectively). There will be a slight interruption in service while the devices register and receive firmware updates. When the upgrade is complete, reboot the servers. When Cisco CallManager Servers C and G are rebooted wait 5 to 10 minutes for the devices to fail-back.

Step 5 The final step of the upgrade process is to reboot all of the servers in the cluster, which synchronizes the DBConnection values in the registry. Start by rebooting the Publisher (Server A). Next, reboot the TFTP server (F). Once the reboot is complete, reboot the primary Cisco CallManagers (Servers B and E). When the servers come back online, wait 5 to 10 minutes to allow the devices to begin the fail-back process. Next, reboot the primary Cisco CallManager Servers C and G and wait until the servers come back online. Finally, reboot backup Cisco CallManagers (Servers D and H). The cluster upgrade is now complete.

The following are the general rules of upgrading a Cisco CallManager cluster:

- Always upgrade the Publisher and standalone TFTP server (if it exists) first.
- Upgrade the backup Cisco CallManagers second.
- Upgrade the primary Cisco CallManagers last.
- Make sure that when the SA and Administrator passwords are set that they are the same for all servers in the cluster.

Summary

Cisco CallManager on supported hardware platforms (MCS, Compaq, or IBM) provides a network business communications system for high-quality telephony over IP networks.

Be sure to read and follow the *Installing Cisco CallManager Release 3.1* document that comes with the MCS servers:

> http://www.cisco.com/univercd/cc/td/doc/product/voice/c_callmg/3_1/install/xtocid18583.

Building a Cisco IP Telephony network starts with planning and then installation. Proper planning and a good, clean installation are the keys to a successful CIPT deployment.

The Cisco IP Telephony Backup utility can be used to backup and restore the Cisco CallManager database. Use this tool to backup the servers and refer to your company's policy on how backups are conducted.

Remember the rules for upgrading a cluster:

- Always upgrade the Publisher and standalone TFTP server (if exists) first.
- Upgrade the backup Cisco CallManagers second.
- Upgrade the primary Cisco CallManagers last.
- Make sure that when the SA and Administrator passwords are set and that they are the same for all servers in the cluster.

Post-Test

How well do you think you know this information? The post-test is designed to help you gauge your knowledge about this chapter. Of the 10 questions, if you answer one to three questions correctly, we recommend that you re-read this chapter. If you answer four to seven questions correctly, we recommend that you review those sections that you need to know more about. If you answer 8 to 10 questions correctly, you probably understand this information well enough to move on to the next chapter. You can find the answers to the post-test for this chapter in Appendix B, "Answers to Chapter Pre-Test and Post-Test Questions."

1 When you are installing a cluster of Cisco CallManagers, what must be the first server installed?

   a. Subscriber

   b. Primary

   c. Call Processor

   d. Publisher

2 List at least five pieces of information you should have prior to starting an installation.

____

____

____

3 List at least two recommended post-installation tasks.

____

____

____

4 If you choose a network directory as the backup destination, how should you configure the directory in Windows 2000?

____

____

____

5 List at least five unnecessary services to be set to manual and stopped on all servers in a cluster, and list the two additional services to be set to manual and disabled on Subscriber servers.

____

____

____

6 When you are upgrading a cluster, which server in the cluster must you upgrade first?

____

____

____

This question is worth four points. Describe the four steps used to upgrade a Cisco CallManager cluster of three servers supporting 2500 users.

____

____

____

7

8

9

10

PART III

Advanced Configuration

Upon completing this chapter you will be able to do the following tasks:

- Given a list of best practices for security in a Cisco AVVID environment, correctly define each security best practice.
- Given a list of QoS tools, correctly define each tool in the list.
- Given a Cisco IP Phone, power the phone one of three ways.
- Given a Catalyst switch that supports 802.1q trunking, configure two VLANs per port: one for data and the other for voice traffic.
- Given a Cisco IOS gateway, configure dial peer information so that it can interoperate in a CIPT network.

CHAPTER 9

LAN Infrastructure for Cisco IP Telephony

To ensure successful implementation of Cisco IP Telephony solutions, you must first consider your LAN infrastructure. Before adding voice to your network, your data network must be configured properly. The concepts and implementation techniques discussed in this chapter can be used regardless of whether you have a headquarters with tens of thousands of users or a small branch with fewer than a hundred users. However, the size of the network determines the actual components and platforms you will select and the details that determine the scalability, availability, and functionality of your network. This chapter covers the following topics.

- Abbreviations
- Security
- Importance of QoS
- QoS Tools
- Connecting the Cisco IP Phone
- Configuring IOS H.323 and MGCP Gateways

Pre-Test

Do you already know this information? The pre-test is designed to help you gauge your knowledge about this chapter. Of the 10 questions, if you answer one to three questions correctly, we recommend that you read this chapter. If you answer four to seven questions correctly, we recommend that you skim through this chapter, reading those sections that you need to know more about. If you answer 8 to 10 questions correctly, you probably understand this information well enough to skip this chapter. You can find the answers to these questions in Appendix B, "Answers to Chapter Pre-Test and Post-Test Questions."

1 What are some of the basics when considering security for Cisco AVVID?

____

____

____

2 In building a secure Cisco AVVID network, what are some things you should design into the network?

____

____

____

3 What are two network factors that affect voice quality?

____

____

____

4 What are the three categories of QoS tools?

____

____

____

5 What are the available power schemes supported by Cisco IP Phones?

____

____

____

6 Prior to sending power down the line, which process is used between the Catalyst inline power switch and a Cisco IP Phone?

____

____

____

7 What are the three phone design IP addressing options?

____

____

____

8 When configuring the voice VLANs on an interface or a port on a Catalyst switch, which commands are used for the following switches?

a. Catalyst 4000 and 6000: ______________________________

b. Catalyst 3524, 2900XL: ______________________________

9 On an H.323 gateway, how does the router know which **dial-peer** statement that has the same destination pattern to use first? Or which command in the dial peer configuration is used to identify which dial peer is used first when the dial peers have the same destination pattern?

____

____

____

10 Assume you are running a version of Cisco IOS Software that supports MGCP on a gateway. What is the command that enables the MGCP application on that gateway?

____

____

____

Abbreviations

This section defines the abbreviations used in this chapter. For more information about terms and abbreviations used in this chapter refer to the *IP Telephony Network Glossary* at the following URL:

www.cisco.com/univercd/cc/td/doc/product/voice/evbugl4.htm.

Table 9-1 provides the abbreviation and the complete term.

Table 9-1 *Abbreviations and Complete Terms Used in This Chapter*

Abbreviation	Complete Term
AAA	Authentication, Authorization, and Accounting
AF31	Assured Forwarding 31
BPDU	bridge protocol data unit
BW	bandwidth
CatOS	Catalyst Operating System
CBAC	context-based access control

continues

Table 9-1 *Abbreviations and Complete Terms Used in This Chapter (Continued)*

Abbreviation	Complete Term
CEF	Cisco Express Forwarding
CoS	Class of Service (*Newton's Telecom Dictionary*, 17th Edition, 3rd Definition)
CRTP	Compressed Real-time Transport Protocol
DA	destination address
DC	direct current
DES	Data Encryption Standard
DoS	denial of service
DSCP	Differentiated Services Code Point
DSP	digital signal processor
EF	explicit forwarding
EIGRP	Enhanced Interior Gateway Routing Protocol
FCS	frame check sequence
FTP	File Transfer Protocol
FXO	foreign exchange office
FXS	foreign exchange station
HSRP	Hot Standby Router Protocol
HTTP	Hypertext Transfer Protocol
IDS	Intrusion Detection System
IIS	Internet Information Services
IOS	Internetworking Operating System
IPv4	Internet Protocol version 4
ITU-T	International Telecommunication Union Telecommunication Standardization Sector
LFI	Link Fragmentation and Interleave
LLQ	low-latency queuing
MAC	media access control
MGCP	Media Gateway Control Protocol
MLPPP	multilink point-to-point protocol
ms	milliseconds
NAT	Network Address Translation

Table 9-1 *Abbreviations and Complete Terms Used in This Chapter (Continued)*

Abbreviation	Complete Term
OSPF	Open Shortest Path First
POTS	plain old telephone service
PQ	priority queue
QoS	quality of service
RCP	Remote Copy Protocol
RFC	Request For Comments
RR	round robin
RSH	Remote Shell Protocol
RTP	Real-time Transport Protocol
SA	source address
SMTP	Simple Mail Transfer Protocol
SNMP	Simple Network Management Protocol
TACACS+	Terminal Access Controller Access Control System Plus
TCP	Transmission Control Protocol
ToS	type of service
TX	transmission
UDP	User Datagram Protocol
UPS	uninterruptible power supply
VAD	voice activity detection
VLAN	virtual LAN
VoIP	Voice over IP
VTP	Virtual Terminal Protocol
WAN	wide-area network
WRED	weighted random early detection
WRR	weighted round robin

Voice Security

Voice security is a much more sensitive topic than data security. Users often expect that all voice communications are confidential, even when they don't have the same expectations of an e-mail containing the same information.

You can prevent the vast majority of attempts to compromise the security of a Cisco AVVID network by implementing reasonable security measures on network infrastructure components and servers. This section is will not go into depth about all the security measures to be implemented for voice, but it will bring to light some concerns.

The sections that follow discuss the following issues:

- **Infrastructure Security Best Practices**—Explains general network infrastructure security that is not specific to Cisco AVVID solutions. Most network administrators will benefit from some of these security guidelines. However, those very experienced with security concerns may wish to skip directly to the following section.
- **Securing CallManager Servers**—Discusses Cisco AVVID-specific network security issues.

For information about security in a Cisco AVVID environment go to the following link:

www.cisco.com/univercd/cc/td/doc/product/voice/ip_tele/solution/4_design.htm#xtocid503062

Protecting Network Elements

In protecting the network and network elements, consider the following topics discussed in this chapter:

- Maintain physical device security.
- Use card readers and video surveillance in all data centers and wiring closets.
- Restrict Telnet access.
- Use TACACS+/RADIUS for all devices.

The sections that follow address specific security precautions relevant to specific network elements.

Maintaining Physical Device Security

As with most computing devices, Cisco routers, switches, servers, and other infrastructure components are not designed to provide protection against penetration or destruction by an attacker with direct physical access. You must take reasonable steps to prevent physical access by unauthorized personnel.

Common precautions include restricting access to wiring closets and data centers to staff that are considered "trusted"; generally, such staff already have direct or indirect logical access to the devices being protected and therefore gain no advantage from physical access. In data centers where untrusted staff may be present, use separate locking cabinets for individual items or racks that have more stringent security requirements.

When using keyed or electronic locks on doors, be sure to consider any facilities, security, and janitorial staff that may have the ability or clearance to bypass the locks.

Authentication, Authorization, and Accounting

AAA brings together a centralized administration of accounts and password information. Use all or parts of the topics in this section to help secure your network.

RADIUS and TACACS+

Using either RADIUS or TACACS+ to perform per-user AAA functions is vital to enhance security and accountability for infrastructure devices. These features enable centralized administration of account and password information, eliminating per-device maintenance efforts (and errors) when trusted users are added or removed. You can also use CiscoSecure ACS to require strong passwords, password aging, and intrusion lockout.

Access and Enable Passwords

When using RADIUS or TACACS+ for AAA, each user has their own password for accessing network devices. After gaining access to the device, another password (either a per-user or network-wide password) is necessary to enable commands that enable configuration changes or viewing of sensitive configuration information. Cisco does not recommend setting users to default to level 15 (enabled) as it reduces the number of security safeguards. This section relates to such network infrastructure devices as routers and switches, not Cisco CallManagers.

Choosing Good Passwords

Ideally, you should use one-time password systems such as SofToken, SecurID, or DES Gold Cards to prevent an attacker from reusing trusted users' passwords. However, many organizations consider these tools to be cumbersome or too expensive. If you use normal passwords, they should be chosen so that individuals cannot guess them by trying one or more dictionary words in sequence or by observing the user type. For maximum security, you should include numbers or punctuation symbols, as well as mixed case letters. All passwords should be changed periodically.

Secure Cisco IOS Software Devices

This section pertains to the Cisco IOS devices and discusses how to secure them with some example configurations.

Restrict Virtual Console Access

Limiting virtual console access to the IP address range(s) of operations staff and network management hosts is a useful method to prevent unauthorized users from accessing network devices, even if a password is discovered.

Example 9-1 defines an access list that permits only hosts on network 192.89.55.0 to connect to the virtual console.

Example 9-1 *Access List to Restrict Virtual Console Access*

```
access-list 12 permit 192.89.55.0  0.0.0.255
line vty 0 4
 access-class 12 in
```

For more information, refer to the following Web site:

www.cisco.com/univercd/cc/td/doc/product/software/ios120/12cgcr/np1_r/1rprt2/1rip.htm

Restrict SNMP Access

Nearly all of the information viewable or configurable via virtual console can also be accessed via SNMP. Because an SNMP community is essentially a password that does not require a user name, you must restrict this method of access as completely as possible. Only those hosts with a verified need to perform SNMP writes should have full access.

Example 9-2 defines an access list that permits only hosts on network 192.89.55.0 to perform SNMP reads with the community foobar and only the host 192.89.55.132 to perform SNMP writes with the community foobaz.

Example 9-2 *Access List to Restrict SNMP Access*

```
access-list 12 permit 192.89.55.0  0.0.0.255
snmp-server community foobar RO 12
access-list 13 permit host 192.89.55.132
snmp-server community foobaz RW 13
```

For more information, refer to the following Web site:

www.cisco.com/univercd/cc/td/doc/product/software/ios120/12cgcr/fun_r/frprt3/frmonitr.htm#xt ocid1998360

Enable Session Timeouts

Sometimes operations staff can become distracted or be called away from their systems while logged in to network devices. Automatically disconnecting idle users helps prevent accidental access by unauthorized users.

The configuration in Example 9-3 sets the real and virtual consoles to automatically disconnect the user after five minutes of inactivity.

Example 9-3 *Setting the Real and Virtual Consoles to Time Out*

```
line con 0
 session-timeout 5
line vty 0 4
 session-timeout 5
```

For more information, refer to the following Web site:

www.cisco.com/univercd/cc/td/doc/product/software/ios120/12cgcr/dial_r/drprt1/drtermop.htm# 4907

Encrypt Configured Passwords

Some passwords, such as those for dialup links or local users, must be stored in the device's configuration file. Encrypting the passwords stored in the configuration file makes it difficult for a casual observer to determine or remember these passwords if they come into possession of the configuration file.

To enable encryption of static passwords in the configuration file, enter the following command:

```
service password-encryption
```

For more information, refer to the following Web site:

www.cisco.com/univercd/cc/td/doc/product/software/ios120/12cgcr/secur_r/srprt5/srpass.htm#4 899

Disable Minor Host Services

By default, several services are enabled that allow an attacker to more easily consume device resources, indirectly attack other hosts, or gain information about operations staff currently accessing the network devices. You can disable these services to prevent malicious use of these services or the information they may provide.

The configuration in Example 9-4 disables these services.

Example 9-4 *Disabling Minor Host Services*

```
no service tcp-small-servers
no service udp-small-servers
no service finger
```

For more information, refer to the following Web site:

www.cisco.com/univercd/cc/td/doc/product/software/ios120/12cgcr/fun_r/frprt3/frgenral.htm

Disable or Restrict the HTTP Server

Web configuration is disabled on most platforms; however, novice network administrators often enable it. If HTTP configuration is not necessary, you should disable it entirely. If disabling the service is not feasible, restrict HTTP access to management addresses.

The configuration in Example 9-5 defines an access list that permits only hosts on network 192.89.55.0 to connect to the HTTP server.

Example 9-5 *Restricting HTTP Server Access with Access Lists*

```
access-list 12 permit 192.89.55.0  0.0.0.255
ip http access-class 12
```

For more information, refer to the following Web site:

www.cisco.com/univercd/cc/td/doc/product/software/ios120/12cgcr/fun_r/frprt1/frui.htm

Disable Forwarding of Directed Broadcasts

Directed broadcasts are Unicast packets that are addressed to another subnet's broadcast address. While forwarding these packets provides limited diagnostic value, there is a significant risk of becoming an amplifier in various types of DoS attacks. Cisco IOS Software Release 12.0 and later disables directed broadcasts by default, but they should be manually disabled on all prior versions.

The configuration in Example 9-6 disables directed broadcasts on interfaces Ethernet0/0 and Serial1/1.

Example 9-6 *Disabling Directed Broadcasts on Specific Interfaces*

```
interface Ethernet 0/0
 no ip directed-broadcast
interface Serial 1/1
 no ip directed-broadcast
```

For more information, refer to the following Web site:

www.cisco.com/univercd/cc/td/doc/product/software/ios120/12cgcr/np1_r/1rprt2/1ripadr.htm

Disable Forwarding of Source-Routed Packets

Source-routed packets contain additional hop-by-hop routing information that can supersede what is present in routing tables. Although it was initially intended as a diagnostic tool for network operators, it is not very valuable and can be used to exploit security vulnerabilities. You should disable it on all routers.

To disable forwarding of packets containing source-route information, enter the following command:

```
no ip source-route
```

For more information, refer to the following Web site:

www.cisco.com/univercd/cc/td/doc/product/software/ios120/12cgcr/np1_r/1rprt2/1rip.htm

Disable RCP and RSH Services

Use the Berkeley rcp command to copy files to a device and the rsh command to execute commands without logging in. However, be aware that these services have extremely weak authentication and should not be enabled unless there is no other option (such as SSH support in Cisco IOS Software Release 12.1T).

To disable RCP and RSH services, enter the following commands:

```
no ip rcmd rcp-enable
no ip rcmd rsh-enable
```

For more information, refer to the following Web site:

www.cisco.com/univercd/cc/td/doc/product/software/ios120/12cgcr/fun_r/frprt2/fraddfun.htm

Enable Neighbor Authentication

Most common networking protocols provide a means for neighbors to authenticate each other to ensure that unauthorized devices are not enabled to affect the stability or security of the network. These authentication mechanisms prevent casual attempts at disrupting proper operation but should not be expected to stop a determined attacker.

Secure Devices Running the Catalyst Operating System

This section pertains to the CatOS devices and discusses how to secure them with some example configurations.

Restrict Virtual Console and SNMP Access

You can limit virtual console and SNMP access to the IP address range(s) of operations staff and network management to prevent unauthorized users from accessing network devices, even if a password is discovered.

To define an access list that permits only hosts on network 192.89.55.0 to connect to the virtual console or make SNMP requests, enter the following command:

```
set ip permit 192.89.55.0 255.255.255.0 all
```

For more information, refer to the following Web site:

www.cisco.com/univercd/cc/td/doc/product/lan/cat6000/sw_5_5/cnfg_gd/ip_perm.htm

Configure SNMP Communities

Nearly all the information that you can view or configure via a virtual console can also be accessed via SNMP. Because an SNMP community is essentially a password that does not require a username, it is essential that this method of access is restricted as completely as possible. You should enable access only to those hosts with a verified need to perform SNMP writes.

Example 9-7 defines RO, RW, and RWA communities of foo, bar, and baz, respectively:

Example 9-7 *Configuring SNMP Communities*

```
set snmp community read-only foo
set snmp community read-write bar
set snmp community read-write-all baz
```

For more information, refer to the following Web site:

www.cisco.com/univercd/cc/td/doc/product/lan/cat6000/sw_5_5/cnfg_gd/snmp.htm

Enable Session Timeouts

Sometimes operations' staff members are distracted or are called away from their systems while logged in to network devices. Automatically disconnecting idle users helps prevent accidental access by unauthorized users.

To set the switch to automatically disconnect the user after five minutes of inactivity, enter the following command:

```
set logout 5
```

For more information, refer to the following Web site:

www.cisco.com/univercd/cc/td/doc/product/lan/cat6000/sw_5_5/cmd_refr/set_m_pi.htm

Enable Port Security

Many infrastructure attacks require that the attacker assume the MAC address of the victim's device or replace the victim's device with a counterfeit clone. Enabling port security on Catalyst switches automatically causes the switch to disable any port that changes MAC addresses or uses another port's MAC address. You should use this feature with caution in environments where laptops or other devices are commonly connected to different switch ports.

NOTE Applying port security limits your ability to do administrator-free moves, adds, and changes, and phones are dedicated to a single port.

The configuration in Example 9-8 enables port security on ports 2/1 through 2/48 and configures the disablement to last 30 minutes after each violation.

Example 9-8 *Enabling Port Security and Setting Port Disablement*

```
set port security 2/1-48 enable
set port security 2/1-48 shutdown 30
```

For more information, refer to the following Web site:

www.cisco.com/univercd/cc/td/doc/product/lan/cat6000/sw_5_5/cnfg_gd/sec_port.htm

Enable VTP Server Authentication

VTP servers are capable of adding or removing VLANs within a VTP domain. Because you can potentially configure any Catalyst switch as a VTP server for a given domain, setting a password on the intended VTP servers is the only way to prevent an attacker from sending false **vtp** commands.

To enable the password foobar for the current VTP domain, enter the following command:

```
set vtp passwd foobar
```

For more information, refer to the following Web site:

www.cisco.com/univercd/cc/td/doc/product/lan/cat6000/sw_5_5/cnfg_gd/vtp.htm

Secure Network Design

In building a secure network, design the following in the network:

- Place all VoIP devices on "voice-only" VLANs. Use separate RFC 1918 addresses for VoIP devices (For more information about RFC 1918 visit www.ietf.org/rfc/rfc1918.txt).
- Place a firewall between a Cisco CallManager cluster and all other devices.
- Use sensors to monitor for attacks.

The sections that follow address specific network design precautions to enhance voice security.

IOS Security/Firewall Features: Anti-Spoofing Filters

Many widespread DoS attacks rely on the ability of the attacker to send packets with forged (spoofed) source addresses, which makes tracking the true source of the attack very difficult. To help prevent your site from sourcing these types of attacks, you should block any outbound packets outside of your own address space.

Many network attacks rely on an attacker falsifying, or "spoofing," the source addresses of IP datagrams. Some attacks do not rely on spoofing to work at all, and other attacks are much harder to trace if the attacker can use somebody else's address instead of his or her own. Therefore, it's valuable for network administrators to prevent spoofing wherever feasible.

Anti-spoofing should be done at every point in the network where it's practical, but is usually both easiest and most effective at the borders between large address blocks or between domains of network administration. It's usually impractical to do anti-spoofing on every router in a network because of the difficulty of determining which source addresses may legitimately appear on any given interface.

If you're an Internet service provider (ISP), you may find that effective anti-spoofing, together with other effective security measures, causes expensive, annoying problem subscribers to take their business to other providers. ISPs should be especially careful to apply anti-spoofing controls at dialup pools and other end-user connection points (see also RFC 2267).

Administrators of corporate firewalls or perimeter routers sometimes install anti-spoofing measures to prevent hosts on the Internet from assuming the addresses of internal hosts, but don't take steps to prevent internal hosts from assuming the addresses of hosts on the Internet. It's a far better idea to try to prevent spoofing in both directions. There are at least three good reasons for doing anti-spoofing in both directions at an organizational firewall:

- Internal users will be less tempted to try launching network attacks and less likely to succeed if they do try.
- Accidentally misconfigured internal hosts will be less likely to cause trouble for remote sites (and therefore less likely to generate angry telephone calls or damage your organization's reputation).
- Outside crackers often break into networks as launching pads for further attacks. These crackers may be less interested in a network with outgoing spoofing protection.

Anti-Spoofing with Access Lists

It's not practical to give a simple list of commands that will provide appropriate spoofing protection; access list configuration depends too much on the individual network. However, the basic goal is simple: to discard packets that arrive on interfaces that are not viable paths from the supposed source addresses of those packets. For example, on a two-interface router connecting a corporate network to the Internet, any datagram that arrives on the Internet interface, but whose source address field claims that it came from a machine on the corporate network, should be discarded.

Similarly, any datagram arriving on the interface connected to the corporate network, but whose source address field claims that it came from a machine outside the corporate network, should be discarded. If CPU resources allow it, anti-spoofing should be applied on any interface where it's feasible to determine what traffic may legitimately arrive.

ISPs carrying transit traffic may have limited opportunities to configure anti-spoofing access lists, but such an ISP can usually at least filter outside traffic that claims to originate within the ISP's own address space.

In general, anti-spoofing filters must be built with input access lists; that is, packets must be filtered at the interfaces through which they arrive at the router, not at the interfaces through which they leave the router. This is configured with the **ip access-group list in** interface configuration command. It's possible to do anti-spoofing using output access lists in some two-port configurations, but input lists are usually easier to understand even in those cases. Furthermore, an input list protects the router itself from spoofing attacks; whereas, an output list protects only devices "behind" the router.

When anti-spoofing access lists exist, they should always reject datagrams with broadcast or multicast source addresses and datagrams with the reserved loopback address as a source address. It's usually also appropriate for an anti-spoofing access list to filter out all ICMP redirects, regardless of source or destination address. Appropriate commands are

```
access-list number deny icmp any any redirect
access-list number deny ip 127.0.0.0 0.255.255.255 any
access-list number deny ip 224.0.0.0 31.255.255.255 any
access-list number deny ip host 0.0.0.0 any
```

Note that the fourth command will filter out packets from many BOOTP/DHCP clients and, therefore, is not appropriate in all environments.

Anti-Spoofing with RPF Checks

In almost all Cisco IOS software versions that support Cisco Express Forwarding (CEF), it's possible to have the router check the source address of any packet against the interface through which the packet entered the router. If the input interface isn't a feasible path to the source address according to the routing table, the packet will be dropped.

This works only when routing is symmetric. If the network is designed in such a way that traffic from host A to host B may normally take a different path than traffic from host B to host A, the check will always fail and communication between the two hosts will be impossible. This sort of asymmetric routing is common in the Internet core. You should make sure that your network doesn't use asymmetric routing before enabling this feature.

This feature is known as a reverse path forwarding (RPF) check and is enabled with the command **ip verify unicast rpf**. It is available in Cisco IOS Software Release 11.1CC, 11.1CT, 11.2GS, and all 12.0 and later versions but requires that CEF be enabled in order to be effective.

IOS Security/Firewall Features: TCP Intercept

TCP Intercept is an IOS feature that is specifically designed to protect vulnerable hosts from SYN attacks. These attacks abuse a common flaw in TCP implementations to render the host temporarily unable to accept incoming connections.

SYN flood attacks, also known as *TCP flood* or *half-open connections* attacks, are common DoS attacks perpetrated against IP servers. SYN flood attacks are perpetrated as follows:

1. The attacker spoofs a nonexistent source IP address or IP addresses and floods the target with SYN packets pretending to come from the spoofed host(s).
2. SYN packets to a host are the first step in the three-way handshake of a TCP-type connection; therefore, the target responds as expected with SYN-ACK packets destined to the spoofed host or hosts.

3 Because these SYN-ACK packets are sent to hosts that do not exist, the target sits and waits for the corresponding ACK packets that never show up. This causes the target to overflow its port buffer with embryonic or half-open connections and stop responding to legitimate requests.

In version 5.2 of the PIX OS the TCP Intercept feature was added to the SYN Flood Guard. The TCP Intercept feature improves the embryonic connection response of the PIX Firewall. When the number of embryonic connections exceeds the configured threshold, PIX Firewall intercepts and proxies new connections.

IOS Security/Firewall Features: CBAC

Cisco's CBAC extends the functionality of a traditional access list to include firewall functionality by monitoring stateful connections.

You can find more information in the Cisco Press book *Cisco Secure PIX Firewalls* (ISBN: 1-58705-035-8) and from the following Web site:

www.cisco.com/univercd/cc/td/doc/product/software/ios120/12cgcr/secur_c/scprt3/sccbac.htm

IOS Security/Firewall Features: Intrusion Detection

The Cisco IDS is the industry's first real-time network intrusion detection system that protects the network perimeter, extranets, and the increasingly vulnerable internal network. The system uses sensors, which are high-speed network appliances, to analyze individual packets to detect suspicious activity. If the data stream in a network exhibits unauthorized activity or a network attack, the sensors detect the misuse in real time, forward alarms to an administrator, and remove the offender from the network. Many Cisco routers, switches, and the Cisco PIX Firewall also have IDS functionality.

Cisco IOS Security/Firewall Features: Private Address Space and NAT

Using RFC 1918 private address space (10/8, 172.16/12, and 192.168/16) for the internal parts of a corporate network is often considered a security measure because these addresses cannot be directly reached from the public Internet.

While using private addressing is useful in many instances for other purposes, it is not ideal. Any penetration from the public Internet is likely to first step through one of the company's exposed Internet servers that can usually reach private addresses behind a NAT or proxy device.

Also, because most attacks come from inside attackers, the ability to hide devices from attackers on the Internet does not pose any protection against the most significant threat.

Intersubnet Connectivity

When configuring voice and data VLANs, keep in mind which components need to access each other in the same subnet and also access other parts of your network. Figure 9-1 shows three VLANs: one for the Cisco CallManager cluster, one for data, and one for voice.

Figure 9-1 shows that for intersubnet connectivity, you want to provide the devices in the voice VLAN access to other devices in the voice VLAN. However, you want to block voice VLAN access to the data VLAN and the Cisco CallManager VLAN.

Figure 9-1 *Common Intersubnet Connectivity Policy*

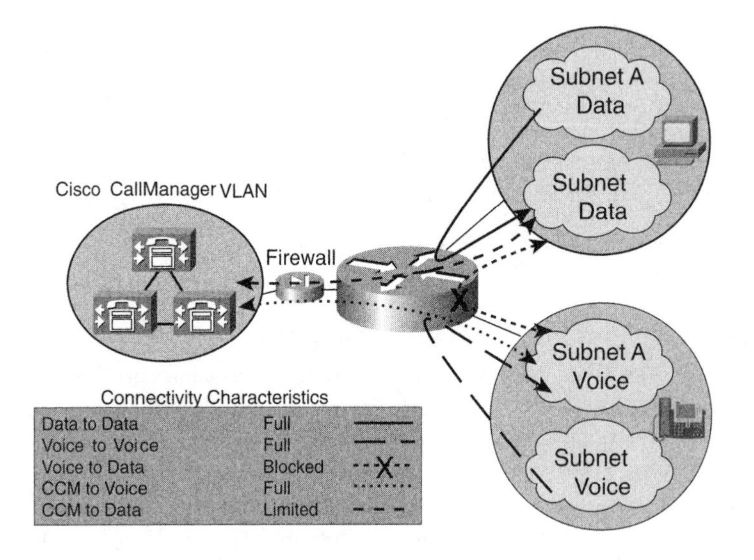

As for the data VLAN, you want to enable devices in the data VLAN to access other devices on the data VLAN and permit only certain devices access to the Cisco CallManager VLAN, and you do not want the data VLAN to access the voice VLAN. If you are using Cisco IP SoftPhone in your CIPT network, you will need to enable your data VLANs access the voice VLANs.

The Cisco CallManager VLAN should be able to have full access to the voice VLAN and limited access to the data subnet.

Protecting Cisco CallManager

A Cisco CallManager server is built from several distinct components (Windows 2000 Server, SQL Server, and Internet Information Server), and security for each of these components must be addressed separately. The sections that follow focus on addressing security issues on a Microsoft Windows 2000 platform.

Install All Available Security Patches Approved by Cisco Systems

Microsoft operating systems, particularly Windows NT and 2000, are subject to intense scrutiny by the cracking community.

When a new vulnerability is discovered, Microsoft responds to security vulnerabilities with patches, usually within days. Due to the frequency with which bugs are discovered, it is essential that an administrator make sure that all possible security patches *approved* by Cisco Systems are applied to all Microsoft products in a timely fashion. Installing patches that have not been approved by Cisco Systems could have a detrimental effect on your Cisco CallManager.

Stop Unnecessary Service

One of the fundamental principles of securing a server is that each service running will expose potential security vulnerabilities. Because securing every service is difficult and time-consuming, it is a logical task to disable all services that are not mandatory, even if those services aren't immediately known to have any security vulnerabilities.

Unless otherwise needed on the system, all of the following services should be stopped and set to Manual Start status:

- Alerter Service
- Computer Browser
- Distributed File System
- DHCP Client
- FTP Publishing Service

Disk Security

FAT-style file systems do not provide any inherent means to restrict access to specific users or encrypt entire disks. Because unauthorized users being able to see or alter privileged information could compromise the security of Cisco CallManager, you must use NTFS. This is the default file system for new installations of Cisco CallManager.

Sanitize Accounts

There is no logical reason for unknown users to be logging into the Cisco CallManager system. Disable the **Guest** account and remove any users from the **Guests** group. This is done via the Users and Passwords feature in the Windows Control Panel (**Start > Settings > Control Panel > Users and Passwords**).

Many attacks rely on trying to blindly execute commands as Administrator; changing the name of the Administrator account to **CallmgrAdmin** or some other logical name helps prevent these attacks from working properly even if the system is compromised. This is done via the Security Options in the Local Security Policy.

Many attacks require access as a non-privileged user on the system. You can prevent these attacks by allowing only administrators to log on to the CallManager locally. This is done via User Rights Assignments in the Local Security Policy.

Because Administrator accounts are privileged, it is essential to ensure that unauthorized users cannot guess the passwords for these accounts. Requiring that all local passwords on the Cisco CallManager system to be at least eight characters long prevents casual observation of the password. In addition an automatic lockout after five incorrect attempts helps prevent an attacker from finding a correct password by entering random guesses. These tasks can be performed via Local Security Settings in the Local Security Policy.

For a step-by step approach to securing a Cisco AVVID network go to the following URL:

www.cisco.com/univercd/cc/td/doc/product/voice/ip_tele/solution/4_design.htm#xtocid503063

At the URL you will find tasks on the following:

- Securing Microsoft IIS
- Microsoft SQL Server
- Setting Up a Secure SQL Server 7.0 Installation

The Importance of QoS

This section discusses the importance of QoS in a CIPT network.

Table 9-2 compares data with voice and the importance for QoS on a voice network as discussed in the text that follows.

Table 9-2 *Converged Data and Voice—Opposite Needs/Behavior*

Data	Voice
Bursty	Smooth
Greedy	Benign
Drop insensitive	Drop sensitive
Delay insensitive	Delay sensitive

Two factors affect voice quality: lost packets (jitter) and delayed packets (latency). Packet loss causes voice clipping and skips. Current Cisco DSP algorithms can correct for up to 30 ms of lost voice. Cisco VoIP technology uses 20 ms samples of voice payload per VoIP packet. One of the many design goals for a CIPT network is to ensure zero percent packet loss for VoIP traffic. If packet loss happens in spite of a good design, the DSP can conceal a single packet drop and minimize any noticeable voice quality degradation.

Packet delay can cause either voice quality degradation, due to the end-to-end voice latency or packet loss, if the delay is variable. If the end-to-end voice latency becomes too long (250 ms, for example), the conversation begins to sound like two parties talking on a CB radio. If the delay is variable, there is a risk of jitter buffer overruns at the receiving end. Eliminating drops and delays is even more imperative when including fax and modem traffic over IP. By examining the causes of packet loss and delay, you can understand why QoS is needed in all areas of enterprise networks.

Packet Loss

Network congestion can lead to packet drops and variable packet delays. Voice packet drops from network congestion are usually caused by full transmit buffers on interfaces somewhere in the network. As links or connections approach 100 percent utilization, the queues servicing that connection will fill. When a queue fills, packets attempting to enter the full queue are discarded. This can occur on a campus Ethernet switch as easily as in a service provider's Frame Relay network.

Because network congestion is typically sporadic, delays from congestion tend to vary. These variable delays are caused by long wait times from the egress interface queue or large serialization delays. See the next section for a discussion of both of these issues.

Delay and Jitter

Delay is the amount of time it takes a packet to reach the receiving endpoint after being transmitted from the sending endpoint. This time period is called the end-to-end delay and can be divided into two areas: fixed network delay and variable network delay. *Jitter* is the delta, or difference, in the total end-to-end delay values of two voice packets in the voice flow.

Figure 9-2 shows the delay budget goal of voice.

Figure 9-2 *Cumulative Transmission Path Delay: Delay Budget Goal < 150 ms*

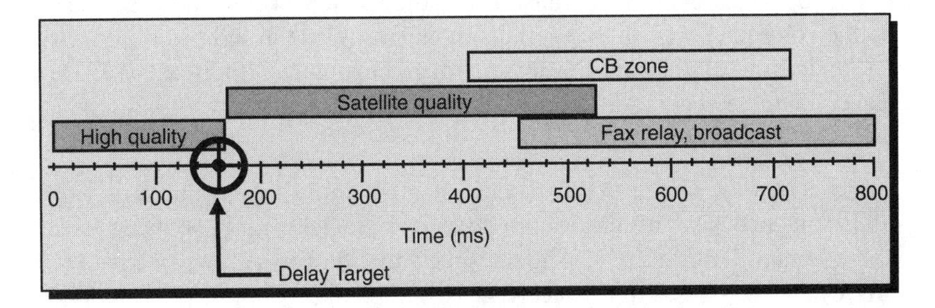

ITU's G.114 "Recommendation"= 0–150 ms 1-Way Delay

You should examine fixed network delay during the initial design of the VoIP network. As Figure 9-2 cites, the International Telecommunication Union Telecommunication Standardization Sector (ITU-T) standard G.114 states that a 150 ms one-way delay budget is acceptable for high voice quality. The Cisco Technical Marketing Team has shown that there is a negligible difference in voice quality scores using networks built with 200 ms delay budgets. Examples of fixed network delay include the propagation delay of signals between the sending and receiving endpoints, voice encoding delay, and the voice packetization time for various VoIP codecs. Propagation delay calculations work out to almost 6.3 μs/km. The G.729A codec, for example, has a 25 ms encoding delay value (two 10 ms frames + 5 ms look-ahead) and an additional 20 ms of packetization delay.

Figure 9-3 shows serialization delays on network interfaces.

Figure 9-3 *Large Packets on Slow Links*

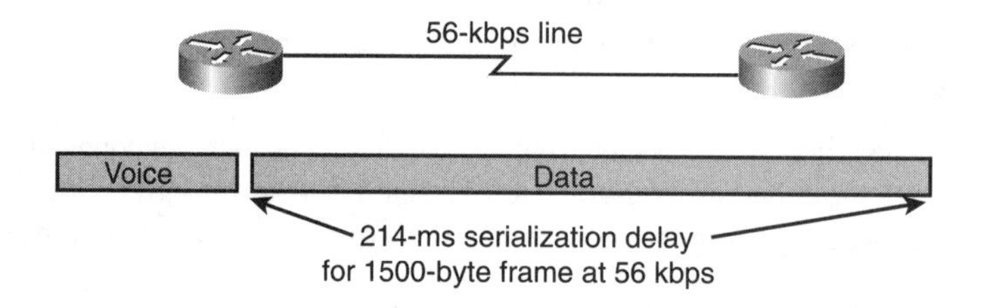

Large packets "freeze out" voice—results in jitter

Congested egress queues and serialization delays on network interfaces can cause variable packet delays. Without Priority or LLQ, queuing delay times equal serialization delay times as link utilization approaches 100 percent. Serialization delay is a constant function of link speed and packet size. The larger the packet and the slower the link clocking speed, the greater the serialization delay. While this is a known ratio, it can be considered variable because a larger data packet can enter the egress queue before a voice packet at any time or not at all. If the voice packet must wait for the data packet to serialize, the delay incurred by the voice packet is its own serialization delay plus the serialization delay of the data packet in front of it. Using Cisco Link Fragmentaion and Interleave (LFI) techniques, you can configure serialization delay to ensure that the delay value will not be larger than a configured maximum.

Because network congestion can occur at any point in time within a network, buffers can fill instantaneously. This can lead to a difference in delay times between packets in the same voice stream. This difference (jitter) is the variation between when a packet is expected to arrive and when it is actually received. To compensate for these delay deltas between voice packets in a conversation, VoIP endpoints use jitter buffers to turn these delay variations into a constant value so that voice can be played out smoothly. A jitter buffer temporarily holds the packets prior to voice playout in order to smooth out the variations in packet delay values.

Cisco VoIP endpoints use domain-specific part algorithms, which have an adaptive jitter buffer between 20–50 ms. The actual size of the buffer varies between 20–50 ms based on the expected voice packet network delay. These algorithms examine the timestamps in the RTP header of the voice packets, calculate the expected delay, and adjust the jitter buffer size accordingly. When this adaptive jitter buffer is configured, a 10 ms portion of "extra" buffer is configured for variable packet delays. For example, if a stream of packets enters the jitter buffer with RTP timestamps indicating 23 ms of encountered network jitter, the jitter buffer for the receiving VoIP endpoint will be sized at a maximum of 33 ms. If a packet's jitter is greater than 10 ms above the expected 23 ms delay variation (23 + 10 = 33 ms of dynamically allocated adaptive jitter buffer space), the packet is dropped.

Table 9-3 shows the resulting delay for the corresponding link speed and packet size.

Table 9-3 *Serialization Delay as a Function of Link Speed and Packet Size*

	Packet Size					
Link Speed	**64 Bytes**	**128 Bytes**	**256 Bytes**	**512 Bytes**	**1024 Bytes**	**1500 Bytes**
56 kbps	9 ms	18 ms	36 ms	72 ms	144 ms	214 ms
64 kbps	8 ms	16 ms	32 ms	64 ms	128 ms	187 ms
128 kbps	4 ms	8 ms	16 ms	32 ms	64 ms	93 ms
256 kbps	2 ms	4 ms	8 ms	16 ms	32 ms	46 ms
512 kbps	1 ms	2 ms	4 ms	8 ms	16 ms	23 ms
768 kbps	.640 ms	1.28 ms	2.56 ms	5.12 ms	10.24 ms	15 ms

QoS Tools

Voice quality is only as good as the quality of your weakest network link. Packet loss, delay, and delay variation all contribute to degraded voice quality. Also, because instantaneous buffer congestion can occur at any time in any portion of the network, network quality is an end-to-end design issue. The QoS tools discussed in this section are a set of mechanisms to increase voice quality on data networks by decreasing dropped voice packets during times of network congestion and minimizing both the fixed and variable delays encountered in a given voice connection.

The QoS tools can be separated into three categories and are described as follows:

- Classification
- Queuing
- Network provisioning

The sections that follow describe the QoS tools in more detail.

Classification QoS Tool

Classification means marking a packet or flow with a specific priority. This marking establishes a trust boundary that must be enforced.

Figure 9-4 shows what is in a packet based on the type of QoS tags used.

Figure 9-4 *Classify at Layer 3 or Layer 2*

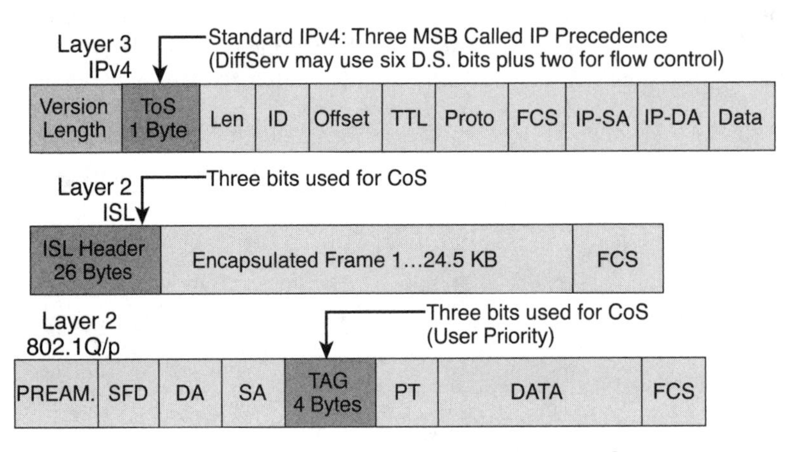

Classification should take place at the network edge, typically in the wiring closet or within the Cisco IP Phones or voice endpoints. Packets can be marked as important by using Layer 2 CoS settings in the User Priority bits of the 802.1p portion of the 802.1Q header or the IP Precedence/DSCP bits in the ToS byte in the IPv4 header. All Cisco IP Phone RTP packets should be tagged with a values of CoS=5 for the Layer 2 802.1p settings and IP Precedence=5 for Layer 3 settings. Additionally, all Call Control packets should be tagged with a Layer 2 CoS value of 3 and a Layer 3 ToS of 3.

NOTE By default, the Cisco IP Phones set CoS and ToS values to 5.

The previous example uses IP Precedence to mark traffic as a transitional step until all IP devices support the DiffServ Code Point. Ideally, all Cisco VoIP endpoints will use a DSCP value of EF for the RTP voice bearer flows and DSCP equals AF31 for VoIP Control traffic.

Table 9-4 shows the IP Precedence and the DSCP used per Layer 2 CoS.

Table 9-4 *CoS Values*

Classification Layer 2 CoS	IP Precedence	DSCP
CoS 0	Routine (IP Precedence 0)	0–7
CoS 1	Priority (IP Precedence 1)	8–15
CoS 2	Immediate (IP Precedence 2)	16–23
CoS 3	Flash (IP Precedence 3)	24–31
CoS 4	Flash-override (IP Precedence 4)	32–39
CoS 5	Critical (IP Precedence 5)	40–47
CoS 6	Internet (IP Precedence 6)	48–55
CoS 7	Network (IP Precedence 7)	56–63

Queuing as a QoS Tool

Figure 9-5 shows the before and after results of using a queuing mechanism to interleave voice between large data packets.

Figure 9-5 *Slow Link Efficiency Tools*

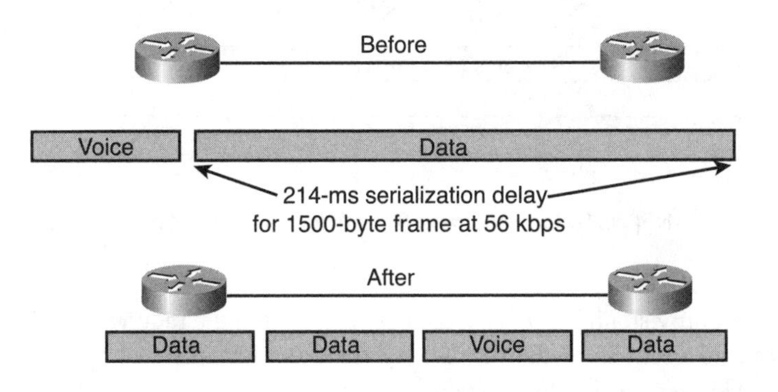

Table 9-5 lists the queuing mechanisms, based on media type, for interleaving voice between large data packets

Table 9-5 *Queuing Mechanisms by Media Type*

Media	Queuing Mechanism
Point-to-point links	MLPPP with Fragmentation and Interleave
Frame Relay	FRF.12
ATM	MLPPP over ATM
ATM/Frame Relay Interworking	MLPPP over ATM and Frame Relay

Based on classification, queuing denotes assigning a packet or flow to one of multiple queues for appropriate treatment in the network.

When data, voice, and video are placed in the same queue, packet loss and variable delay are much more likely to occur. By using multiple queues on the egress interfaces and separating voice into a different queue than data, network behavior becomes much more predictable. Queuing is addressed in all sections of this chapter because buffers can reach capacity in any portion of the network.

Network Provisioning as a QoS Tool

Figure 9-6 shows network provisioning and the recommended calculations for bandwidth provisioning.

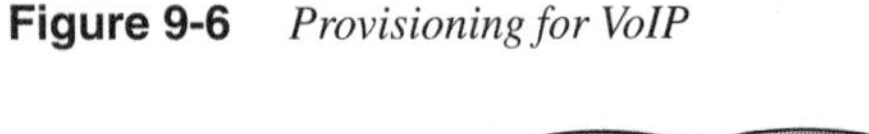

Figure 9-6 *Provisioning for VoIP*

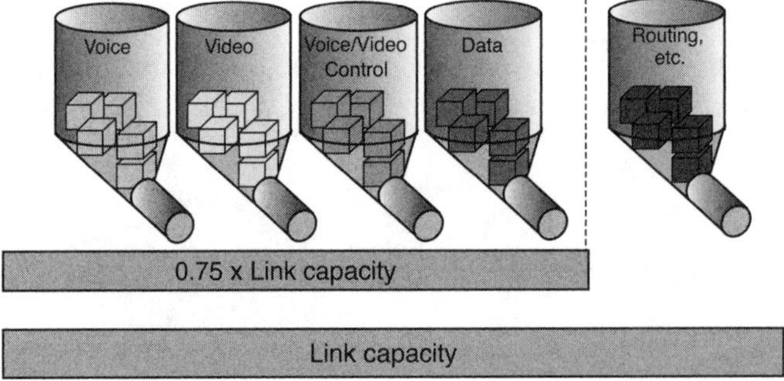

Link capacity=(Min BW for voice and Min BW for video and Min BW for data) / 0.75

Network provisioning entails accurately calculating the required bandwidth needed for voice conversations, all data traffic, any video applications, and necessary link management overhead, such as routing protocols.

When calculating the required amount of bandwidth for running voice over the WAN, it is important to remember that all the combined application traffic (voice, video, and data) should equal only 75 percent of the provisioned bandwidth. The remaining 25 percent is used for overflow and administrative overhead, such as routing protocols.

Traffic Shaping as a QoS tool

Generic Traffic Shaping (GTS) constrains traffic on an outbound interface to a specific bit rate (token bucket approach). Thus, traffic adhering to a particular profile can be shaped to meet downstream requirements.

GTS applies on an interface basis, and works with a variety of L2 technologies, (for example, Frame Relay, ATM, SMDS, and Ethernet) and can be used as an access list to select the traffic to be shaped.

GTS can be set up to adapt dynamically to available bandwidth on FR subinterface using BECN integration or simply to shape to a pre-specified rate. GTS is available in Cisco IOS Software Release 11.2 and later. Traffic shaping should only be used on the data traffic and not on the VoIP traffic.

Tables 9-6 and 9-7 highlight using a G.729 codec in VoIP. Table 9-6 gives values without Layer 2 headers, while Table 9-7 shows the effect of the overhead that the different Layer 2 protocols add.

Table 9-6 *VoIP Bandwidth Requirements*

Codec	Sampling Rate	Voice Payload in Bytes	Packets per Second	IP Bandwidth per Conversation
G.711	20 ms	160	50	80 kbps
G.711	30 ms	240	33	53 kbps
G.729a	20 ms	20	50	24 kbps
G.729a	30 ms	30	33	16 kbps

Table 9-7 *Effect of Overhead on Bandwidth Requirements Based on Layer 2 Protocols*

Code	Ethernet (14 Bytes of Header)	PPP (6 Bytes of Header)	ATM (53 Bytes Cells with a 48-Byte Payload)	Frame Relay (4 Bytes of Header)
G.711 at 50 pps	85.6 kbps	82.4 kbps	106 kbps	81.6 kbps
G.711 at 33 pps	56.5 kbps	54.4 kbps	70 kbps	54 kbps
G.729a at 50 pps	29.6 kbps	26.4 kbps	42.4 kbps	25.6 kbps
G.729a at 33 pps	19.5 kbps	17.4 kbps	28 kbps	17 kbps

You must consider bandwidth requirements and provisioning for that bandwidth when using VoIP. Bandwidth consumption is a matter of the type of codec being used, the voice payload size, and the packets per second.

Table 9-8 shows the bandwidth calculations using cRTP for a VoIP packet.

Table 9-8 *cRTP VoIP Bandwidth Calculations*

Codec	PPP (6 Bytes of Header)	ATM (53 Bytes Cells with a 48-Byte Payload)	Frame Relay (4 Bytes of Header)
G.711 at 50 pps	68 kbps	N/A	67 kbps
G.711 at 33 pps	44 kbps	N/A	44 kbps
G.729a at 50 pps	12 kbps	N/A	11.2 kbps
G.729a at 33 pps	8 kbps	N/A	7.4 kbps

As you can see, the G.729 codec has dropped to 12 kbps rather than 26.4 kbps from the implementation without cRTP.

NOTE cRTP, or RTP Header Compression, is a method for making the VoIP packet headers smaller to regain some of the lost bandwidth. cRTP takes the 40-byte IP/UDP/RTP header on a VoIP packet and compresses it to 2–4 bytes per packet, yielding approximately 11.2 kbps of bandwidth for a G.729 encoded call with cRTP.

Connecting the Cisco IP Phone in a CIPT Network

Figure 9-7 shows a typical highly available enterprise network.

Figure 9-7 *Typical Enterprise Network*

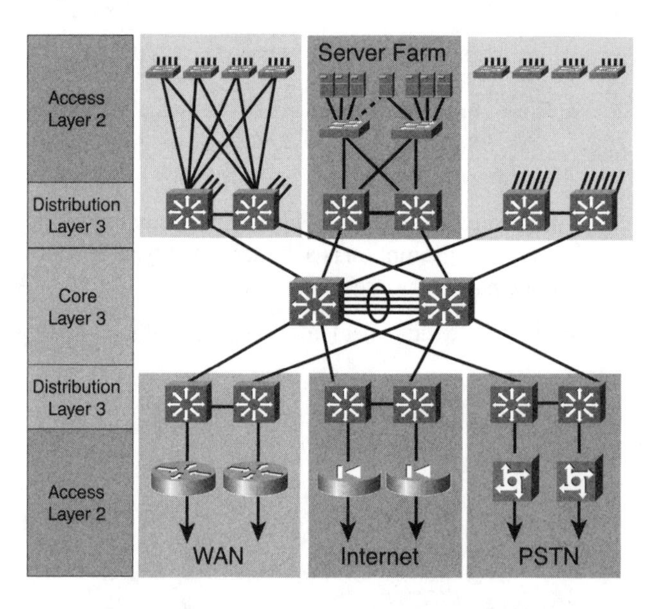

Building an end-to-end IP telephony system requires an IP infrastructure based on Layer 2 and Layer 3 switches and routers with switched connections to the desktop. Network designers must ensure that the endpoints are connected using switched 10/100 Ethernet ports, as illustrated in the preceding figure.

The following sections discuss high availability, single wire power options, IP addressing, and reducing campus congestion.

High Availability

The distributed architecture of a CIPT solution provides the inherent availability that is a prerequisite for voice networking. CIPT solutions are also inherently scalable, enabling seamless provisioning of additional capacity for infrastructure, services, and applications.

In the world of converged networking and in contrast to the world of the PBX, availability is designed into a distributed system rather than into a box. Redundancy is available in the individual hardware components for services such as power and supervisor modules. Network redundancy, however, is achieved with a combination of hardware, software, and intelligent network design practices.

Network redundancy is achieved at many levels as demonstrated in Figure 9-7. A physical connection exists from the edge devices where Cisco IP Phones and computers are attached to two spatially diverse aggregation devices. In the event that an aggregation device fails or connectivity is lost for any reason (such as a broken fiber or a power outage), failover of traffic to the other device is possible. By provisioning clusters of Cisco CallManager servers to provide resilient call control, other servers can pick up the load if any device within the cluster fails.

Advanced Layer 3 protocols such as HSRP and fast converging routing protocols, such as OSPF and EIGRP, can be used to provide optimum network layer convergence around failures.

Advanced tools are also available for the MAC layer (Layer 2). Tunable Spanning Tree parameters and the capability to supply a per-VLAN Spanning Tree (PVST) enable fast convergence. Value-added features such as Uplink-Fast and Backbone-Fast enable intelligently designed networks to further optimize network convergence.

High availability of the underlying network plays a major role in ensuring a successful deployment. This translates into redundancy, resiliency, and fast convergence.

Single Wire and Power Options in a CIPT Network

Figure 9-8 shows the Cisco IP Phone connected to the switch and to the data device (computer or workstation) via the switched Ethernet port on the Cisco IP Phone. This is the most common connectivity option and aids in rapid deployment with minimal modifications to the existing environment. This arrangement has the advantage of using a single port on the switch to provide connectivity to both devices. Also, no changes to the cabling plant are required if the phone is line powered. The disadvantage is that the computer also loses connectivity if the Cisco IP Phone goes down.

Figure 9-8 *Singe Wire and Power Options*

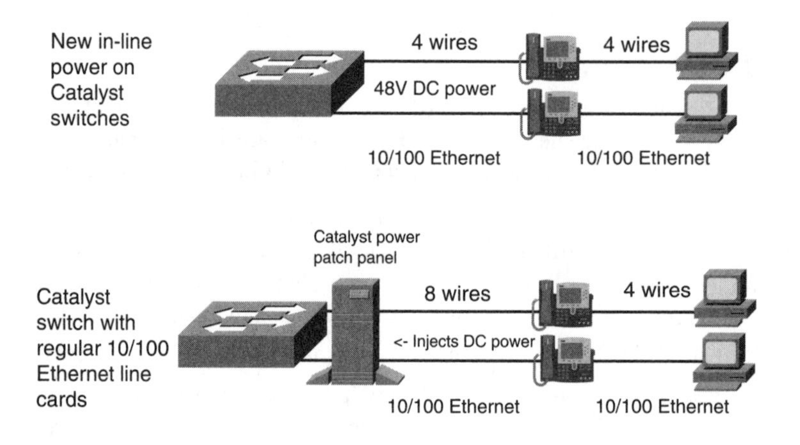

Power to Cisco IP Phones

Cisco IP Phones support a variety of power options. This section discusses each of the three available power schemes:

- Inline power
- External patch panel power
- Wall power

Inline Power

The advantage of inline power is that it does not require a local power outlet. It also permits centralization of power management facilities or UPS devices

With the inline power method, pairs 1 and 2 (pins 1, 2, 3, and 6) of the four pairs in a Category 5 or Category 6 cable are used to transmit power (6.3W) from the switch. This method of supplying power is sometimes called phantom power because the power signals travel over the same two pairs used to transmit Ethernet signals. The power signals are completely transparent to the Ethernet signals and do not interfere with their operation.

The inline method of supplying power requires the new power-enabled line card for the switch. This mechanism is currently available in the following Cisco Catalyst systems:

- Catalyst 6000 Family switches with minimum Cisco CatOS Release 5.5 or higher.
- Catalyst 4000 Family switches. (Catalyst 4006 with Power Entry Module and Auxiliary Power Shelf. Requires minimum of two power supplies to power 240 ports.) Minimum Cisco CatOS Release 6.1 or higher.
- Catalyst 3524-PWR (standalone 24-port 10/100 two gigabit uplinks). Minimum Cisco IOS Software Release 12.0(5)XU or higher.
- Catalyst 4224 Access Gateway switch supplies 24 ports of inline power.

Before the Catalyst switch applies power, it first tests for the presence of a Cisco IP Phone. By first testing for the unique characteristics of the Cisco IP Phone and then applying power (using a low current limit and for a limited time), the Catalyst switch avoids damage to other types of 10/100 Ethernet terminating devices.

Establishing Power to the Cisco IP Phone

To establish power to the Cisco IP Phone, the power-enabled Catalyst switch performs the following steps:

1. The switch performs phone discovery by sending modified Fast Link Pulses (FLPs) down the wire to the Cisco IP Phone. In its unpowered state, the Cisco IP Phone loops these FLPs word back to the switch.

   When the switch receives this FLP word, it knows that the device connected is a Cisco IP Phone and it is safe to deliver power to the device. This behavior is exhibited only by Cisco IP Phones so that other devices connected to the switch port are safe from receiving current. This hardware polling is done by the system at fixed intervals on a port-by-port basis until a LINK signal is seen or the system has been configured not to apply inline power to that port.

2. When the switch uses phone discovery to find an Cisco IP Phone, it applies power to the device. The Cisco IP Phone powers up, energizing the relay and removing the loopback (normally closed relay becomes open) between transmit and receive pairs. It also begins sending Fast Ethernet Fast Link Pulses to attempt to auto-negotiate Ethernet speed and duplex with the switch. From this point, the Cisco IP Phone functions as a normal 10/100 Ethernet device.

   If the Fast Ethernet FLPs are received within five seconds, the Catalyst switch concludes that the attached device is a Cisco IP Phone and it maintains the power feed. Otherwise, power is removed and the discovery process is restarted.

3 Once the Cisco IP Phone is powered and responding, the phone discovery mechanism enters a steady state. If the phone is removed or the link is interrupted, the discovery mechanism starts again. The port is checked for a link every five seconds if a phone has been discovered and power has been applied to the port.

   The advantage of this mechanism is that power is supplied to the phone by the switch just as it is in a traditional telephony environment. In some installations, it is entirely possible that only two pairs have been terminated out of the four available for the data run between the wiring closet and the desktop location. In such cases the inline power method can enable customers to deploy IP telephony by using the existing cable plant without any modification.

The inline power method requires Catalyst software Release 5.5 for Catalyst 6000, Cisco CatOS 6.1 or higher for Catalyst 4000, and Cisco IOS Software Release 12.0(5)XU or higher for Catalyst 3524-PWR. These software releases support all the necessary commands to enable the switch to deliver power through the power-enabled line card. You also have the option of explicitly not providing power through the line card, but the auto-detection feature has the capability of determining whether an attached phone requires power or not.

Configuring the Inline Power Mode

The inline power mode can be configured on each port on the switch using the one of the following commands:

- For Cisco CatOS:

```
set port inlinepower mod/port {auto | off}
```

- For native Cisco IOS Software:

```
Switch(config-if)# power inline {auto | never}
```

  The two modes are defined as follows:

  — **auto**—The supervisor engine tells the port to supply power to the phone only if it has discovered the phone using the phone discovery mechanism. This is the default behavior.

  — **off**—The supervisor engine instructs the port not to apply power, even if it can and if it knows that there is a connected Cisco IP Phone device.

If the **set port inlinepower** command executes successfully, the system displays a message similar to

```
Inline power for port 7/1 set to auto
```

If the **set port inlinepower** command does not execute successfully, the system displays a message similar to

```
Failed to set the inline power for port 7/1
```

You can configure the default power allocation using the following command:

```
set inlinepower defaultallocation value
```

This command specifies how much power (in watts) to apply on a per-port basis. The default value of 7W is good for any currently available or planned Cisco IP Phone model. The phone has the intelligence to report to the switch how much power it actually needs (using Cisco Discovery Protocol), and the switch can adjust the delivered power accordingly, but under some circumstances you might want to reconfigure the default allocation.

If the **set inlinepower defaultallocation** command executes successfully, the system displays a message similar to

```
Default Inline Power allocation per port: 10.0 Watts (0.24 amps @42V)
```

If the **set inlinepower defaultallocation** command does not execute successfully, the system displays the following error message:

```
Default port inline power should be in the range of 2000..12500 (mW)
```

You can display the details on the actual power consumed by using the following command:

```
show port inlinepower {mod | mod/port}
```

Example 9-9 shows a sample display from the **show port inlinepower** command.

Example 9-9 **show port inlinepower** *Command Output*

```
Default Inline Power allocation per port: 12.500 Watts (0.29 Amps @42V)
Total inline power drawn by module 7:  37.80 Watts (0.90 Amps @42V)y
module 5:  37.80 Watts ( 0.90)
Port  InlinePowered   Power   Allocated   Admin   Oper   Detected mWatt  mA  42V
-----         -----   ------      --------     -----      --------
 7/1      auto      off        no          0        0
 7/2      auto      on         yes         12600    300
 7/3      auto      faulty     yes         12600    300
 7/4      auto      deny       yes         0        0
 7/5      on        deny       yes         0        0
 7/6      on        off        no          0        0
 7/7      off       off        no          0        0
```

Power Consumption Cisco IP Phone model 7960 consumes 6.3W. Depending on the number of phones attached or planned, the system should be equipped with a 1300W power supply, the 2500W power supply, or the 4000W power supply.

NOTE The new power supply for the Cisco Catalyst 6000 family switches needs 220V to deliver 2500W of power. When powered with 110V, it delivers only 1300W. In addition, the power supply needs 20A regardless of whether it is plugged into 110V or 220V.

For more information about power, you want to check the Phone Power Calculator at

wwwin.cisco.com/ent/voice/xls/cat6000_power_calculator.xls

Error and Status Messages You can configure the system to send syslog messages that indicate any deviations from the norm. These messages include the following deviations:

- Not enough power available:

```
5SYS-3-PORT_NOPOWERAVAIL:Device on port 5/12 will remain unpowered
```

- Link did not come up after powering up the port:

```
%SYS-3-PORT_DEVICENOLINK:Device on port 5/26 powered but no link up
```

- Faulty port power:

```
%SYS-6-PORT_INLINEPWRFLTY:Port 5/7 reporting inline power as faulty
```

Power status can also be displayed on a per-port basis using the **show port status** command. The command displays the following values:

- **On**—Power is being supplied by the port.
- **Off**—Power is not being supplied by the port.
- **Power-deny**—System does not have enough power, so the port does not supply power.

Dual Supervisors When the system is using dual supervisors, power management per port and phone status are synchronized between the active and standby supervisor. This is done on an ongoing basis and is triggered with any change to the power allocation or phone status. The usefulness and functioning of the high availability features are unaffected by the use of inline power.

Power Protection Cisco recommends that backup power be used for a higher degree of redundancy and availability.

The following factors contribute to power/environment availability:

- Environmental cooling and temperature control
- Equipment BTU and determination
- Environmental conditioning provisioning processes
- Power provisioning process to ensure circuit wattage availability and circuit redundancy for redundant equipment and power supplies
- Equipment surge protection
- UPS battery backup systems
- Generator systems
- SNMP or other remote management processes for UPS systems
- Geographic location of equipment where lightening strikes, floods, earthquakes, severe weather, tornadoes, or snow/ice/hail storms can affect power reliability

- Power cabling above ground
- Construction near or within equipment facility
- Power cabling infrastructure conformance to NEC and IEEE wiring standards for safety and ground control

For more information about power redundancy check with studies done by the APC corporation (www.apcc.com).

Ports and Power Supplies Table 9-9 shows the number of Cisco IP Phones that can be supported with the 1050W, 1300W, and 2500W power-enabled line cards on a Cisco Catalyst 6509 with the Policy Feature Card (PFC).

Table 9-9 *Power Supply and Phones*

Cisco IP Phones Supported with Power-Enabled Line Cards Power Supply	Cisco IP Phones Supported at 6.3W per Phone
1050W	60 Cisco IP Phones
1300W	96 Cisco IP Phones (two modules)
2500W	240 Cisco IP Phones (five modules)

External Patch Panel Power

If the switch does not have a power-enabled line card or one is not available for the switch being used, then a Cisco power patch panel can be used. The power patch panel can be inserted in the wiring closet between the Ethernet switch and the Cisco IP Phone.

The patch panel has a 250W power supply and it accommodates 48 ports and is capable of supplying power to each of the 48 ports at 6.3W per Cisco IP Phone. Cisco recommends a UPS for backup in the event of a power failure.

This arrangement of applying power to the phone uses all four pairs in the Category 5 cable. Unlike the inline method, Ethernet pairs do not carry power signals. Rather, the remaining pairs of Category 5 cable are used for delivering power from the patch panel.

Pairs 1 and 2 from the switch are patched straight through to pairs 1 and 2 coming from the phone. Pairs 3 and 4 from the phone terminate at the patch panel (Ethernet does not use pairs 3 and 4), and power is applied across them to power the phone. The actual conductors used are pins 4 and 5 (pair 3) and pins 7 and 8 (pair 4) for power and ground return. This means that all four pairs in the Category 5 cable need to be terminated at the user's desk and in the wiring closet.

The Cisco power patch panel operates in discovery mode:

- The field-programmable gate array (FPGA) starts at port 1 and sends a 347-kHz loopback tone. The FPGA samples port 1 for 50 ms to detect the loopback tone. Within this 50 ms, 16 transitions must be received to verify that it is the correct frequency.
- If the FPGA verifies that this received signal is the correct one, power is enabled to the port. If either nothing is received or is the wrong frequency, then power is not enabled to the port (or disabled on the port if it was previously powered up).
- The FPGA continuously cycles through the ports repeating the above two steps for each port. Each port is polled for 50 ms every 600 ms to ensure there is still a device attached.
- Because there are no relays involved and the only purpose of these pairs is for power, you can see that this method is considerably more simple than the previous method. The biggest drawback is that it requires that all four pairs of cable be run.

Wall Power

The last option is to power the Cisco IP Phone from a local transformer module plugged into a nearby outlet (maximum of three meters).

IP Addressing for Cisco IP Phones

Figure 9-9 shows the IP addressing deployment options and the recommended method for IP addressing.

Figure 9-9 *IP Addressing Deployment Options*

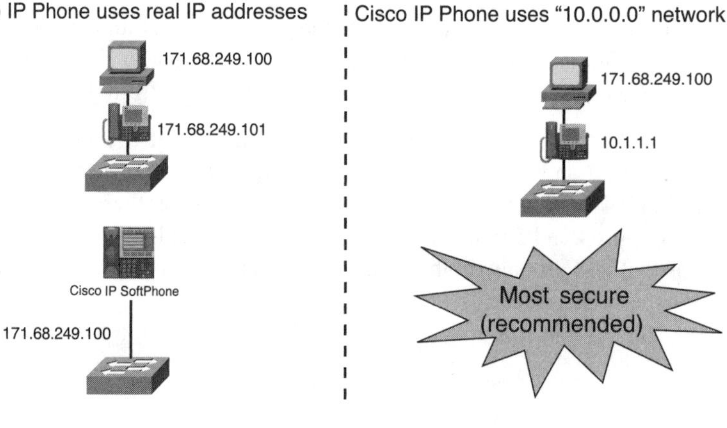

There are three phone IP addressing options:

- Create a new subnet and use that for Cisco IP Phones in a different IP address space (registered or RFC 1918 address space).
- Provide an IP address in the same subnet as the existing data device (PC or workstation).
- Start a new subnet in the existing IP address space (doing so may require you to re-create the entire IP addressing plan for the organization).

Figure 9-10 shows the configuration on the switch that can provide automatic subnet placement for voice and data devices using the same port.

Figure 9-10 *Automatic Subnet Placement*

Catalyst multiservice port provides
automatic phone VLAN configuration

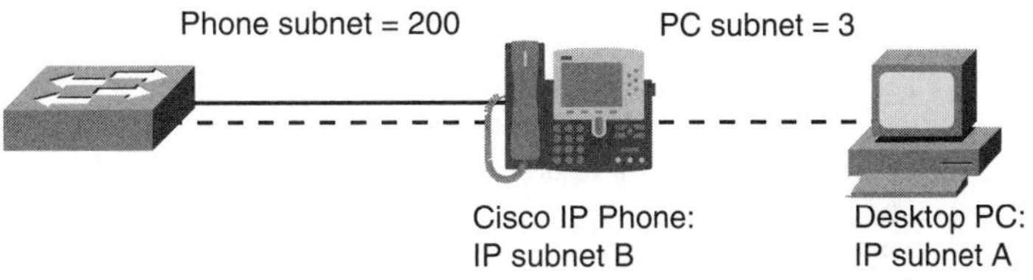

The benefits of using automatic subnet placement are as follows:

- No end-user intervention required.
- Provides the benefits of VLAN technology for the phone.
- Preserves existing IP address structure.
- Uses standards-based 802.1Q technology between switch and phone.

You can implement all these options using either DHCP or static configuration. Adding Cisco IP Phones can potentially double the organization's need for IP address space. While this may not be an issue in some enterprises, others may not have the available address space in particular subnets or even throughout the enterprise. Because of the IP address space concerns, as well as the requirement of separation between the voice and data networks for administrative and QoS reasons, Cisco recommends you create a new subnet for the Cisco IP Phones.

NOTE Using a separate subnet (and potentially a separate IP address space) may not be an option for some small branch offices.

To create voice and data subnets using Catalyst 2948G, 2980G, 4000, and 6000 switches, use the **set port auxiliaryvlan** CatOS command as demonstrated in Example 9-10.

Example 9-10 *Creating Voice and Data Subnets for Catalyst 2948G, 2980G, 4000, and 6000 Switches*

```
cat6k-access> (enable) set vlan 10 name 10.1.10.0_data
cat6k-access> (enable) set vlan 110 name 10.1.110.0_voice
cat6k-access> (enable) set vlan 10 5/1-48
cat6k-access> (enable) set port auxiliaryvlan 5/1-48 110

cat4k> (enable) set vlan 11 name 10.1.11.0_data
cat4k> (enable) set vlan 111 name 10.1.111.0_voice
cat4k> (enable) set vlan 11 2/1-48
cat4k> (enable) set port auxiliaryvlan 2/1-48 111
```

For Catalyst 3500 and 2900XL series switches, you create voice and data subnets using a different set of commands as demonstrated in Example 9-11.

Example 9-11 *Creating Voice and Data Subnets for Catalyst 3500 and 2900XL Switches*

```
interface FastEthernet0/1
     switchport trunk encapsulation dot1q
     switchport trunk native vlan 12
     switchport mode trunk
     switchport voice vlan 112
     spanning-tree portfast
```

You can configure the VLAN to match the subnet address for easier troubleshooting.

Classification and Queuing on the Cisco IP Phone

An integral part of the Cisco network design architecture has always been classifying (marking) traffic as close to the edge of the network as possible. When connecting the Cisco IP Phone using a single cable model, the phone is now the edge of the managed network. Therefore, the Cisco IP Phone can and should classify traffic flows.

Figure 9-11 shows Layer 2 and 3 traffic classification to make voice important traffic.

Three user priority bits in the 802.1p portion of the 802.1Q header are used for signaling Layer 2 CoS information. By default, all VoIP RTP bearer flows from the Cisco IP Phone are set to a Layer 2 CoS value of 5 and a Layer 3 IP Precedence value of 5. Using IP Precedence is a transitional step as all Cisco VoIP devices will eventually migrate to the DSCP for Layer 3 classification. At that time, Cisco VoIP endpoints using DSCP, instead of IP Precedence, will use a DSCP value of 46, or Expedited Forwarding (EF). These CoS and ToS values are significant when examining how classification and queuing works both within an Cisco IP Phone and in an enterprise network.

Figure 9-11 *Layer 2 and Layer 3 Traffic Classification*

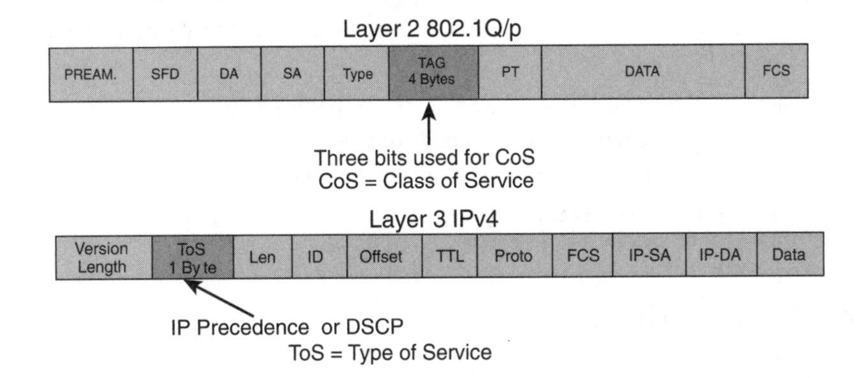

At the heart of a Cisco IP Phone is a three-port 10/100 switch. One port, P3, is an internal port used for connecting the actual voice electronics in the phone. Port P2 is used to connect a daisy-chained PC and Port P1 is used to uplink to the wiring closet Ethernet switch. Each port has four queues with a single threshold (4Q1T) configuration. One of these queues, Queue 0, is a high PQ for all BPDU and CoS=5 traffic. These queues are all serviced in a round-robin (RR) fashion with a timer used on the high PQ. If this timer expires while the queue scheduler is servicing the other queues, the scheduler automatically moves back to the high PQ and empties its buffer, ensuring voice quality.

Because the Cisco IP Phone's high PQ is accessible to any Layer 2 CoS=5 traffic, it is critical to make sure the PC connected to the Cisco IP Phone's access port is not classifying traffic. Cisco recommends you extend the Ethernet switch's trust boundary to the Cisco IP Phone, but not beyond.

Extending the Classification Trust Boundary to the Phone

For the Catalyst 6000 switch, use the **set port trust-ext** command in CatOS 5.5 to instruct the Cisco IP Phone to mark all VoIP frames as CoS=5 and all data traffic from the attached PC as CoS=0 (this is done by CDP). Once you configure the phone to manipulate the CoS value, you also need to configure the line card to accept the Cisco IP Phone's CoS. The best way to accomplish this is to configure an ACL to trust all CoS classification on Ethernet ports in the Auxiliary VLAN. See the following CLI commands for these switches:

```
cat6k-access>(enable)set port qos 5/1-48 trust-ext untrusted
->default
```

Currently the Catalyst 2948G, 2980G, and 4000 switches, do not offer the **set port qos** [*mod/port*] **trust** *trust-ext* commands. Therefore, these switches must rely on the default configuration of the Cisco IP Phone, which uses CoS=5 for all VoIP streams and reclassifies CoS on all PC traffic to "0."

When connecting Cisco IP Phones to Catalyst 3500s and 2900XLs using the single cable model, the same functionality is needed. To configure the Cisco IP Phone to not trust the CoS settings from the PC, use the following commands:

```
interface FastEthernet0/1
switchport priority extend cos 0
```

Reducing Campus Congestion

This section discusses campus congestion in a CIPT network.

Figure 9-12 shows how QoS is done in the Cisco IP Phone.

Figure 9-12 *Campus QoS—IP Phone: Giving Priority to Voice*

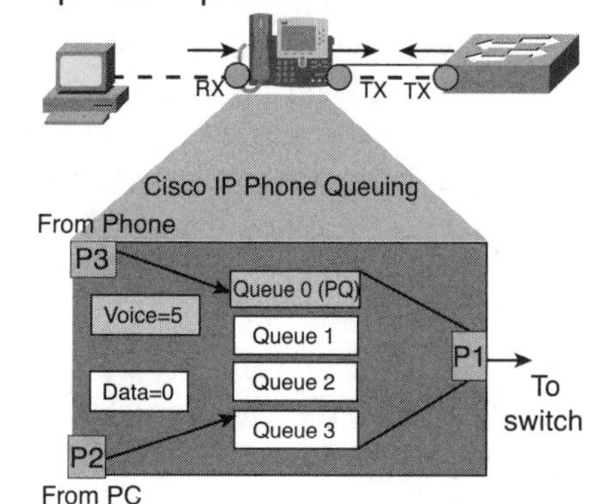

- The CoS/ToS settings for all voice traffic from the IP Phone will be set to 5 (High Priority).
- The CoS settings for all traffic from the PC will be re-classified to 0 (Low Priority) in the IP Phone.

Until recently, the conventional wisdom was that QoS would never be an issue in the enterprise campus due to the bursty nature of network traffic and the capability of buffer overflow. However, we now know that buffering, not bandwidth, is the primary issue in the campus. Therefore, QoS tools are required to manage these buffers to minimize loss, delay, and delay variation.

Transmit buffers have a tendency to fill to capacity in high-speed campus networks due to the bursty nature of data networks combining with the high volume of smaller TCP packets. Bursty is similar to spiking such as when the CPU spikes to open an application. If an output buffer fills, ingress interfaces are not able to place new flow traffic into the output buffer. Once the ingress buffer fills (which can happen very quickly), packet drops will occur. These drops will typically be more than a single packet in any given flow. Current Cisco DSP algorithms can correct for 30 ms of lost voice. Cisco VoIP technology uses 20 ms samples of voice payload per VoIP packet, which means that only a single voice RTP packet can be lost during any given time period. If two successive voice packets are lost, voice quality will degrade. You should always design a CIPT network for zero percent packet loss and if there is packet loss the phone DSPs can conceal most of it without much notice to voice degradation.

VoIP traffic is sensitive to delay and drop. As long as a campus uses Gigabit Ethernet trunks, which have extremely fast serialization times, delay should never be a factor regardless of the size of the queue buffer. However, drops will adversely affect voice quality in the campus. Using multiple queues on transmit interfaces is the only way to eliminate the potential of drops caused by buffers operating at 100 percent capacity. By separating voice and video into their own queues, flows are never dropped at the ingress interface if data flows fill up the data transmit buffer.

NOTE It is critical to remember to verify that Flow Control is disabled when enabling QoS (multiple queues) on Catalyst switches. Flow Control will interfere with the configured queuing behavior by acting on the ports before queuing is activated. Flow Control is disabled by default.

Figure 9-13 shows how multiple queues operate within Cisco switches (Catalyst 2900XL, 3500, 4000, 6000) where the following is true:

- Output buffers can reach 100 percent in campus networks resulting in dropped voice packets.
- QoS required when there is a possibility of congestion in buffers.
- Multiple queues are the only way to guarantee voice quality.

The scheduler process can use a variety of methods to service each of these transmit queues. The easiest method is a RR algorithm, which services queue 1 through queue N in a sequential manner. While not robust, this is an extremely simple and efficient method that can be used for branch office and wiring closet switches. Distribution layer switches use a WRR algorithm, so higher priority traffic is given a scheduling weight. Another option is to combine RR or WRR scheduling with priority scheduling for delay-and-drop-sensitive applications. This uses a PQ, which is always serviced first when there are packets in the queue. If there are no frames in the PQ, then the additional queues are scheduled using RR or WRR.

Figure 9-13 *Campus QoS: Giving Priority to Voice Requirement—Switches with Multiple Queues*

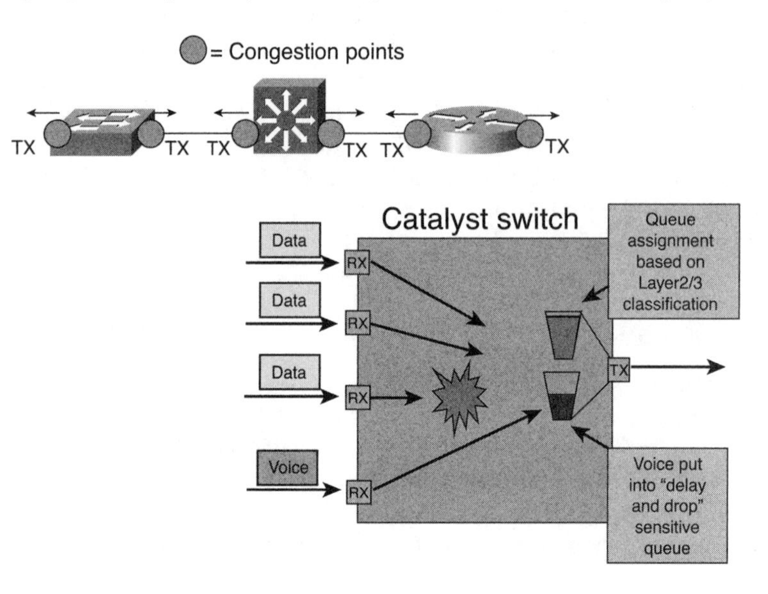

An important consideration is how many queues are actually needed on transmit interfaces in the campus:

- Should you add a queue to wiring closet switches for each CoS value?
- Should you add eight queues to the Distribution layer switches?
- Should you add a queue for each of the 64 DSCP values?

Remembering that each port has a finite amount of buffer memory is important. A single queue will have access to all the memory addresses in the buffer. As soon as you add a second queue, the finite buffer amount is split into two portions, one for each queue. All frames entering the switch not classified for entry into the newly created second queue are now contending for a much smaller portion of buffered memory registers. Therefore, during periods of high traffic, the buffer will fill and frames will be dropped at the ingress interface. Considering that the vast majority of network traffic is TCP-based (comprising TCP ACKs (40 bytes), TCP SYN/ACKs (44 bytes) and 512–1024-byte TCP application traffic (SMTP, HTTP, FTP)), dropping a packet results in a resend. In other words, dropped packets within TCP-oriented networks increase network congestion. Queuing should be used cautiously and only when particular drop-and-delay-sensitive priority traffic is traversing the network.

Two queues are adequate for wiring closet switches where buffer management is less critical. Whether these queues are serviced in an RR, WRR, or Priority Queuing manner is less critical because the scheduler process is extremely fast when compared to the aggregate amount of traffic.

Distribution switches require much more complex buffer management because of the flow aggregation occurring at this layer. Not only do you need PQs, but you also need thresholds within the standard queues.

Cisco chose to use multiple thresholds within queues instead of continually increasing the number of interface queues. Each time a queue is configured and allocated, all the memory buffers associated with that queue can only be used by frames meeting the queue entrance criteria. For example, we will assume that a Catalyst 4000 10/100 Ethernet port has two queues configured: one for VoIP (VoIP bearer and control traffic) and the default queue which is used for HTTP, e-mail, FTP, logins, NT Shares, and NFS. The 128 KB queue is split into a 7:1 transmit and receive ratio. The TX buffer memory is then further separated into high-and-low priority partitions in a 4:1 ratio.

If the default traffic (the Web, e-mail, and file shares) begins to congest the default queue, which is only 24 KB, then packets begin dropping at the ingress interfaces regardless of whether or not the VoIP control traffic is using any of its queue buffers. The dropped packets of the TCP-oriented applications cause each of these applications to re-send the data again, aggravating the congested condition of the network. If this same scenario were configured with a single queue but multiple thresholds used for congestion avoidance, then the default traffic would share the entire buffer space with the VoIP control traffic. Only during periods of congestion when the entire buffer memory approaches saturation, would the lower priority traffic (HTTP and e-mail) be dropped.

This is not to say that multiple queues are not a vital component in Cisco AVVID networks. As discussed earlier, the VoIP bearer streams *must* use a separate queue to eliminate the adverse affects of drops and delays to voice quality. However, every single CoS or DSCP value should not get its own queue because the small size of the resulting default queue will cause many TCP re-sends and actually increase congestion.

The VoIP bearer channel is also a poor candidate for queue congestion avoidance algorithms like weighted random early detection (WRED). Queue thresholding uses the WRED algorithm to manage queue congestion when a preset threshold value is set. WRED works by monitoring buffer congestion and discarding TCP packets if the congestion begins to increase. The result of the drop is that the sending endpoint detects the dropped traffic and slows the TCP sending rate by adjusting the window size.

A WRED drop threshold is the percentage of buffer utilization at which traffic with a specified CoS value is dropped, leaving the buffer available for traffic with higher-priority CoS values. The key is the word "random" in the algorithm name. Even with weighting configured, WRED can still discard packets in any flow; it is just statistically more likely to drop from the lower CoS thresholds.

Voice traffic also uses UDP and dropping a single UDP packet will not result in a slowing down of traffic as TCP does as mentioned in the preceding paragraph.

Configuring IOS H.323 and MGCP Gateways

In both the IOS H.323 and MGCP gateways, some internal configurations need to be performed to interoperate in a CIPT network. The Cisco IOS H.323 gateways work like a relay race for dial plan information. In a relay race, a series of runners travel a specified length passing a baton to each new runner on their team and then the final runner continues the race to the finish. In a similar manner, Cisco CallManager sends the dialed digits to the router. The router examines those digits and matches them to its dial plan information, and then sends it out the correct port.

The MGCP gateway is like a hurdle race. The runner knows where the finish is but has to leap over hurdles to get to the finish line. Cisco CallManager receives and matches the dialed digits and knows the call's endpoint destination but has to go through an MGCP gateway. The MGCP gateway is just a hurdle the call needs to go over to reach the endpoint.

Not only do you need to configure these gateways for dial plan information but also for Cisco CallManager redundancy.

A Cisco IOS H.323 gateway needs some basic dial peer configurations to interoperate in a CIPT network. For use with an FXS or FXO port, you need to configure a **dial-peer** statement using POTS as the technology. For calls from the gateway to the Cisco CallManager cluster, you need to configure a **dial-peer** statement using VoIP.

Examples 9-12 and 9-13 show two examples of **dial-peer** statements used in an H.323 gateway in a CIPT network.

Example 9-12 *Cisco IOS H.323 Dial Plan Configuration Using POTS*

```
dial-peer voice 11 pots
  destination-pattern 11601
  port 1/0/0
```

Example 9-13 *Cisco IOS H.323 Dial Plan Configuration Using VoIP*

```
dial-peer voice 110 voip
  destination-pattern 110..
  voice-class h323 1
  session target ipv4:172.16.10.3
```

Example 9-12 shows that the voice will use POTS, either an FXS or FXO. So if that router gets a call or dialed digits that match "11601" it routes that call out port 1/0/0.

Example 9-13 shows that the voice uses VoIP. If the router gets a call or dialed digits that match "110.." (the dots or periods represent any digit in the range of 0 to 9), then it sends the call or dialed digits to the session target with an IP address of 172.16.10.3

(Cisco CallManager). The **voice-class** statement sets up how the router handles Cisco CallManager redundancy.

Example 9-14 shows how the **voice-class** statement is used to configure Cisco CallManager redundancy in an H.323 gateway. Looking at the bottom of the configuration, we see how the **voice-class h323 1** is configured.

Example 9-14 *H.323v2 Cisco CallManager Redundancy CLI Commands*

```
dial-peer voice 101 voip
  destination-pattern 1111
  session target ipv4:10.1.1.101
  preference 0
  voice-class h323 1 dtmf-relay h245-alphanumeric
dial-peer voice 102 voip
  destination-pattern 1111
  session target ipv4:10.1.1.102
  preference 1
  voice-class h323 1
  dtmf-relay h245-alphanumeric
voice-class h323 1
  h225 timeout tcp establish 5
```

The line that shows **h225 timeout tcp establish 5** determines how long (in seconds) the router waits for a TCP timeout before failing over to the next **dial-peer** statement. So based on the configuration in Example 9-14, the router will wait five seconds for a TCP timeout and then use the next configured **dial-peer** statement.

The **dial-peer** statements in this example look very similar. The similarity is that both dial peers are using the same destination pattern and the same **voice-class** statement. The two dial peers have different session targets: one goes to a Cisco CallManager server and the other goes to another Cisco CallManager Server.

How does the router know which **dial-peer** statement to use first? The **preference** statement in the dial peer configuration determines which **dial-peer** statement to use first. The **dial-peer** statement with the preference set to 0 will be used first, then 1, then 2, and so on.

NOTE Cisco recommends that you not configure VAD on dial peers.

Example 9-15 shows part of a running configuration on an MGCP gateway.

Example 9-15 *MGCP Running Configuration*

```
mgcp
mgcp call-agent 172.20.71.30
mgcp dtmf-relay codec all mode out-of-band
mgcp sdp simple
!
ccm-manager switchback immediate
ccm-manager redundant-host 172.20.71.26 172.20.71.47
ccm-manager mgcp
!
voice-port 1/1/1
!
dial-peer voice 4 pots
 application MGCPAPP
 port 1/1/1
```

Table 9-10 defines the configuration commands used in Example 9-15.

Table 9-10 *MGCP Command Definitions*

Command	Definition
mgcp	Enables the MGCP application on the gateway.
mgcp call-agent	Defines the IP address of the primary Cisco CallManager.
mgcp dtmf-relay codec all mode out-of-band	States the gateway will send DTMF tones out of band.
mgcp sdp simple	States the gateway will use simple desktop protocol as simple. Only choice is simple.
ccm-manager switchback immediate	Determines the method used to failback to the primary Cisco CallManager. There are four choices: **graceful**—Gateway will wait until there are no active calls and then switchback (recommended). **immediate**—Gateways goes immediately back to the primary Cisco CallManager, whether or not there are calls active. **schedule-time**—Enables you to configure a time for switchback within the next 24 hours format (hh:mm). **uptime-delay**—Gateways will switchback during some delay in up time delay.
ccm-manager redundant-host *ip address*	Lists the secondary and tertiary Cisco CallManagers.
ccm-manager mgcp	Lets Cisco CallManager know the gateway is using MGCP.

There are a few similarities with the MGCP configuration in Example 9-15 and the H.323 gateway configuration in Example 9-14. Both configurations use **dial-peer** statements and associate the POTS **dial-peer** statement with a port number. But the MGCP gateway configuration of dial peers does not have a destination pattern, it has a statement **application MGCPAPP**. This configuration lets Cisco CallManager know that this port is using the MGCP application on this gateway and that for calls to go out this port, it checks with Cisco CallManager for dial plan information.

There are quite a few **show** commands on each gateway for checking status. More information about **status** and **show** commands can be found in the CVOICE course.

Summary

QoS tools can be separated into three categories and are described as follows:

- Classification
- Queuing
- Network provisioning

Cisco Catalyst switches (Catalyst 4000, 4224, 6000, 3524, and 2900XL) support dual VLANs per port to provide a voice and data subnet and enable a single wire to the desktop. This enables the administrator to provide priority to voice traffic.

Cisco IOS gateways that are H.323 gateways in a CIPT network need to be configured to process their part of the call leg either going in or out of a CIPT network.

Cisco IOS gateways that are MGCP gateways in a CIPT network need to be configured and let Cisco CallManager know that it is using the MGCP application for dial plan information.

Post-Test

How well do you think you know this information? The post-test is designed to help you gauge your knowledge about this chapter. Of the 10 questions, if you answer one to three questions correctly, we recommend that you re-read this chapter. If you answer four to seven questions correctly, we recommend that you review those sections that you need to know more about. If you answer 8 to 10 questions correctly, you probably understand this information well enough to move on to the next chapter. You can find the answers to these questions in Appendix B, "Answers to Chapter Pre-Test and Post-Test Questions."

1 What are some of the basics when considering security for Cisco AVVID?

____

____

____

2 In building a secure Cisco AVVID network, what are some things you should design into the network?

____

____

____

3 What are two network factors that affect voice quality?

____

____

____

4 What are the three categories of QoS tools?

____

____

____

5 What are the available power schemes supported by Cisco IP Phones?

____

____

____

6 Prior to sending power down the line, which process is used between the Catalyst inline power switch and a Cisco IP Phone?

____

____

____

7 What are the three phone design IP addressing options?

____

____

____

8 When configuring the voice VLANs on an interface or a port on a Catalyst switch, which commands are used for the following switches?

a. Catalyst 4000 and 6000:

b. Catalyst 3524, 2900XL:

9 On an H.323 gateway, how does the router know which **dial-peer** statement that has the same destination pattern to use first? Or which command in the dial peer configuration is used to identify which dial peer is used first when the dial peers have the same destination pattern?

____

____

____

10 Assume you are running a version of Cisco IOS Software that supports MGCP on a gateway. What is the command that enables the MGCP application on that gateway?

____

____

____

Upon completion of this chapter you should be able to do the following tasks:

- Given the three device requirements for call preservation, correctly define each device requirement.
- Given a Cisco IP Telephony device, identify the switchover algorithm and disconnect supervision supported by that device.

CHAPTER 10

Call Preservation

This chapter discusses the call preservation feature of Cisco CallManager. In short, call preservation allows calls to remain active on devices (Cisco IP Phones and gateways) when their primary Cisco CallManager is unreachable or fails. This chapter describes how to select gateway devices to ensure calls that need to stay active continue to stay active if their primary Cisco CallManager is unreachable or fails. This chapter discusses the following topics:

- Abbreviations
- Distributed Call Processing
- Service Requirements for Call Preservation
- Call Preservation Example
- PSTN Gateways Controlled by MGCP
- Call Detail Records (CDRs)

Pre-Test

Do you already know this information? The pre-test is designed to help you gauge your knowledge about this chapter. Of the 12 questions, if you answer one to four questions correctly, we recommend that you read this chapter. If you answer five to eight questions correctly, we recommend that you skim through this chapter, reading those sections that you need to know more about. If you answer 9 to 12 questions correctly, you probably understand this information well enough to skip this chapter. You can find the answers to these questions in Appendix B, "Answers to Chapter Pre-Test and Post-Test Questions."

1 List the three device requirements for supporting call preservation.

____

____

____

2 List the three protocols used by Cisco CallManager to establish calls between devices.

____

____

____

3 Cisco CallManager servers in a cluster use what type of links to establish calls between devices registered to different Cisco CallManager servers?

____

____

____

4 What are the three disconnect supervision mechanisms devices must provide for any media connections to be preserved during system failure?

____

____

____

5 Define or give an example of the "end user release" disconnect supervision.

____

____

____

Complete the missing information marked by questions 6–12 in the following table:

Devices	Switchover Algorithm	Disconnect Supervision
Cisco IP Phones • 12-button series • 30-button series • 79*XX* series	6.	7.
CFB/MTP • Software services • Transcoder	8.	9.
MGCP Gateways • DT24+ (T1-PRI/CAS) • E1/T1-PRI • T1-CAS • 24 port FXS • IOS platforms (FXO/FXS, T1-PRI/CAS)	10.	11. 12.

Abbreviations

This section defines the abbreviations used in this chapter. For more information about terms and abbreviations used in this chapter refer to the *IP Telephony Network Glossary* at the following URL:

www.cisco.com/univercd/cc/td/doc/product/voice/evbugl4.htm

Table 10-1 provides a list of abbreviations with the corresponding complete terms.

Table 10-1 *Frequently Used Abbreviations in This Chapter*

Abbreviations	Definitions
CAS	channel associated signaling
CCM	Cisco CallManager
CFB	conference bridge
FXO	foreign exchange office
FXS	foreign exchange station
ICMP	Internet Control Message Protocol
Link OOS	Link out of service
MGCP	Media Gateway Control Protocol

continues

Table 10-1 *Frequently Used Abbreviations in This Chapter (Continued)*

Abbreviations	Definitions
MSF	media streaming failure
MTP	media termination point
PRI	Primary Rate Interface
PSTN	public switched telephone network
QC	quiet clear
SCCP	Skinny Client Control Protocol

Distributed Call Processing

In a Cisco CallManager cluster, the call processing for devices is distributed among the Cisco CallManager servers doing call processing. This is known as distributed call processing. Call preservation is based on call signaling and call connections. Different elements in a Cisco IP Telephony network communicate to establish calls and connections between devices. In a Cisco IP Telephony network, a call refers to the signaling type setup. A connection refers to call control.

Figure 10-1 shows calls and connections that are part of Cisco CallManager signaling.

Figure 10-1 *Cisco CallManager Signaling*

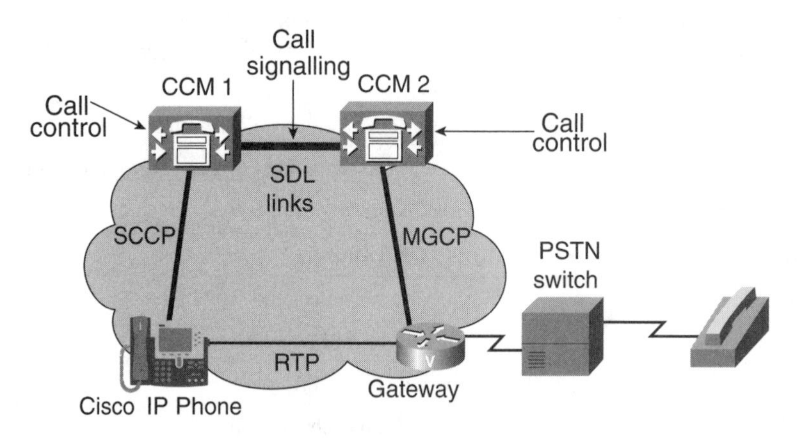

Cisco CallManager establishes calls between devices via several Internet protocols (SCCP, MGCP, H.323). Cisco CallManager servers in a cluster use signal distribution layer (SDL) links to establish calls between devices registered to different Cisco CallManager servers. Media connections between devices in a call stream sessions directly between the devices using RTP.

If an error occurs that affects the signaling path, which is controlled by a Cisco CallManager server, the affected calls can no longer be controlled by the same Cisco CallManager. Some common types of errors are Cisco CallManager server failure and network failure.

The RTP streaming sessions can be preserved as long as there are no network errors affecting communication between the devices.

Service Requirements for Call Preservation

The three device requirements for supporting call preservation are

- Active connection maintenance
- Disconnect supervision
- Switchover algorithm

Active Connection Maintenance

When a device is in an active streaming mode, it is considered an active connection.

When a device detects that its link to its active Cisco CallManager server is out of service (Cisco CallManager server failure or network failure), the device maintains active media connections (streams) as shown in Figure 10-2. An active media connection is one where both send and receive RTP ports are active.

Figure 10-2 *Active Connection*

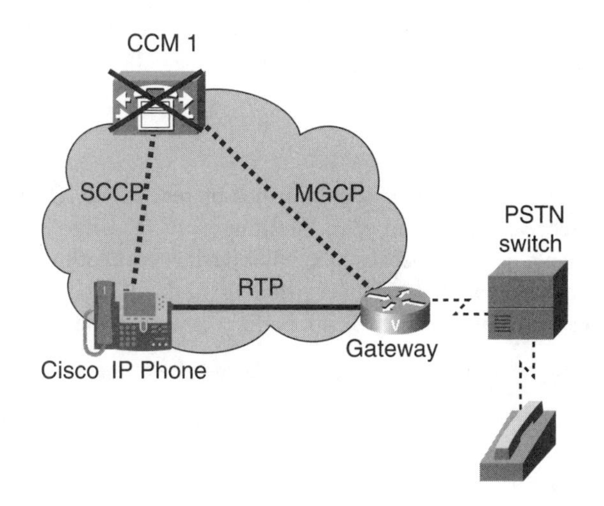

Disconnect Supervision

Devices must provide at least one of the following disconnect supervision mechanisms for any media connections that are preserved during system failure:

- End user release
- Timed
- MSF

End User Release

The end user release disconnect supervision is where the device can detect when the user of the connection hangs up. For instance, Cisco IP Phones provide end user release disconnect supervision when there are changes in the hook status, either on-hook or off-hook.

MGCP gateways can support several interfaces to the PSTN and most PSTN interfaces provide end user release disconnect supervision. Only FXO (loop start) interfaces lack this capability from Cisco CallManager's perspective.

Timed

Timed disconnect supervisions are used by devices when there is no end user that can initiate the release of a connection. The device releases preserved streams (closes RTP ports) after a finite period of time. The period of time is device- and application-specific. For instance, any interface that cannot provide other forms of disconnect supervision on their interface, that is, voice mail systems, applications, and so on, must use timed disconnect supervisions.

MSF

MSF disconnect supervision is accomplished by reacting to ICMP port unreachable indications received as a result of transmitting media packets. The device sends an ICMP port unreachable message because the other device terminating the RTP stream has disconnected the call and is no longer listening for that RTP stream. The device detects that any preserved streaming session is not being received by the destination.

When the MSF condition is detected, the device closes the affected RTP ports.

Switchover Algorithms

The switchover algorithms decide when a device will fail over to the secondary Cisco CallManager when the device detects that its link is out of service to the primary Cisco CallManager. The three switchover algorithms are

- **Graceful**—Delay registration with any available Cisco CallManager until all active streaming sessions are stopped in the device.
- **Immediate**—Immediately switchover to an available Cisco CallManager and communicate the status of preserved connections to this new Cisco CallManager, such that the release of the preserved connections can be managed.
- **Timed**—After timed interval (the default setting is three minutes after a Cisco CallManager failure), any calls in progress through a gateway are dropped by the system and the gateway re-homes to the secondary Cisco CallManager.

Table 10-2 shows the supported device information for disconnect supervision and switchover algorithm.

Table 10-2 *Switchover Algorithm and Disconnect Supervision for Devices*

Devices	Switchover Algorithm	Disconnect Supervision
Cisco IP Phones	Graceful	End user
12-button series		
30-button series		
79*XX* series		
CFB/MTP	Immediate	MSF
Software services		
Transcoder		
MGCP Gateways	Immediate	End user (where applicable)
DT24+ (T1-PRI/CAS)		MSF
E1/T1-PRI		
T1-CAS		
24-port FXS		
IOS platforms (FXO/FXS, T1-PRI/CAS)		

Call Preservation Example

Figure 10-3 illustrates how processes in the Call Control layer and Media Control layer of Cisco CallManager 1 realize that the terminating device control process is in Cisco CallManager 2 and registers for link signals.

Figure 10-3 *Cisco CallManager to Cisco CallManager Call*

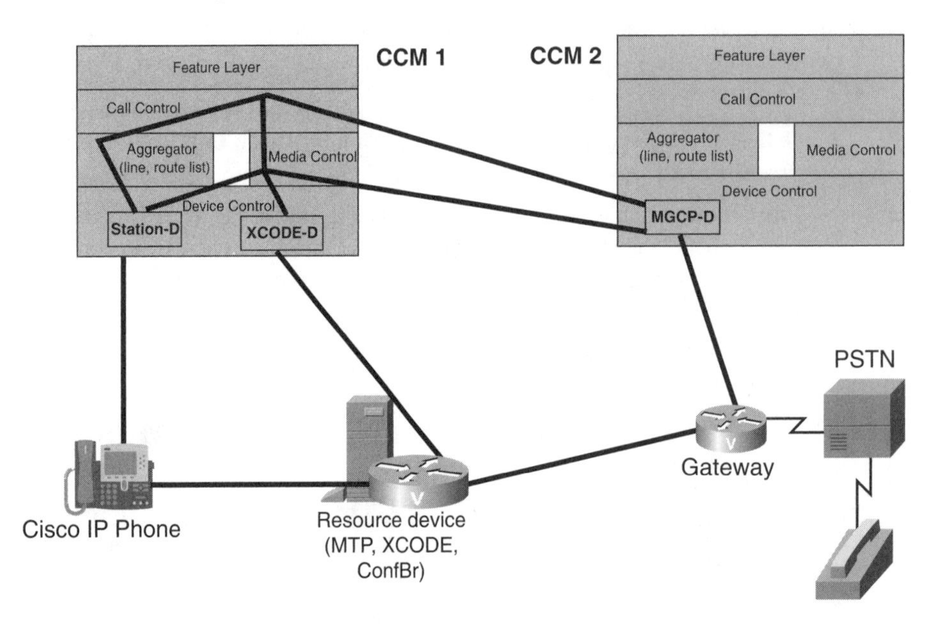

Similarly, the device control process (MGCP-D) recognizes that a call is being originated in Cisco CallManager 1 and registers for link signals to Cisco CallManager 1.

When a Cisco CallManager fails, the surviving Cisco CallManager(s) detect the other Cisco CallManager failure as shown in Figure 10-4.

Each upper layer initiates release of any affected call. If the call is in active state, a QC flag is included in the release signals as shown in Figure 10-5. The release sequence proceeds as normal until the Device Control layer processes receive the release signals. The effect is that all entities in layers above the Device Control layer that are associated with the affected call are freed.

Normally during a release sequence, the Device Control process sends signals to the device to stop the streaming sessions and close associated RTP ports. If the QC flag is received and the call is known to be active, the Device Control process refrains from sending such signals, leaving the media intact, and marks the associated call preserved.

This information is used to handle release sequences for the surviving side of the preserved call.

If the device supports MSF disconnect supervision, a start MSF signal is sent to the device for each preserved connection.

Figure 10-4 *Cisco CallManager Failure*

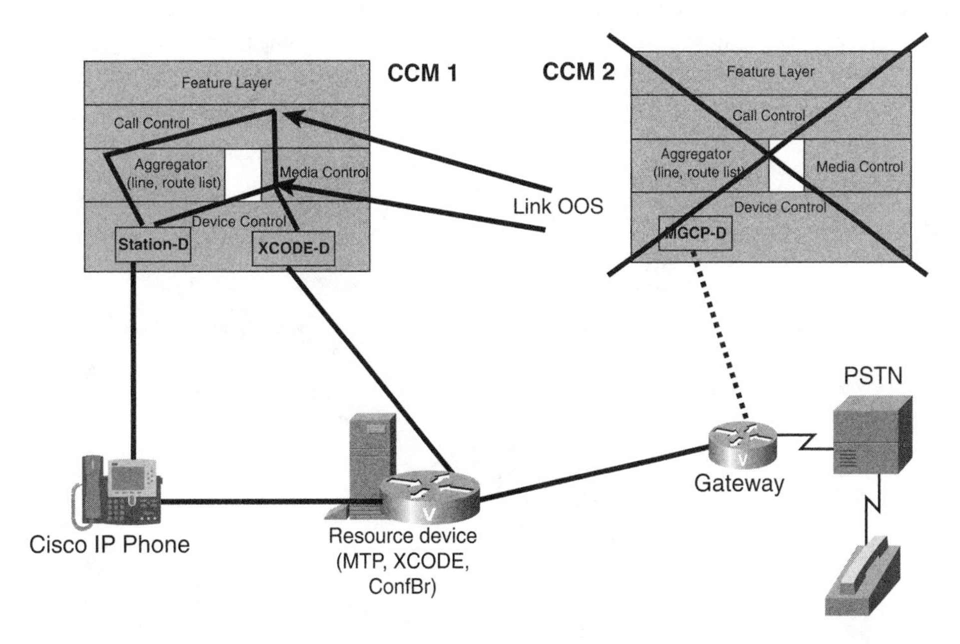

Figure 10-5 *Quiet Clearing*

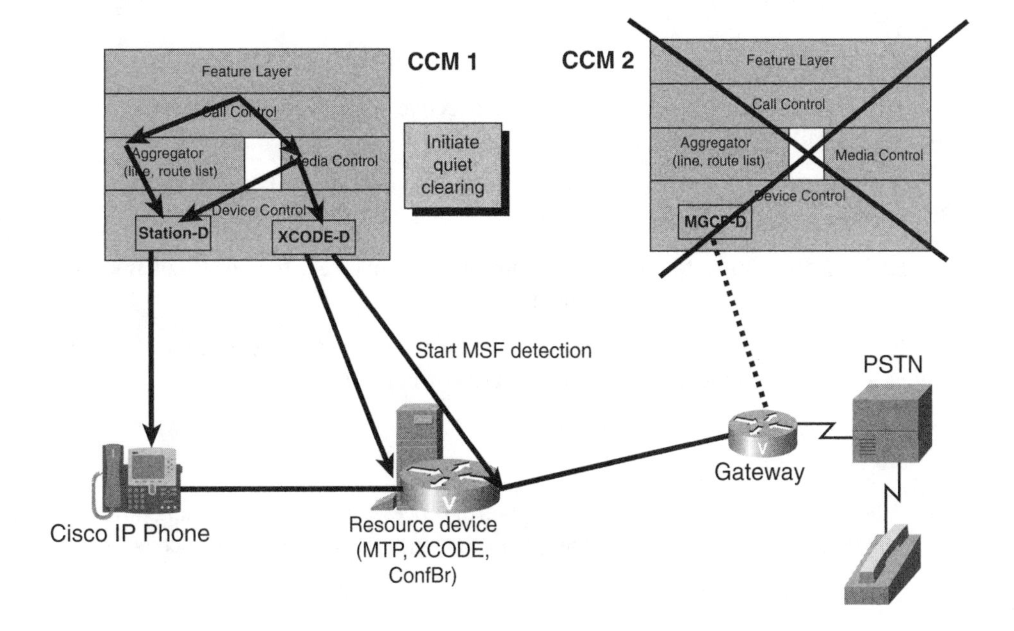

Devices that perform switchover immediately can register with a Cisco CallManager server while involved in active connections as demonstrated in Figure 10-6. Cisco CallManager must restore call state information for these preserved connections to manage their release during call clearing. A Device Control process is created to manage the registering device.

Figure 10-6 *Registration*

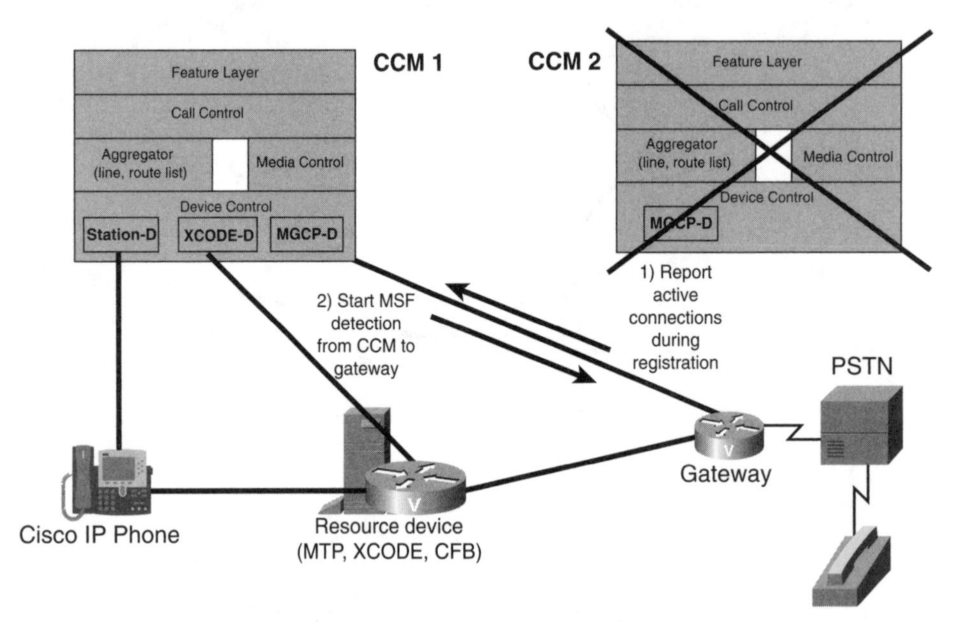

The device communicates the status of any active connections to the new Cisco CallManager server.

If an MGCP gateway supports MSF disconnect supervision, a start MSF signal is sent to the device for each preserved connection.

Each surviving or restored Device Control process manages the release of any connections in the associated device independently.

Based on the type of disconnect supervision supported in the device, the Device Control process receives one of the two signals as shown in Figure 10-7:

- An end user release signal (Release)
- An MSF signal

When either is received, the Device Control process has enough call state information to satisfy the release sequence to the releasing device, based on the protocol supported.

Figure 10-7 *Disconnect Supervision*

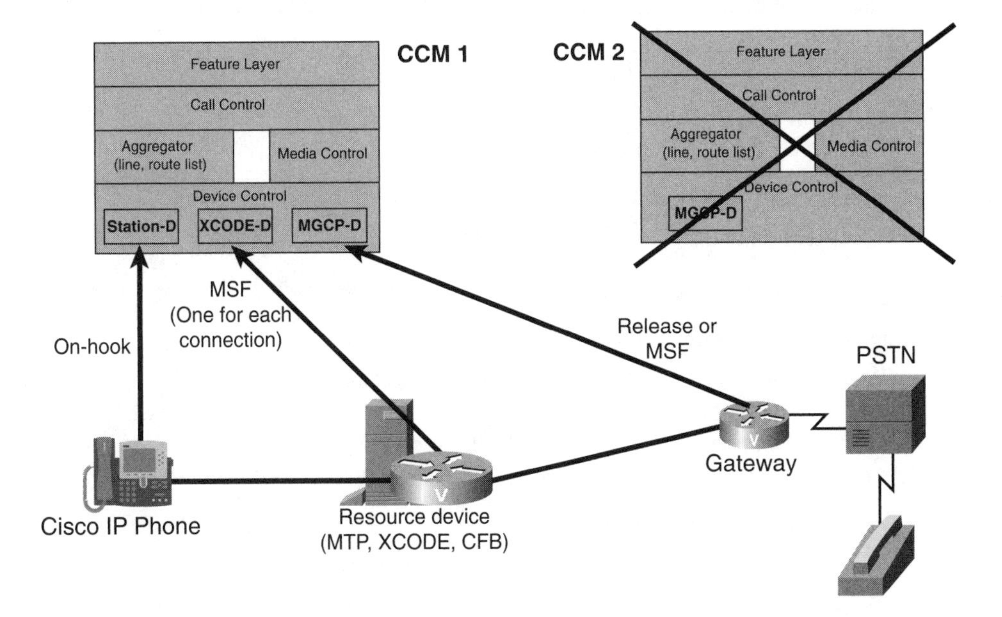

There is no linkage between the Device Control processes that are involved in the preserved call.

Each end is released autonomously.

PSTN Gateways Controlled by MGCP

A PSTN gateway device typically manages more than one voice channel simultaneously, especially in cases where the PSTN connection is a T1 or E1 interface. This is to say that several active calls are likely to exist through a gateway at any point in time. Therefore, when a failure condition occurs, it is very desirable to maintain these calls. However, it is as desirable to make the inactive channels on the gateway available to another Cisco CallManager to allow new call setups. This is accomplished by the MGCP gateway immediately registering with a lower order Cisco CallManager upon failure conditions. Therefore, for *switchover*, the Immediate option is implemented. Upon registration, Cisco CallManager requests the status (whether active or inactive) of the gateway's endpoints (channels). MGCP provides this mechanism in the sequence of Audit Endpoint messages.

When the MGCP Audit Endpoint message, which requests the connection identifier, is received, the MGCP gateway returns an Audit Endpoint Res message, which provides the connection identifier of any connection maintained on the associated endpoint.

Cisco CallManager uses the connection identifiers to avoid reusing the channels and deleting the connections in the gateway when the associated calls are released.

End user disconnect supervision from the gateways, where possible, enables existing connections to be effectively maintained. It is different based on the PSTN interface type and the manner in which the interfaces are controlled by Cisco CallManager. The two types of Cisco CallManager control for MGCP gateways are

- MGCP Call Control Signaling
- PRI-Backhaul

MGCP Call Control Signaling

For an endpoint of a PSTN interface that is controlled by a Cisco CallManager using MGCP Call Control signaling, only a connection that is fully active (in SEND/RECEIVE mode) is maintained by the gateway during switchover. A connection on this endpoint that exists in any other mode is considered in a transient state and is not maintained.

Cisco IOS Software gateways that support FXS, FXO, and T1-CAS PSTN interfaces use MGCP signaling for control of the ports.

End user disconnect supervision is impossible for FXO ports using loop start. Therefore, all supported platforms provide MSF disconnect supervision for loop start interfaces.

PRI-Backhaul

Several PSTN interfaces can be supported using Q.931 Call Control signaling for call control and MGCP as the media control protocol. This is known as PRI-Backhaul. The VG200 (T1 and E1 interfaces) supports PRI-Backhaul

Similarly, Cisco Catalyst 6000 modules support PRI-Backhaul under the following devices:

- Catalyst 6000 T-1 and DT24+
- Catalyst 6000 E-1 and DE30+
- Catalyst 6000 FXS analog interface module

In this configuration, Cisco CallManager uses MGCP commands to create, modify, and delete media connections in the gateway as calls are setup and released over PRI.

For an endpoint of a PSTN interface that is controlled by a Cisco CallManager using PRI-Backhaul Call Control signaling, any existing connection, regardless of mode, is maintained by the gateway during switchover.

Cisco CallManager takes the responsibility of managing the existing connections to satisfy the release sequences with the other end of PRI when this type of gateway registers.

To manage PRI calls, Cisco CallManager needs the call reference value. The active Cisco CallManager requests the gateway to send information (which includes the Q.931 call reference value) for each connection by initiating the Audit Connection sequence.

For a connection that exists on an endpoint of a PSTN interface that is controlled by a Cisco CallManager using PRI-Backhaul Call Control signaling, the gateway returns the requested connection information that was included in the Create Connection command (explicitly the Q.931 call reference value) in the Audit Connection ACK message.

CDRs

Cisco CallManager generates CDRs for calls. No CDR generation is provided for those calls whose CDR information is collected in the failed Cisco CallManager. This is true for all affected calls, regardless of whether the call is preserved.

For calls that are affected by a Cisco CallManager failure or device communication failure, but whose CDR information is being collected in a surviving Cisco CallManager, the CDR information is generated when the calls are internally released as a result of the failure detection, the Quiet Clear mechanism. For these calls, there is no record of the portion of the call that continued after the system failure occurred.

Summary

The three device requirements for supporting call preservation are

- Active connection maintenance
- Disconnect supervision
- Switchover algorithm

Devices must provide at least one of the following disconnect supervision mechanisms for any media connections that are preserved during system failure:

- End user release
- Timed
- MSF

The switchover algorithms decide when a device will fail over to the secondary Cisco CallManager when the device detects that its link is out of service to the primary Cisco CallManager. The three switchover algorithms are

- **Graceful**—Delay registration with any available Cisco CallManager until all active streaming sessions are stopped in the device.

- **Immediate**—Immediately switchover to an available Cisco CallManager and communicate the status of preserved connections to this new Cisco CallManager, such that the release of the preserved connections can be managed.
- **Timed**—After timed interval (the default setting is three minutes after a Cisco CallManager failure), any calls in progress through a gateway are dropped by the system and the gateway re-homes to the secondary Cisco CallManager.

Survivable Remote Site Telephony (SRST) is configured on supported IOS platforms and provides limited call processing to remote branch offices in a Centralized deployment model.

Post-Test

How well do you think you know this information? The post-test is designed to help you gauge your knowledge about this chapter. Of the 12 questions, if you answer one to four questions correctly, we recommend that you reread this chapter. If you answer five to eight questions correctly, we recommend that you review the sections that you need to know more about. If you answer 9 to 12 questions correctly, you probably understand this information well enough to move on to the next chapter. You can find the answers to these questions in Appendix B, "Answers to Chapter Pre-Test and Post-Test Questions."

1 List the three device requirements for supporting call preservation.

___

___

___

2 List the three protocols used by Cisco CallManager to establish calls between devices.

___

___

___

3 Cisco CallManager servers in a cluster use what type of links to establish calls between devices registered to different Cisco CallManager servers?

___

___

___

4 What are the three disconnect supervision mechanisms devices must provide for any media connections to be preserved during system failure?

____

____

____

5 Define or give an example of the "end user release" disconnect supervision.

____

____

____

Complete the missing information marked by questions 6–12 in the following table:

Devices	Switchover Algorithm	Disconnect Supervision
Cisco IP Phones • 12-button series • 30-button series • 79*XX* series	6.	7.
CFB/MTP • Software services • Transcoder	8.	9.
MGCP Gateways • DT24+ (T1-PRI/CAS) • E1/T1-PRI • T1-CAS • 24 port FXS • IOS platforms (FXO/FXS, T1-PRI/CAS)	10.	11. 12.

Upon completing this chapter you will be able to do the following tasks:

- Given a component of the media resource manager, correctly define the component listed.
- Given a list of media resources, identify the limits of each media resource in that list.
- Given a music file, construct a diagram that shows the interaction of the Audio Translator, default MOH TFTP server, and the MOH server.

CHAPTER 11

Media Resources

This chapter provides more detail about the media resources available in Cisco CallManager and how the Cisco CallManager architecture layers are involved. Media resources were first introduced in Chapter 4, "Cisco CallManager Administration Service Menu." Media resources include conferencing, MTP, transcoder (X-Code), and music on hold. This chapter covers the following topics:

- Abbreviations
- Media Resource Manager (MRM)
- Media Resource Limits
- Sharing Media Resources in a Cisco CallManager Cluster
- Music on Hold (MOH)

Pre-Test

Do you already know this information? The pre-test is designed to help you gauge your knowledge about this chapter. Of the 10 questions, if you answer one to three questions correctly, we recommend that you read this chapter. If you answer four to seven questions correctly, we recommend that you skim through this chapter, reading those sections that you need to know more about. If you answer 8 to 10 questions correctly, you probably understand this information well enough to skip this chapter. You can find the answers to the pre-test for this chapter in Appendix B, "Answers to Chapter Pre-Test and Post-Test Questions."

1. List the major components the MRM interfaces with.

2. When does call control interface with the MRM?

3. When does the Media layer interface with the MRM?

4. What media resource provides the capability to bridge an incoming RTP stream to an outgoing RTP stream on an H.323v1 gateway?

5. How many full-duplex streams are configurable for software conferences?

6. How many simplex streams are configurable for MOH media resources?

7. How many transcoder resources are available per X-Code device?

8 What are the limits for a hardware conferencing resource on a Catalyst 6000?

____

____

____

9 List at least two reasons for sharing the media resources within a Cisco CallManager cluster.

____

____

____

10 The MOH server is actually a component of which service installed during a Cisco CallManager installation?

____

____

____

Abbreviations

This section defines the abbreviations used in this chapter. For more information about terms and abbreviations used in this chapter refer to the *IP Telephony Network Glossary* at the following URL:

www.cisco.com/univercd/cc/td/doc/product/voice/evbugl4.htm

Table 11-1 provides a list of abbreviations with the corresponding complete terms used frequently in this chapter.

Table 11-1 *Frequently Used Abbreviations with Definitions in This Chapter*

Abbreviation	Complete Term
CCM	Cisco CallManager
ConfBr	conference bridge
MOH	Music on hold
MRG	media resource group
MRM	Media Resource Manager
MRGL	media resource group list
MTP	media termination point

continues

Table 11-1 *Frequently Used Abbreviations with Definitions in This Chapter (Continued)*

Abbreviation	Complete Term
RTP	Real-Time Transport Protocol
SDL	signal distribution layer
TFTP	Trivial File Transport Protocol
X-Code	transcoding

Media Resource Manager (MRM)

Initialization of Cisco CallManager creates an MRM. Each MTP, MOH, transcoder, and ConfBr device defined in the database registers with the MRM. The MRM obtains a list of provisioned devices from the database and constructs and maintains a table to track these resources. The MRM uses this table to validate registered devices. The MRM keeps track of the total devices available in the system, also tracking the devices that have available resources.

When a media device registers, Cisco CallManager creates a controller to control this device. After the device is validated, the system advertises its resources throughout the cluster. This mechanism enables the resource to be shared throughout the cluster.

Resource reservation takes place based on search criteria. The given criteria provides the resource type and the MRGL. Resource deallocation occurs when Cisco CallManager no longer needs the resource. Cisco CallManager updates and synchronizes the resource table after each allocation and deallocation.

The MRM interfaces with the following major components:

- Call control
- Media control
- MTP control
- Unicast bridge control
- MOH control

Call Control

The Call control software component performs call processing, including setup and teardown of connections. The Call control interacts with the Feature layer to provide services like transfer, hold, conference, and so forth. Call control interfaces with the MRM when it needs to locate a resource to set up conference call and MOH features as shown in Figure 11-1.

Figure 11-1 *Call Control*

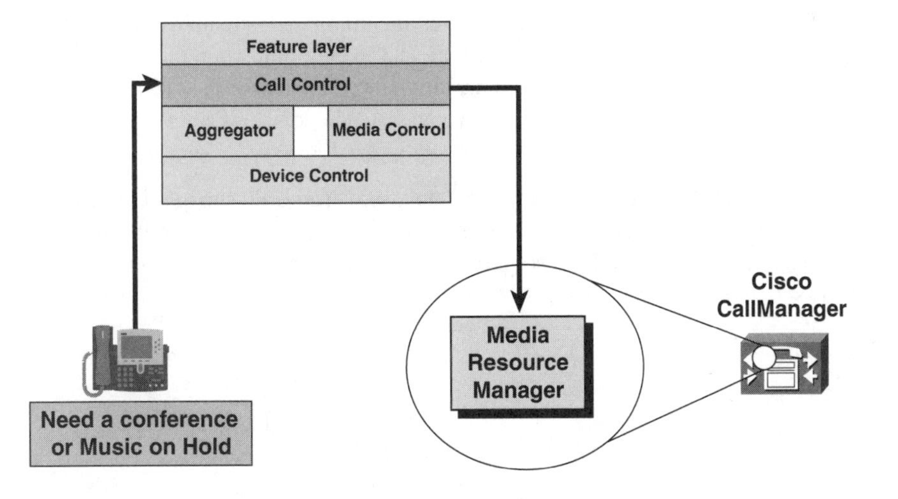

Media Control

The media control software component manages the creation and teardown of media streams for the endpoint. When a request for media to be connected between devices is received, depending on the type of endpoint, media control sets up the proper interface to establish a media stream.

The Media Control layer interfaces with the MRM when it needs to locate a resource to set up an MTP as shown in Figure 11-2.

Figure 11-2 *Media Control*

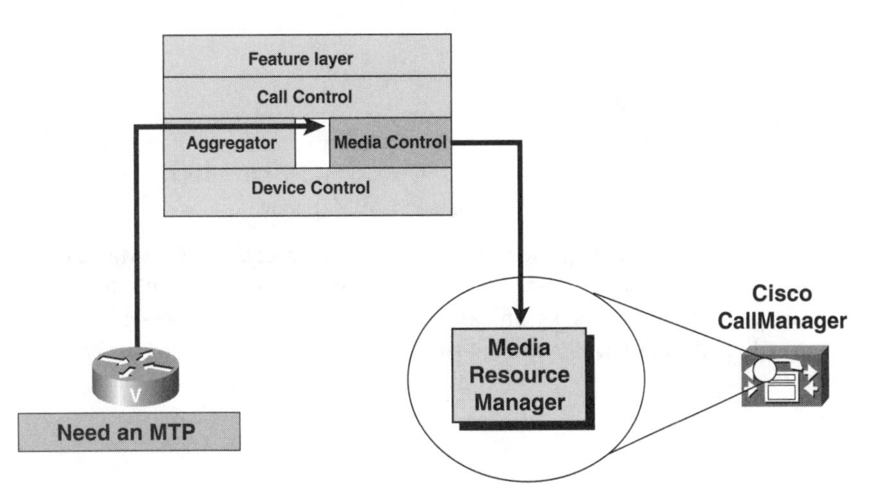

Media Termination Point Control

MTP provides the capability to bridge an incoming RTP stream to an outgoing RTP stream. MTP maintains an RTP session with an H.323 endpoint when the streaming from its connected endpoint stops. MTP currently supports only codec G.711. MTP can also transcode A-law to μ-law. If you have an X-Code device, it can support all codecs if it is being used as an MTP resource.

For each MTP device defined in the database, Cisco CallManager creates an MTP control process. This MTP control process registers with the MRM when it initializes. The MRM keeps track of these MTP resources and advertises their availability throughout the cluster as shown in Figure 11-3.

Figure 11-3 *MTP Control*

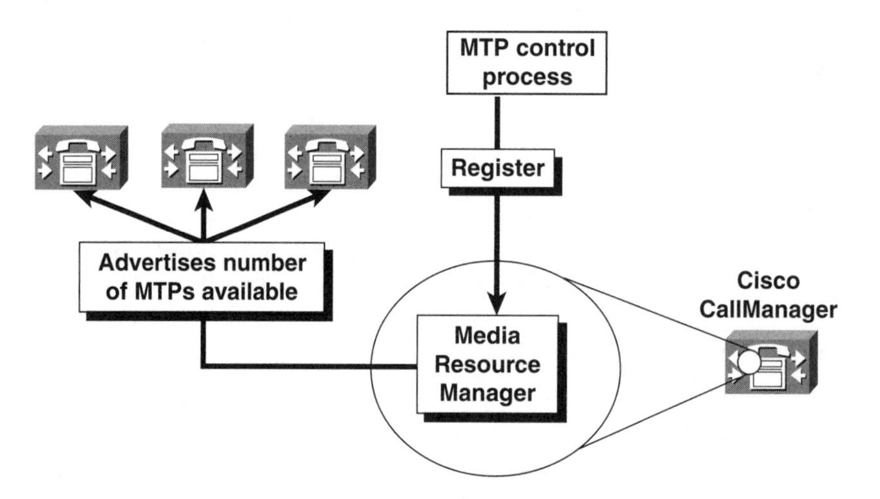

Unicast Bridge Control

A Unicast bridge (ConfBr) provides the capability to mix a set of incoming Unicast streams into a set of composite output streams. Unicast bridge provides resources to implement Ad Hoc and Meet-Me conferencing in Cisco CallManager.

For each Unicast bridge device defined in the database, Cisco CallManager creates a Unicast control process. This Unicast control process registers with the MRM when it initializes. The MRM tracks Unicast stream resources and advertises their availability throughout the cluster. Figure 11-4 shows a device that needs Unicast bridge support.

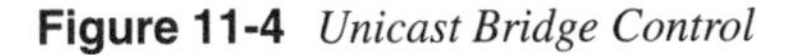
Figure 11-4 *Unicast Bridge Control*

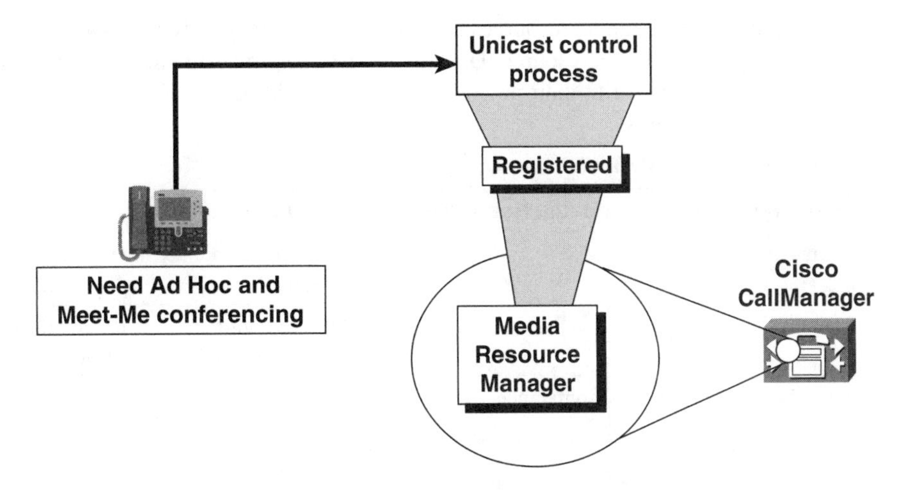

MOH Control

MOH provides the capability to redirect a party on hold to an audio server. For each MOH server device defined in the database, Cisco CallManager creates an MOH control process. This MOH control process registers with the MRM when it initializes. The MRM tracks MOH resources and advertises their availability throughout the cluster. MOH supports both Unicast and Multicast audio sources. Figure 11-5 shows a device that needs an audio source because it is on hold.

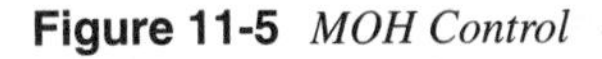
Figure 11-5 *MOH Control*

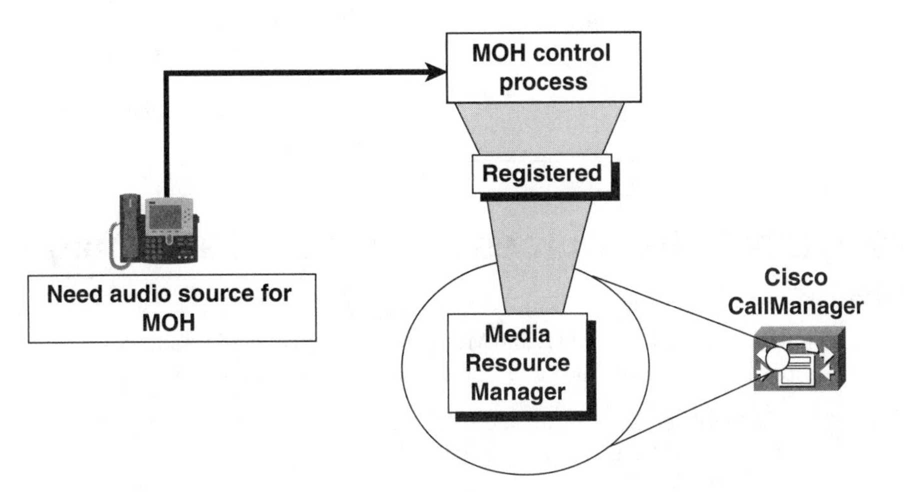

Media Resource Limits

To help manage the media resources available in a cluster, you need to know the limits of each registered media resource. Use this information to design and scale the media resources in the cluster to their potential. Table 11-2 addresses software and hardware media resource limits.

Table 11-2 *Hardware/Software Media Resource Limits*

Hardware/Software	Media Resource	Limitations
Software	MTPs	Up to 128 full-duplex streams are configurable.
		With 128 configured streams, 64 resources are available for MTP application.
	Conference	Up to 128 full-duplex streams are configurable.
		With 128 streams, a software conference media resource can handle 128 users in a single conference.
		With 128 streams, a software conference media resource can handle up to 42 conferencing resources with three users per conference.
	MOH server	Up to 500 simplex streams are configurable.
		With 500 configured streams, up to 500 resources are available for MOH application.
Hardware	Transcoder	48 streams register and provide up to 24 X-Code resources.
	Conference	32 streams register, so the hardware conference media resource can handle 32 users in a single conference or up to 10 conferencing resources with three users per conference.

Sharing Media Resources in a Cisco CallManager Cluster

Conferencing, X-Code, MTP, and MOH are considered media resources. The following are reasons for sharing the media resources within a Cisco CallManager cluster:

- To enable both hardware and software resources to co-exist within a Cisco CallManager and a Cisco CallManager cluster.

- To enable Cisco CallManager to share and access resources available in the cluster.
- To enable Cisco CallManager to perform load distribution within a group of similar resources.
- To enable Cisco CallManager to allocate resources based on user preferences.

Figure 11-6 shows the first reason for being able to enable both hardware and software resources to co-exist. This enables customers to leverage all the media resources within each Cisco CallManager server and fully utilize its digital signal processor (DSP) components that provide hardware conference and X-Code resources.

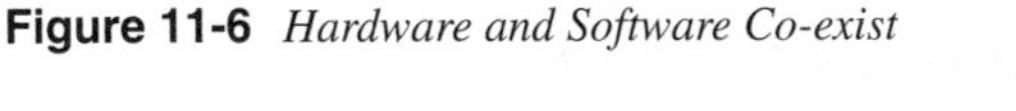

Figure 11-6 *Hardware and Software Co-exist*

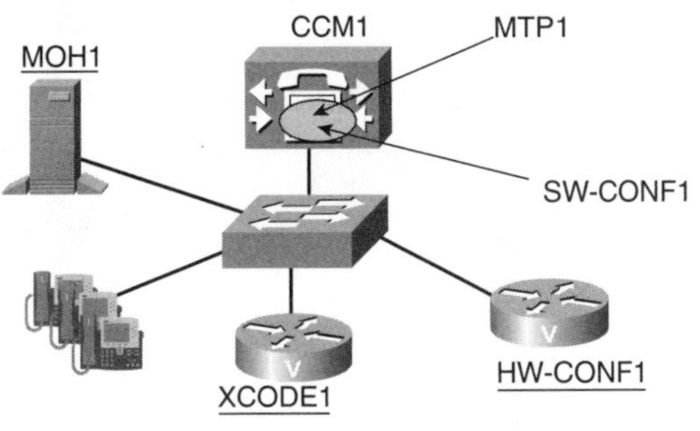

Figure 11-7 shows that a Cisco CallManager node can register all media resources and then share those media resources throughout the cluster. This enables Cisco CallManager nodes in the cluster to use media resources that are not registered to the requesting Cisco CallManager node. This provides for efficiency among Cisco CallManager nodes in the cluster. For example, in the cluster topology in Figure 11-7, CCM2 and CCM3 don't have any resources but they are capable of accessing resources on CCM1.

Sharing media resources enables for load distribution of like media resources. For instance, when you have two MOH servers registered, the MRM is able to alternate between servers to provide MOH as shown in Figure 11-8. First Cisco CallManager will use MOH server 1, then MOH server 2, then MOH server 1, and so on.

The administrator decides the media resource preference for the device. The administrator can also limit the type of media resources to which a device has access. For example, a phone in the lobby can be denied conference resources but allowed to have MOH resources. This can also be very helpful for saving bandwidth for media resources.

Assume that you have two buildings and each building has its own media resource servers as shown in Figure 11-9. It would make sense for the devices in Building 1 to go to the media resource servers in Building 1 first and then go to Building 2 for media resources if Building 1's resources are all used up.

Figure 11-7 *Sharing Media Resources Throughout the Cluster*

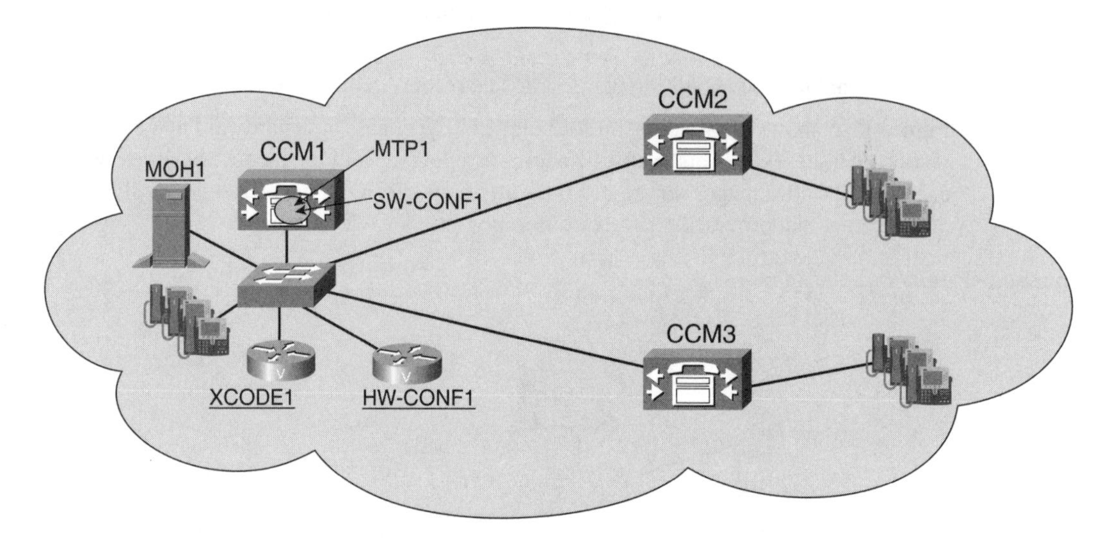

Figure 11-8 *Load Distribution*

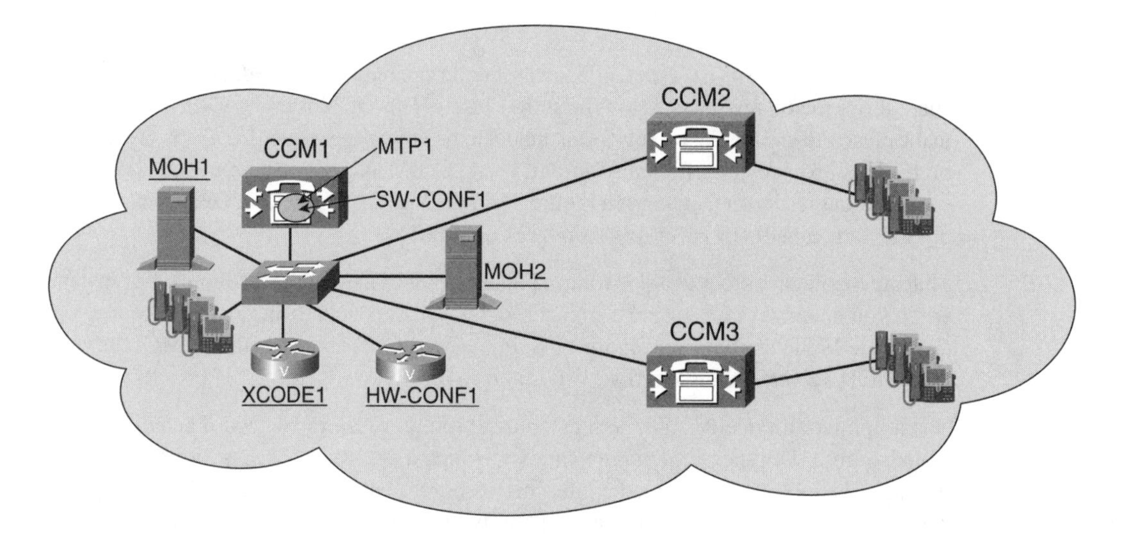

After each device is configured in Cisco CallManager Administration, database replication is used to broadcast its name and its type to the other Cisco CallManager nodes within the cluster, as shown in Figure 11-10.

Figure 11-9 *Allocating Resources/Conserving Bandwidth*

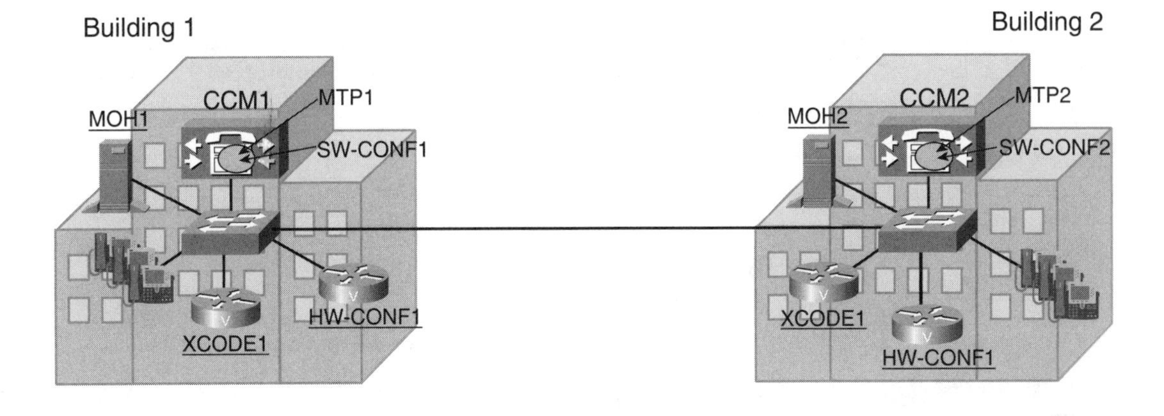

Figure 11-10 *Propagation of Media Resources*

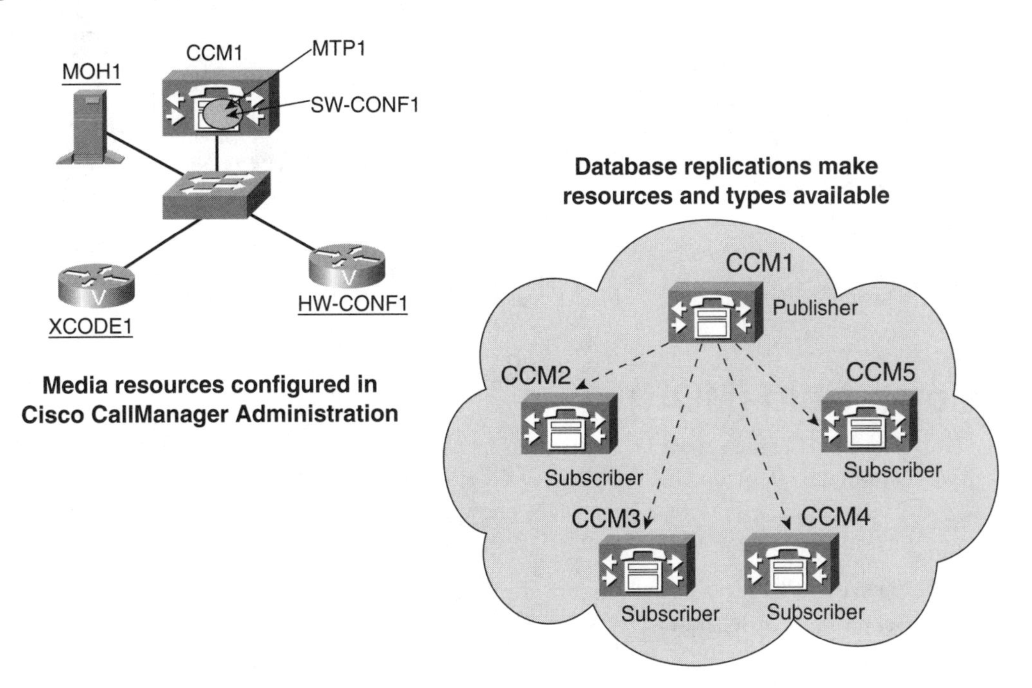

During startup, each Cisco CallManager node gets all the media resource device names and device types from the database.

After each media resource device is registered to a Cisco CallManager node, the device manager in that Cisco CallManager node broadcasts the controller process ID to the other device managers in the cluster, as shown in Figure 11-11.

Figure 11-11 *Device Manager Process ID Broadcasts*

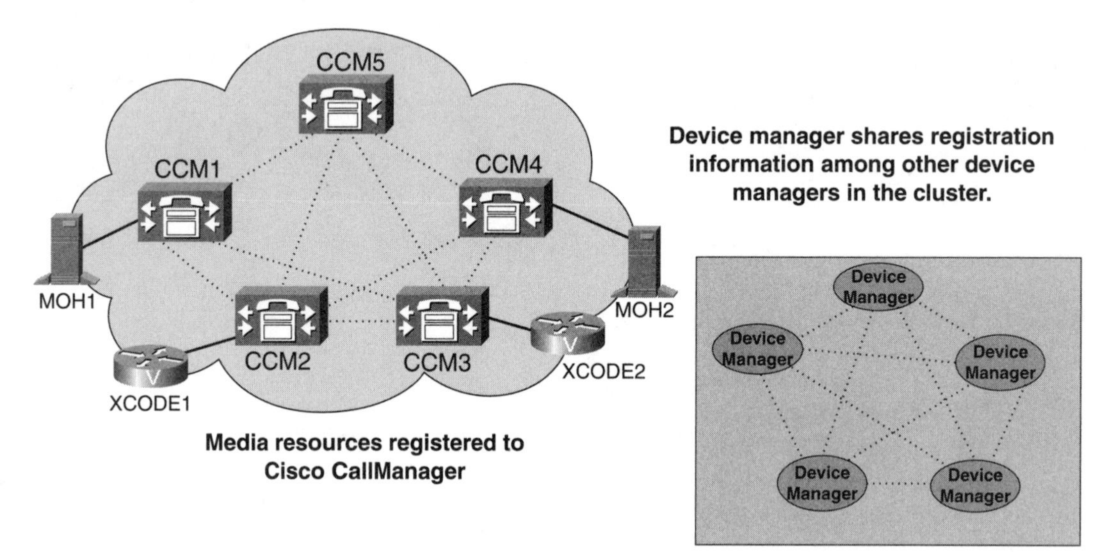

During call processing, the MRM uses this process ID to allocate resources.

Music on Hold (MOH)

The MOH feature provides music to users when they are placed on hold. Figure 11-12 shows the interaction between the Audio Translator, default MOH TFTP server, and the MOH server. Figure 11-12 also shows how you add an audio source file that is processed and utilized by the MOH server.

Each of the large boxes in the figure are components in a cluster; they may reside on a single server or three separate servers.

Figure 11-12 *MOH Interaction*

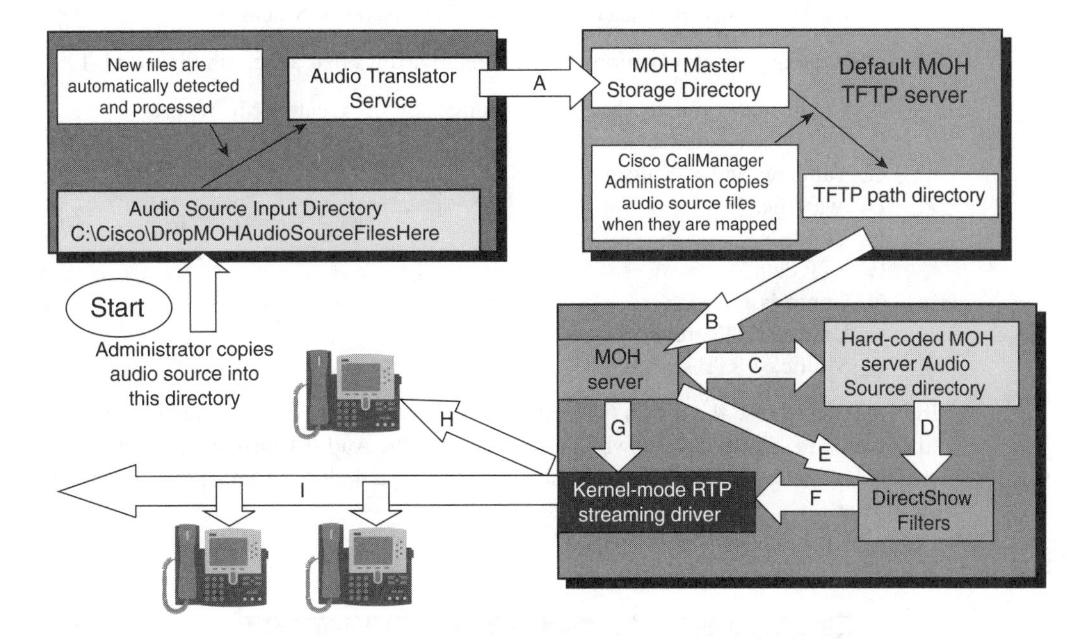

Step 1 First the administrator drops the audio files into the directory path of C:\Cisco\DropMOHAudioSourceFilesHere. Valid input audio files are most standard .wav and .mp3 files. (It takes about 30 seconds to convert a 3 MB .mp3 file.) The file is automatically detected and translated, and the output files, as well as the source files, are then moved into the default MOH TFTP server holding directory. This holding directory will always be whatever the DefaultTFTPMOHFilePath is with \MOH appended.

NOTE You should not use the Audio Translator service during production hours because the service will consume 100 percent of the server's CPU.

When you assign/map the audio source file to an audio source number, the proper audio source files are then copied up one directory to make them available for the MOH servers.

Step 2 The MOH servers download the needed audio source files and then store them in the hard coded directory C:\Program Files\Cisco\MOH.

Step 3 The MOH server then streams the files using DirectShow and the kernel mode RTP driver as needed/requested by Cisco CallManager.

The following are the explanations of the lettered arrows that appear in Figure 11-12:

A. Create audio source files, move original source, create XML report
B. TFTP the needed audio source files
C. Validates and downloads
D. .wav files
E. Controls
F. RTP data
G. Controls
H. Unicast RTP stream
I. Multicast RTP stream

To install MOH, use the install CD making sure the Cisco IP Voice Media Streaming Application is selected, as shown in Figure 11-13. The Audio Translator will also be installed at the same time. Cisco recommends having a dedicated MOH server(s) within the cluster.

Figure 11-13 *Installing MOH*

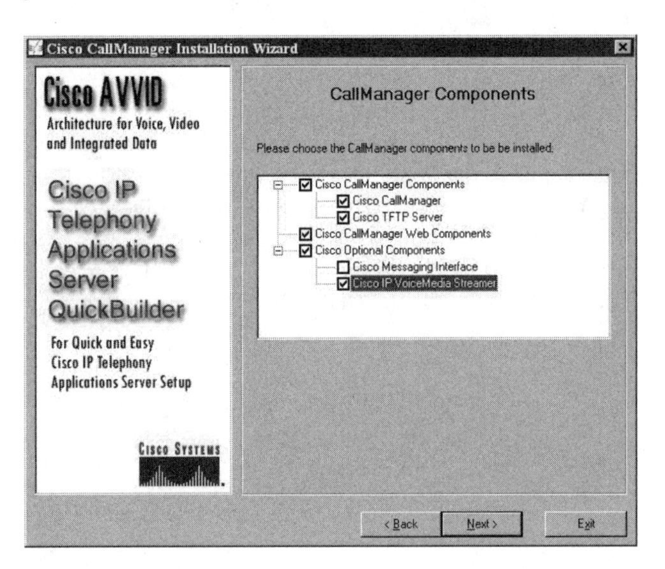

MTP, ConfBr, and MOH devices will automatically be added into the database. The DirectShow filters are automatically installed and registered when the services are registered.

During an installation, a default MOH audio source will be installed and configured if one does not exist. This enables the MOH functionality to work using this default audio source after the install is completed without any other changes.

The device recovery supports up to three Cisco CallManager nodes—this list is read from the device pool assigned to the device.

The Audio Translator (AudioTranslator.exe) runs as an NT service and does not support change notification but will re-read its configuration every five minutes.

CAUTION Cisco recommends that you not install Audio Translator on the same server as Cisco CallManager. When audio files are translated, Cisco CallManager may experience errors or performance degradation. This is because when you are translating files, the DirectX drivers that do the translation consume 100 percent of the CPU causing performance degradation to other processes running on the system.

MOH Server Files

The MOH server is actually a component of the Cisco IP Voice Media Streaming Application (ipvmsapp.exe). The standard device recovery and database change notification is supported.

The kernel mode device driver runs as a non-plug-and-play kernel mode driver (ipvms.sys) and does all of the RTP streaming.

The MOH server supports streaming of up to 50 audio sources and one fixed input source for the G.711 µ-law/A-law, G.729a, and Wideband codecs.

The MOH server uses DirectShow filters to extract the RTP data out of the wave files and pass it to the kernel mode driver.

The physical filenames are

- AudioTranslator.exe—Audio Translator Service.
- Ipvmsapp.exe—ConfBr/MTP/MOH Service (Cisco IP Voice Media Streaming Application).
- Ipvms.sys—Kernel mode device driver.
- IpVMSChangeNotify.dll—Handles change notification (database changes) for the Cisco IP Voice Media Streaming Application service.
- MediaAppPerfMon.dll—Microsoft Performance manager for the Media App service.
- Fxcode.ax—DirectShow filter used by the Audio Translator.
- Ipvmsrend.ax—DirectShow filter used by the Cisco IP Voice Media Streaming Application.
- Mohencode.ax—DirectShow filter used by the Cisco IP Voice Media Streaming Application and Audio Translator.
- Wavdest.ax—DirectShow filter used by the Audio Translator.

Where do you place audio files and then where are the audio files stored in the server waiting to be streamed? The MOH Source Directory and the Default TFTPMOH file path in the Cisco CallManager Administration Service Parameters Configuration page are defined as follows:

- MOHSourceDirectory is the input directory. It points to the directory where the audio source files are placed to be converted and made usable to the MOH server. The install program automatically sets this field. The default setting is C:\Cisco\DropMOHAudioSourceFilesHere.
- DefaultTFTPMOHFilePath is the output UNC path name. This must point to the default MOH TFTP server share. This field is automatically set by the install program; it is also a service-wide setting.

CAUTION Cisco recommends that you not install Audio Translator on the same server as Cisco CallManager. When audio files are translated, Cisco CallManager may experience errors or performance degradation. This is because when you are translating files, the DirectX drivers that do the translation consume 100 percent of the CPU causing performance degradation to other processes running on the system.

Adding/Updating MOH Servers

Figure 11-14 shows how you can add or update an MOH server.

Table 11-3 shows the configurable settings for the MOH Server Configuration page.

Table 11-3 *MOH Server Configuration Settings*

Field	Description
Device Information	
Host Server	Use this required field to choose a host server for the MOH server. To do so, click the drop-down arrow and choose a server from the list displayed. For existing MOH servers, this field is display only.
MOH Server Name	Enter a unique name in this required field for the MOH server. This name can comprise up to 15 characters. Valid characters include letters, numbers, spaces, dashes, dots (periods), and underscores.

Table 11-3 *MOH Server Configuration Settings (Continued)*

Field	Description
Description	Enter a description for the MOH server. This description can comprise up to 50 characters. Ensure Description does not contain ampersand (&), double quotes ("), brackets ([]), less than sign (<), greater than sign(>), or the percent sign (%).
Device Pool	Use this required field to choose a device pool for the MOH server. To do so, click the drop-down arrow and choose a device pool from the list displayed.
Maximum Half Duplex Streams	Enter a number in this required field for the maximum number of half-duplex streams that this MOH server supports. Valid values range from 0 to 500.
Maximum Multicast Connections	Enter a number in this required field for the maximum number of multicast connections that this MOH server supports. Valid values range from 0 to 999999.
Fixed Audio Source Device	Enter the device name of the fixed audio source device. This device serves as the per-server override used if the server has a special sound device installed.
Run Flag	Use this required field to choose a run flag for MOH server. To do so, click the drop-down arrow and choose **Yes** or **No**.
Multicast Audio Source Information	
Enable Multicast Audio Sources on This MOH Server	Check or uncheck this check box to enable or disable multicast of audio sources for this MOH server. If this MOH server belongs to a multicast MRG, a message asks you to enable multicast on this MOH server or to update the specified MRG(s) either by removing this MOH server or by changing the multicast setting of each group listed.

continues

Table 11-3 *MOH Server Configuration Settings (Continued)*

Field	Description
Base Multicast IP Address	If multicast support is needed, enter the base multicast IP address in this field. Valid IP addresses for multicast range from 224.0.1.0 to 239.255.255.255. IP addresses between 224.0.1.0 and 238.255.255.255 fall in the reserved range of IP multicast addresses for public multicast applications. Use of such addresses may interfere with existing Multicast applications on the Internet. Cisco strongly recommends using IP addresses in the range reserved for administratively controlled applications on private networks (239.0.0.0 – 239.255.255.255). Multicast by IP address is preferable in firewall situations.
Base Multicast Port Number	If Multicast support is needed, enter the base Multicast port number in this field. Valid Multicast port numbers include even numbers that range from 16384 to 32767.
Increment Multicast on	Click **Port Number** to increment multicast on port number. Click **IP Address** to increment multicast on IP address.
Selected Multicast Audio Sources (Only audio sources for which the Allow Multicasting check box was checked appear in this listing.)	
No.	**Display only**. This field designates MOH audio stream number associated with a particular multicast audio source. Only audio sources defined as allowing multicasting display.
Audio Source Name	**Display only**. This field designates name of audio source defined as allowing multicasting.
Max Hops	For each multicast audio source, enter the maximum number of router hops through which multicast packets should pass. Valid values range from 1 to 15. Using high values can lead to network saturation. This field is also known as **Time to Live**.

Figure 11-14 *MOH Server Configuration Page in Cisco CallManager Administration*

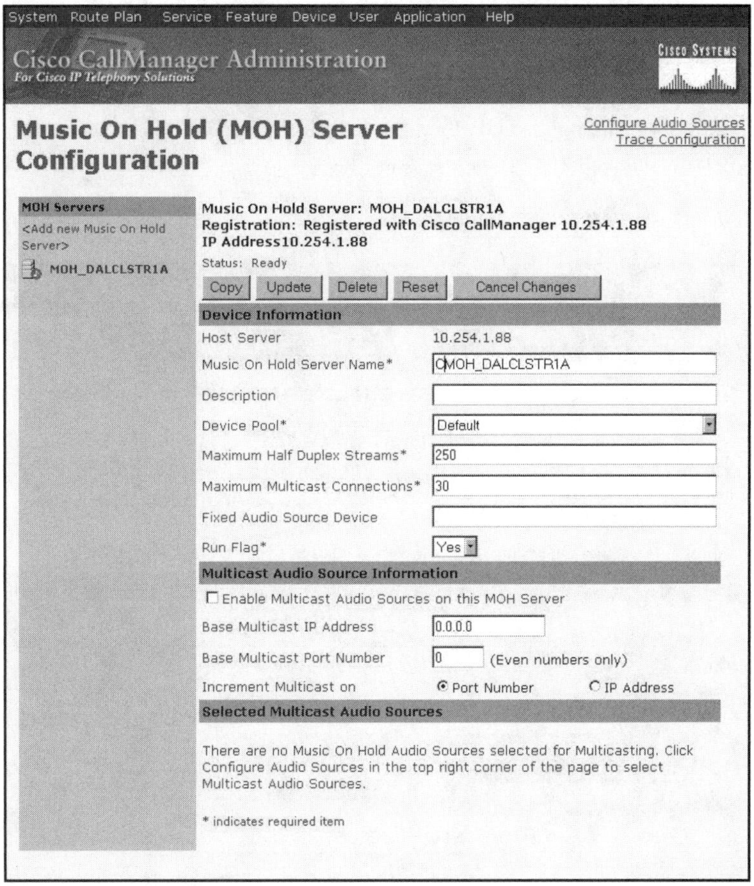

The MOH server has a hard-coded, read-only audio source storage directory—C:\Program Files\Cisco\MOH. Files in this directory should not be changed in any way.

The Fixed Audio Source Device field would normally be left blank; it would only be used if the *Master Fixed Audio Source Device* is not correct for this server. This could happen if a different kind of sound card was installed in just this server.

NOTE The installation program will automatically add the fixed audio source device; you need to add it only if you are manually installing or changing the MOH server.

Setting MOH Service-Wide Settings

To set the MOH service-wide settings, open the Service Parameters Configuration page and select any server on the left side of the screen that has an MOH server installed and then select **Cisco IP Voice Media Streaming** on the left side of the screen. The Service Parameters Configuration screen for MOH is then displayed as in Figure 11-15 on the bottom of that page. The other service parameters on this page are for MTP and Software Conference Bridge.

Figure 11-15 *Service Parameters Configuration for MOH*

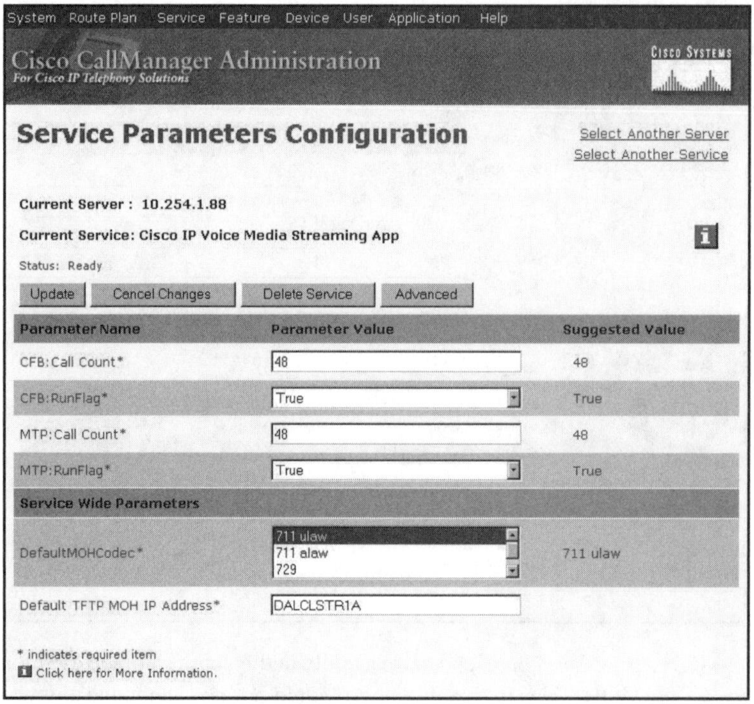

The DefaultMOHCodec field is set to indicate the desired codec(s) supported by the MOH servers in the cluster. This defaults to G.711 μ-law during the install. If you want to enable multiple codecs, hold down the control key to select the codecs you want to enable.

The Default TFTP MOH IP Address is set to the IP address or computer name of the default MOH TFTP server. (This field is automatically set to the server name during the initial installation of Cisco CallManager.)

Adding/Updating MOH Audio Sources

When you add or update all the audio sources except for the fixed audio source, the changes will affect all MOH servers.

All the processed audio sources will show up in the Audio Source File drop-down list.

The Play continuously (repeat) option should always be selected, and the Allow Multicasting flag can be set if Multicast capabilities are needed. If the Play continuously (repeat) option is not selected with Multicast, when the audio file gets to the end, it does not play anymore. If the audio source reaches the end and stops, the administrator will have to stop and start the server to reset the MOH.

Checking Audio File Conversion Status

The File Status window shows you the conversion status and indicates whether the audio file translated correctly or had any errors.

To find the name of the fixed audio source that can be used, use the Control Panel to open the Sounds and Multimedia Properties window. Make sure the proper input source is selected.

To open the Sounds and Multimedia Properties window, open the Control Panel and then select the **Sounds and Multimedia** option. Then select the **Audio** tab. Any Sound Recording device name that shows up in the Preferred device list can be used.

To open the recording Control window, click on the **Volume** button in the Sound Recording section. Make sure that the Line In box, microphone, or CD player is selected.

Configuring the MOH Fixed Audio Source

The fixed audio source name is case sensitive and must be entered exactly as found in the Sounds and Multimedia Properties including spaces or symbols that appear in the name.

If the Multicast check box is selected, remember that if the G.729 codec is enabled it will consume about five to seven percent of your CPU. This setting is global for all MOH servers. If the fixed audio source does not exist on a server it will not be used.

You can override the fixed audio source name on a per-MOH server basis via the MOH Server Configuration page.

The Fixed Audio Source that appears on the left side of the screen in Figure 11-16 will always be available as an option.

The Fixed Audio Source setting is the global setting that affects all MOH servers that have a device by the selected name. If the device does not exist it will be ignored.

Figure 11-16 *MOH Audio Source Configuration Page*

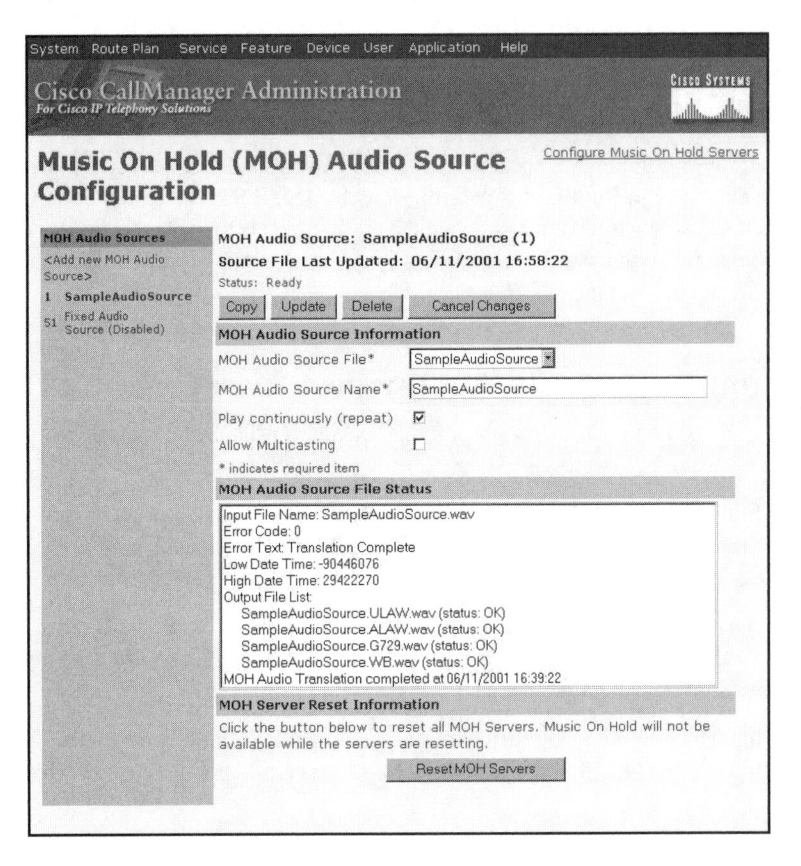

Table 11-4 describes the configurable items on this page.

Table 11-4 *MOH Audio Source Configuration Settings*

Field	Description
MOH Audio Stream Number	Use this required field to choose the stream number for this MOH audio source. To do so, click the drop-down arrow and choose a value from the list displayed. For existing MOH audio sources, this value displays in the MOH Audio Source title.
MOH Audio Source File	Use this required field to choose the file for this MOH audio source. To do so, click the drop-down arrow and choose a value from the list displayed.

Table 11-4 *MOH Audio Source Configuration Settings (Continued)*

Field	Description
MOH Audio Source Name	Enter a unique name in this required field for the MOH audio source. This name can comprise up to 50 characters. Valid characters include letters, numbers, spaces, dashes, dots (periods), and underscores.
Play continuously (repeat)	To specify continuous play of this MOH audio source, check this check box. Cisco recommends checking this check box. If continuous play of an audio source is not specified, only the first party placed on hold, not additional parties, will receive the MOH audio source.
Allow Multicasting	To specify that this MOH audio source allows multicasting, check this check box.
MOH Audio Source File Status	Display only. This pane displays information about the source file for a chosen MOH audio source. For an MOH audio source, the following attributes appear: • Input File Name • Error Code • Error Text • Low Date Time • High Date Time • Output File List (ULAW .wav filename and status, ALAW .wav filename and status, G.729 .wav filename and status, Wideband .wav filename and status) • Time MOH Audio Translation completed
MOH Server Reset Information	To reset all MOH servers, click the **Reset MOH Servers** button. Cisco CallManager makes MOH unavailable while the servers reset.

When the audio source for MOH is selected, it adheres to the following rules for audio source ID selection:

- System administrator (not the end user) defines (or configures) the audio source IDs.
- System administrator selects (or configures) the audio source IDs for the user(s) or device(s).
- The holding party devices decide which audio source ID will apply to the held parties.

- Four levels of prioritized audio. Level four has the highest priority and level one has the lowest. The four levels of prioritized audio are described as follows:
  - Level four is directory/line based (devices which have no line definition, like gateways, do not have this level). The system will select the audio source IDs at this level if defined.
  - If none is defined in level four, the system will search any selected audio source IDs in level three. Level three audio source IDs are device-based.
  - If no level four or level three audio source IDs are selected, the system selects audio source IDs defined in level two, which is device pool-based.
  - If levels two, three, and four have no audio source IDs selected, the final level, level one, will be searched for audio source IDs by the system. The level one audio sources IDs are service-wide service parameters.

The held party devices decide which MRGL will be used for Cisco CallManager to allocate MOH resources. Two levels of prioritized MRGL selection are implemented:

- Level two MRGL (the higher priority) is device-based. The CallManager uses the MRGL in device level two if there is one defined.
- Level one MRGL (the lower priority level) is an optional parameter in the device pool. Cisco CallManager uses the device pool-level MRGL only if there is no MRGL defined in the device level for that device.

NOTE If none of the MRGLs are defined, Cisco CallManager uses the system default resources (the system default resources are ones which are not assigned to established MRGs).

Summary

The MRM interfaces with the following major components:

- Call control
- Media control
- MTP control
- Unicast bridge control
- MOH control

The MRM determines how media resources are allocated within a cluster. This enables for media resources to be shared among devices in the same cluster.

MOH is a media resource that you use to configure and customize the audio streams heard by users on hold.

Post-Test

How well do you think you know this information? The post-test is designed to help you gauge your knowledge about this chapter. Of the 10 questions, if you answer one to three questions correctly, we recommend that you re-read this chapter. If you answer four to seven questions correctly, we recommend that you review those sections that you need to know more about. If you answer 8 to 10 questions correctly, you probably understand this information well enough to move on to the next chapter. You can find the answers to the post-test for this chapter in Appendix B, "Answers to Chapter Pre-Test and Post-Test Questions."

1 List the major components the MRM interfaces with.

2 When does call control interface with the MRM?

3 When does the Media layer interface with the MRM?

4 What media resource provides the capability to bridge an incoming RTP stream to an outgoing RTP stream on an H.323v1 gateway?

5 How many full-duplex streams are configurable for software conferences?

6 How many simplex streams are configurable for MOH media resources?

____

____

____

7 How many transcoder resources are available per X-Code device?

____

____

____

8 What are the limits for a hardware conferencing resource on a Catalyst 6000?

____

____

____

9 List at least two reasons for sharing the media resources within a Cisco CallManager cluster?

____

____

____

10 The MOH server is actually a component of which service installed during a Cisco CallManager installation?

____

____

____

Upon completing this chapter you will be able to do the following tasks:

- Given a list of Cisco IP Telephony (CIPT) characteristics, identify the design characteristics for building a CIPT WAN deployment using distributed and centralized call processing.
- Given a CIPT WAN deployment solution, configure each site to have redundancy using an ISDN or PSTN path.
- Given a CIPT isolated deployment, extend call processing services to the remote branch office.
- Configure a branch office voice gateway router to provide call processing when the IP WAN link is unavailable.

CHAPTER 12

WAN Design Considerations for Cisco IP Telephony Networks

This chapter discusses WAN design considerations for the distributed call processing deployment and centralized call processing deployment models. This chapter discusses the following:

- Abbreviations
- Distributed Call Processing
- Centralized Call Processing
- Survivable Remote Site Telephony

Pre-Test

Do you already know this information? The pre-test is designed to help you gauge your knowledge about this chapter. Of the 10 questions, if you answer one to three questions correctly, we recommend that you read this chapter. If you answer four to seven questions correctly, we recommend that you skim through this chapter, reading those sections that you need to know more about. If you answer 8 to 10 questions correctly, you probably understand this information well enough to skip this chapter. You can find the answers to these questions in Appendix B, "Answers to Chapter Pre-Test and Post-Test Questions."

1 For deployments of the distributed call processing model, what is used to provide CAC?

____

____

____

2 True or False? In a CIPT environment the gatekeeper is designed to work with other Cisco IOS H.323 gateways.

____

____

____

3 For each codec listed below, what bandwidth will the Cisco CallManager deliver to the gatekeeper?

- G.711 ________
- G.729 ________
- G.723 ________

4 What keeps track of the current amount of bandwidth consumed by inter-location voice calls from a given location?

____

____

____

5 What Cisco CallManager route plan feature in a centralized call processing deployment enables for the same access code to be used for PSTN access through a local gateway at multiple sites within the same Cisco CallManager cluster?

____

____

____

6 What is the feature that automatically detects a failure in the network and, using Cisco Simple Network Automated Provisioning capability, initiates a process to intelligently auto-configure the router to provide call processing backup redundancy for the Cisco IP Phones in that office?

____

____

____

7 List at least three of the four supported types of phone calls using SRST:

____

____

____

8 What are the two ways QoS tools ensure voice quality?

____

____

____

9 List at least two of the four caveats that should be considered when deploying locations-based CAC.

____

____

____

10 The maximum bandwidth setting for a zone should take into account the limitation that the WAN link may not be filled with more than what percentage for voice?

____

____

____

Abbreviations

This section defines the abbreviations used in this chapter. For more information about terms and abbreviations used in this chapter refer to the *IP Telephony Network Glossary* at the following URL:

www.cisco.com/univercd/cc/td/doc/product/voice/evbugl4.htm

Table 12-1 provides the abbreviation and the complete term.

Table 12-1 *Abbreviations Used in This Chapter*

Abbreviation	Complete Term
ACF	Admission Confirm
ANI	automatic number identification
ARJ	Admission Reject
ARQ	Admission Request
BRI	Basic Rate Interface
CAC	call admission control
CAS	channel associated signaling

continues

Table 12-1 *Abbreviations Used in This Chapter (Continued)*

Abbreviation	Complete Term
CCM	Cisco CallManager
DA	destination address
DCF	Disengage Confirm
DFW	Dallas/Fort Worth
DID	direct inward dial
DOD	direct outward dial
DRAM	dynamic random-access memory
DRJ	Disengage Reject
DRQ	Disengage Request
DSP	digital signal processor
E&M	Ear and Mouth
FXO	foreign exchange office
FXS	foreign exchange station
GK	gatekeeper
GW	gateway
ICF	Information confirm
IRR	Information request
ISDN	Integrated Services Digital Network
LBR	low bit rate
MCM	Multimedia Conference Manager
MFT	multiflex trunk module
MGCP	Media Gateway Control Protocol
PHL	Philadelphia
POTS	plain old telephone service
PRI	Primary Rate Interface
PSTN	public switched telephone network
QoS	quality of service
RCF	Registration Confirm
RG	route group
RRJ	Registration Reject

Table 12-1 *Abbreviations Used in This Chapter (Continued)*

Abbreviation	Complete Term
RRQ	Registration Request
RTP	Real-time Transport Protocol
SA	Source address
SIP	session initiation protocol
SJ	San Jose
SRST	Survivable Remote Site Telephony
TDM	time-division multiplexing
UCF	Unregistration Confirmation
URJ	Unregistration Reject
URQ	Unregistration Request
VIC	voice interface card
VoIP	Voice over Internet Protocol
VWIC	voice\WAN interface card
WAN	wide-area network

Distributed Call Processing

In a distributed call processing deployment scenario, Cisco CallManagers, voice messaging, and DSP resources are located at each site. This deployment model can support up to 100 sites networked across the IP WAN. Voice calls between sites use the IP WAN as the primary path and the PSTN as the secondary path in the event the IP WAN is down or has insufficient resources to handle additional calls. Whether calls use the IP WAN or the PSTN is transparent to both the calling party and the called party. Figure 12-1 is an example of a distributed call processing deployment model.

Two tools, QoS and CAC, ensure voice quality in two ways:

- By giving voice priority over data.
- By preventing voice from oversubscribing a given WAN link.

Over the WAN link, voice packets need QoS tools for protection from data packets. Although voice packets compete with data packets, they also compete with other voice packets. Protecting voice packets from other voice packets is critical in a distributed call processing model, as shown in Figure 12-2. For more detail about QoS refer to Chapter 9, "LAN Infrastructure for Cisco IP Telephony."

Figure 12-1 *Distributed Call Processing Deployment Model*

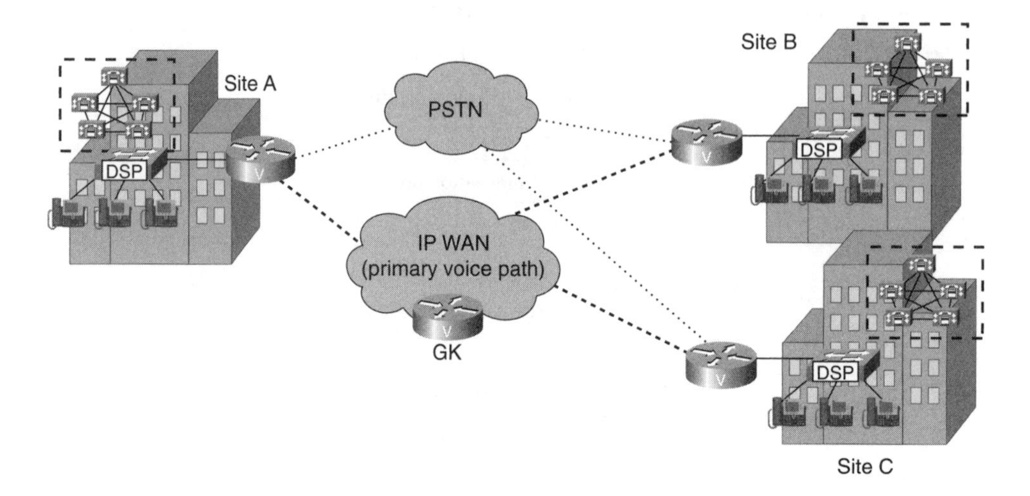

Figure 12-2 *Why CAC?*

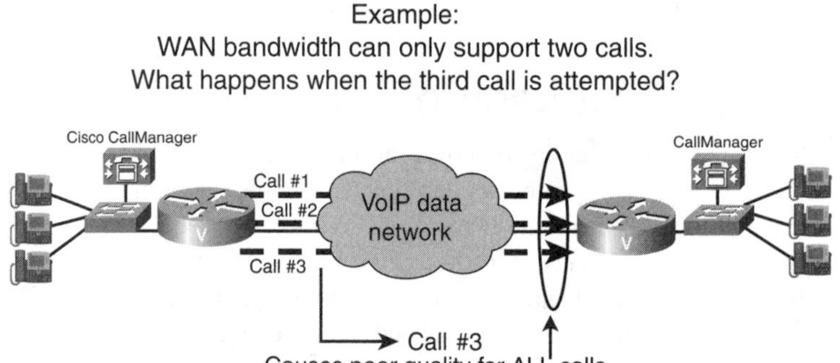

The second task is accomplished by CAC mechanisms. The need for CAC in Cisco AVVID networks is amplified greatly by the fact that all Cisco IP Phones have an open IP path to the WAN; whereas, toll bypass networks could limit the number of physical trunks eligible to initiate calls across the WAN.

For deployments using the distributed call processing model, the H.323 gatekeeper controls CAC. In this design, Cisco CallManager registers with the Cisco IOS gatekeeper, also known as MCM, as a VoIP gateway and Cisco CallManager queries the gatekeeper each time Cisco CallManager wants to make an IP WAN call. The Cisco IOS gatekeeper associates each CallManager with a zone that has specific bandwidth limitations. The Cisco IOS gatekeeper limits the maximum amount of bandwidth consumed by IP WAN voice calls in or out of a zone as shown in Figure 12-3.

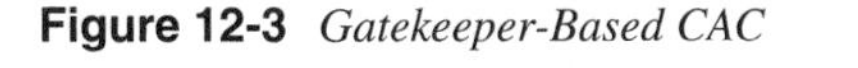

Figure 12-3 *Gatekeeper-Based CAC*

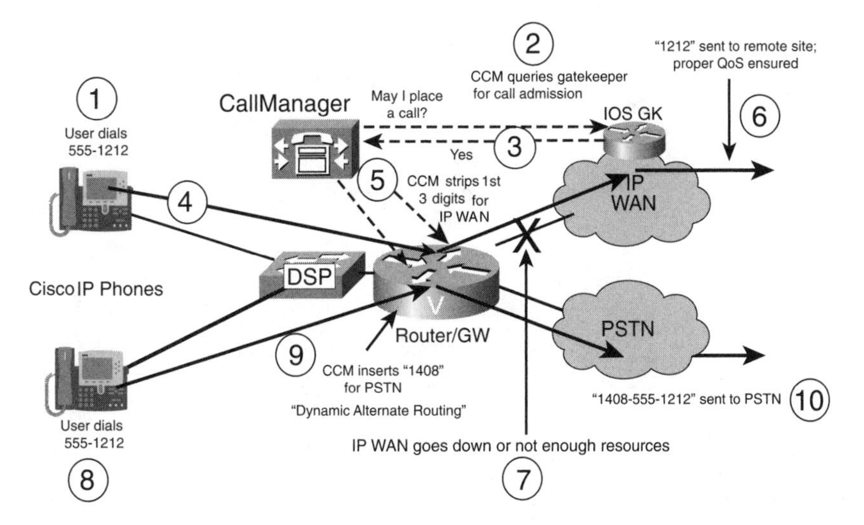

In brief, when Cisco CallManager wants to place an IP WAN call, it first requests permission of the gatekeeper. If the gatekeeper grants the request, the call is placed across the IP WAN. If the gatekeeper denies the request, Cisco CallManager places the call across the secondary path—the PSTN.

This is effectively a call accounting method of providing admission control in which the gatekeeper simply keeps track of the bandwidth the IP WAN calls consume. The maximum bandwidth setting for a zone should take into account the limitation that the WAN link not be filled with more than 75 percent voice, data, and other traffic. The purpose of the calculation is to provide for network overhead.

CAUTION Using the distributed deployment model, Cisco IP Phones are not mobile between sites. Should a Cisco IP Phone register across the WAN, admission control will not operate as designed. This does not work because the phone is registered to a site that is already across the WAN and needs to use the WAN to make calls within its own cluster. It will use bandwidth across the WAN and has no mechanism of notifying the cluster that it is physically at regarding the WAN bandwidth being used.

Understanding Intercluster Trunks

In this model it is important for the dial plan to be tightly coupled with the gatekeeper CAC mechanism. The dial plan ultimately decides when to place a call across the IP WAN and what to do if the gatekeeper rejects it.

When you are using the gatekeeper in a distributed call processing deployment, it is designed to work with intercluster trunks and IOS gateways.

Figure 12-4 shows how intercluster trunks are structured. Each remote Cisco CallManager must be configured as an intercluster trunk gateway in the local cluster. In Figure 12-4, the local cluster is San Jose and the remote cluster is Dallas.

Figure 12-4 *Intercluster Trunks*

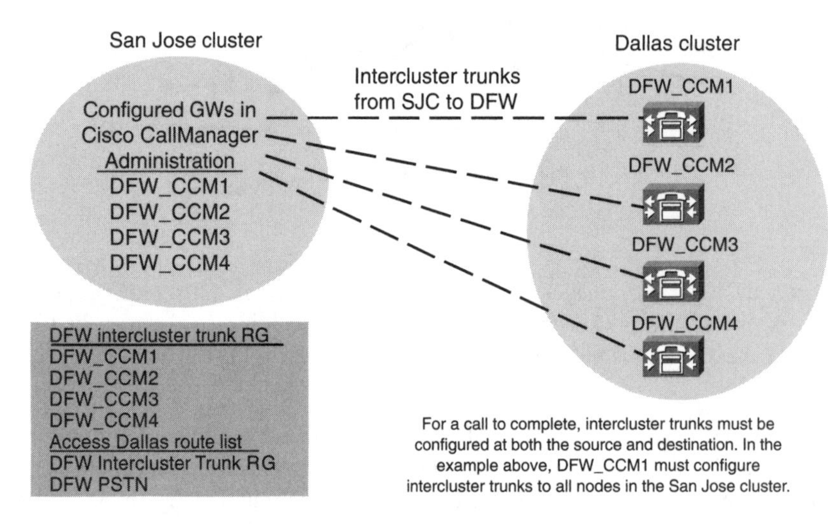

For intercluster trunks to work, the Dallas cluster must configure all San Jose Cisco CallManagers as intercluster trunk gateways. Refer to Chapter 3, "Cisco CallManager Administration Route Plan Menu," about grouping gateway devices into route groups, prioritizing those route groups into a route list, and assigning a route pattern to the route list.

Gatekeeper Features and Configuration

The IOS gatekeeper uses a zone subnet filter to place Cisco CallManager with a zone. This zone subnet filter can be configured on the gatekeeper or on the Gatekeeper Configuration page in Cisco CallManager Administration. (Cisco recommends configuring zones and tech prefix in Cisco CallManager Administration.) Each remote Cisco CallManager must configure across the IP WAN as a gatekeeper-controlled H.323 intercluster trunk. Every gatekeeper-controlled remote Cisco CallManager is a separate VoIP gateway with the gatekeeper. For example, a Cisco CallManager will register as nine VoIP gateways for nine remotes.

Figure 12-5 shows an overview of the registration messages between a Cisco CallManager gateway and the gatekeeper.

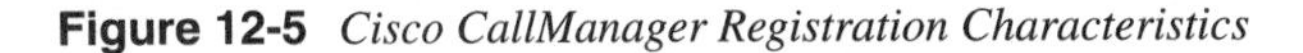

Figure 12-5 *Cisco CallManager Registration Characteristics*

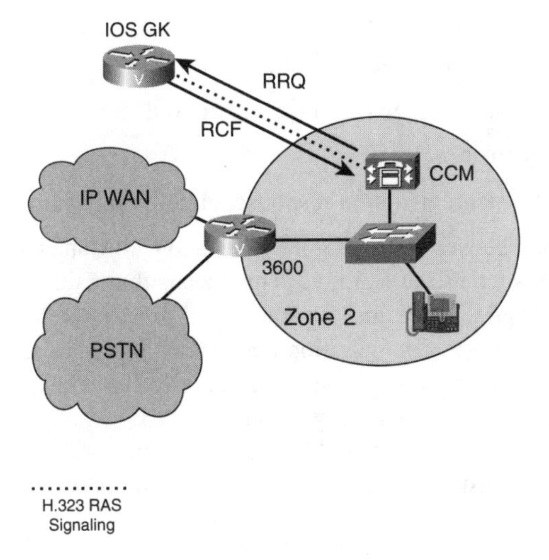

- The RRQ contains an H.323-alias, which is the name of the Cisco CallManager that the gatekeeper process is running on.
- The lightweight KeepAlive RRQ message is shown in Figure 12-5.

Cisco CallManager 3.1 can register E.164 addresses or E.164 address ranges.

Gatekeeper Messages

Cisco CallManager sends a full RRQ every minute by default. This timer is configurable and Cisco CallManager does not use H.323v2 lightweight registration. The gatekeeper then responds with a RCF.

As Cisco CallManager registers with the gatekeeper, the RRQ contains an H.323-alias, which is the server name of the Cisco CallManager. There is also a lightweight KeepAlive RRQ message between the Cisco CallManager and the gatekeeper.

The following lists the messages sent by Cisco CallManager to the gatekeeper:

- RRQ—Registration Request
- ARQ—Admission Request
- DRQ—Disengage Request
- ICF—Information Confirm
- URQ—Unregistration Request

The following lists the messages the gatekeeper sends to Cisco CallManager:

- RCF—Registration Confirm
- RRJ—Registration Reject
- ACF—Admission Confirm
- ARJ—Admission Reject
- DCF/DRJ—Disengage response, either confirmation (DCF) or rejection (DRJ)
- IRR—Information Request
- UCF/URJ—Unregistration response, either confirmation (UCF) or rejection (URJ)

Communication between sites that have a Cisco CallManager node or Cisco CallManager clusters requires the use of H.323v2. This means that a remote Cisco CallManager or Cisco CallManager cluster must be configured as an intercluster H.323 device. Within the Cisco CallManager configuration for an H.323 device, you can configure the device to be gatekeeper-controlled and specify the gatekeeper to be queried. This means that before Cisco CallManager sets up a call with a remote Cisco CallManager it must first send an ARQ with the requested bandwidth to the gatekeeper, as shown in Figure 12-6 (this is using the gatekeeper as just a gatekeeper and not an anonymous device).

Figure 12-6 *Cisco CallManager and Gatekeeper Interaction*

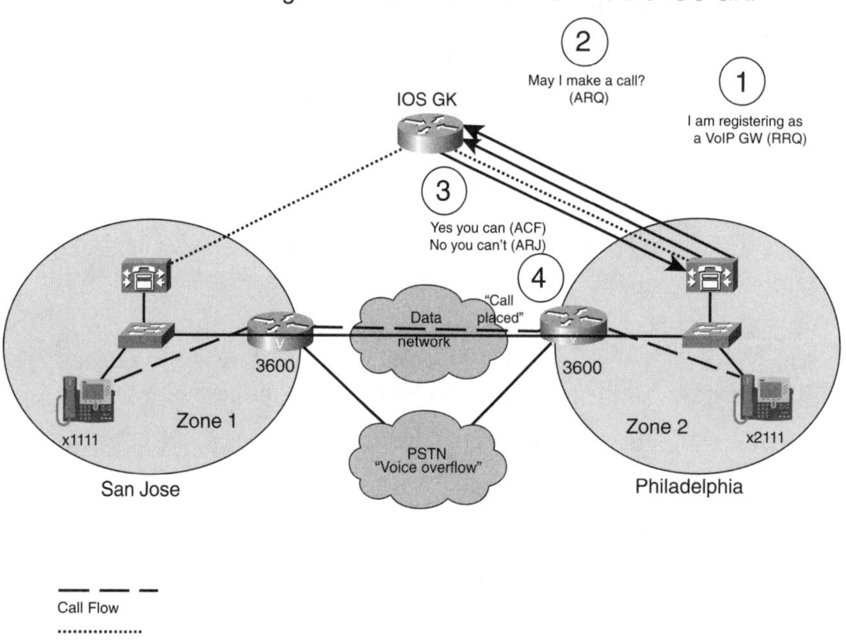

This remote Cisco CallManager is then placed in a route group, which can be associated with various route patterns for IP WAN calls. This route group is configured as the first choice route group in a route pattern or route list; the route group associated with the PSTN is configured as the second choice if the gatekeeper rejected the call. In this way, the call can be transparently routed across the PSTN if the IP WAN is unavailable. For more information about configuring route plans, refer to Chapter 3.

The actual IP WAN bandwidth allocated for a call is defined by putting the remote H.323 Cisco CallManager in a region with which bandwidth can be associated when IP WAN calls must be made to it. Valid codec selections for this are G.711 (80 kbps), G.729 (20 kbps), and G.723 (14 kbps), as shown in Figure 12-7.

Figure 12-7 *ARQ Bandwidth*

Cisco CallManager delivers the following bandwidth requests in ARQ:
G.711—128 kbps
G.723—14 kbps
G.729—20 kbps

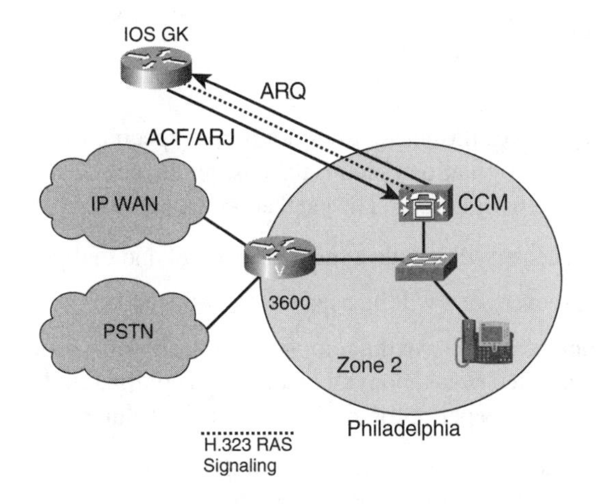

Cisco CallManager uses E.164 in ARQ.

Cisco CallManager delivers the following bandwidth to the gatekeeper:

- G.711 (128 kbps)
- G.729 (20 kbps)
- G.723 (14 kbps)

Gatekeeper as an Anonymous Device

The gatekeeper can be used just as a device that manages bandwidth, or it can be used for dial plan administration and bandwidth admission control, as shown in Figure 12-8.

Figure 12-8 *Anonymous or Not*

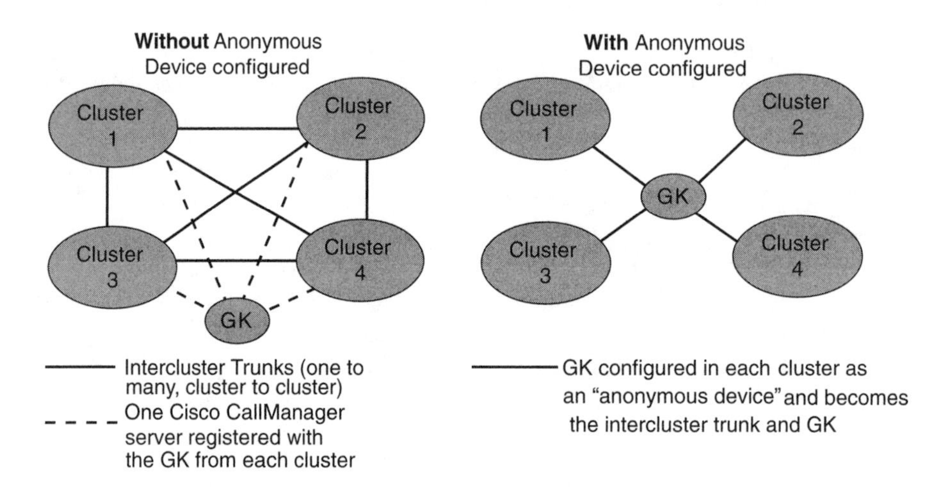

If the remote Cisco CallManager does not have a specific intercluster trunk device configured for the origination Cisco CallManager, then select the anonymous device option in the gatekeeper configuration. The gatekeeper will then have the following characteristics:

- The gatekeeper will handle admission control and dial plan.
- The gatekeeper zone will be used to select the codec.

Cisco CallManagers can have the remote Cisco CallManager intercluster trunks be gatekeeper-controlled or an anonymous device is configured. If an anonymous device is configured, the gatekeeper becomes the intercluster trunk and manages the bandwidth.

Example 12-1 shows a sample gatekeeper configuration. This sample configuration enables only one call at G.729 between zones 1 and 2. If a call of 2111 comes to the gatekeeper, the gatekeeper matches those digits with an IP address in zone 2. If there is enough bandwidth in zone 2, the gatekeeper enables the call to use the IP WAN.

Example 12-1 *Configuring the Gatekeeper*

```
gatekeeper
 zone local zone1 cisco.com
 zone local zone2 cisco.com
 zone prefix zone2 2...
 zone prefix zone1 4...
 gw-type-prefix 1#* default-technology
 bandwidth total zone zone1 20
 bandwidth total zone zone2 20
 bandwidth session zone zone1 20
 bandwidth session zone zone2 20
 no shutdown
```

The command line in the gatekeeper configuration for the tech prefix will look like the following:

```
gw-type-prefix 1#* default-technology
```

In the Gatekeeper Configuration page in Cisco CallManager Administration, add **1#** in the Technology Prefix, as shown in Figure 12-9.

Figure 12-9 *Technology Prefix*

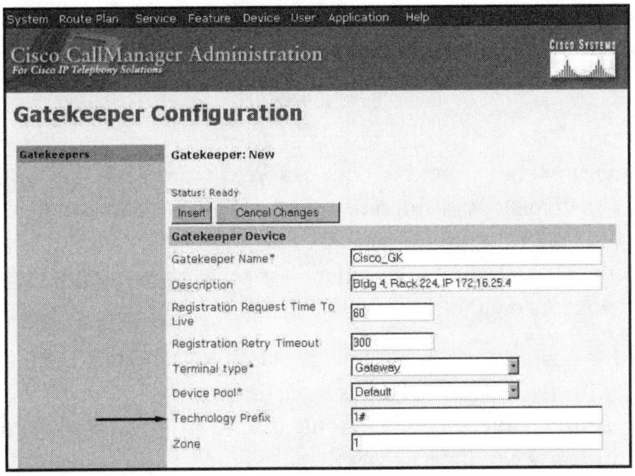

Dial Plan Considerations

In the example depicted in Figure 12-10, users dial five digits for internal calls and seven digits for inter-site calls across the IP WAN. If the IP WAN is down or has insufficient resources, the PSTN is used transparently for inter-site calls. For long-distance calls that will be directed to the PSTN, users dial the access code 9 followed by 1 + area code and seven-digit number. Users dialing local PSTN calls dial 9 plus the seven-digit number. This model also provides gateway redundancy in the event of a gateway or trunk failure to the PSTN. The PSTN gateways are Cisco IOS gateways using H.323.

The goal of this dial plan is to present a consistent dial structure to employees regardless of where the destination is located. The employee can pick up the phone and dial any employee with seven- or five-digit dialing. Figure 12-10 shows an employee is able to dial the San Jose location using only seven digits where calls take the IP WAN as the first choice and the PSTN as the second choice. Thus, users in Philadelphia should be able to dial San Jose users at (408) 526-XXXX by simply dialing 526XXXX.

Figure 12-10 *Dial Plan Goal*

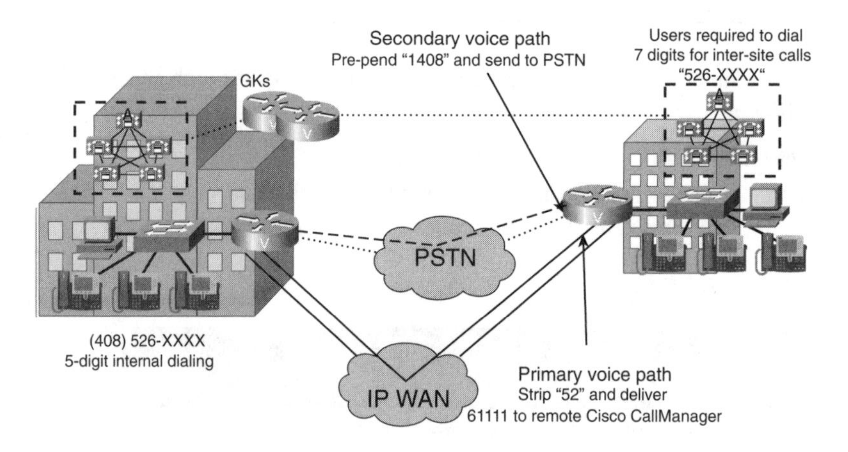

This configuration begins at the route pattern. Figure 12-11 shows the route plan and flow of a call. A route pattern is entered as 52.6XXXX with an assigned route list as SJ. The location of the dot (.) signifies that all digits to the left compose the access code for this route pattern. Also, no digit manipulation is selected or required because each route group needs to invoke its own unique manipulation.

The route list contains two route groups, SJ-IPWAN and PHL-PSTN, listed in order of priority. The SJ-IPWAN route group is listed first and points to the San Jose Cisco CallManager. The digit manipulation specified in route pattern SJ-IPWAN discards the access code (52). This ensures that when the call is sent across the IP WAN, five digits are delivered to the remote Cisco CallManager. This is required because of its internal dial length. The H.323 device associated with the remote Cisco CallManager must be configured to be gatekeeper-controlled to ensure that the gatekeeper is consulted before attempting the call across the IP WAN.

If the gatekeeper rejects the call, the route list uses the next route group, PHL-PSTN. This route group is configured to prepend 1408 to the dialed number to ensure that the call will be recognized by the PSTN and transparently (to the user originating the call) reaches the other end.

Figure 12-11 *Route Plan Configuration*

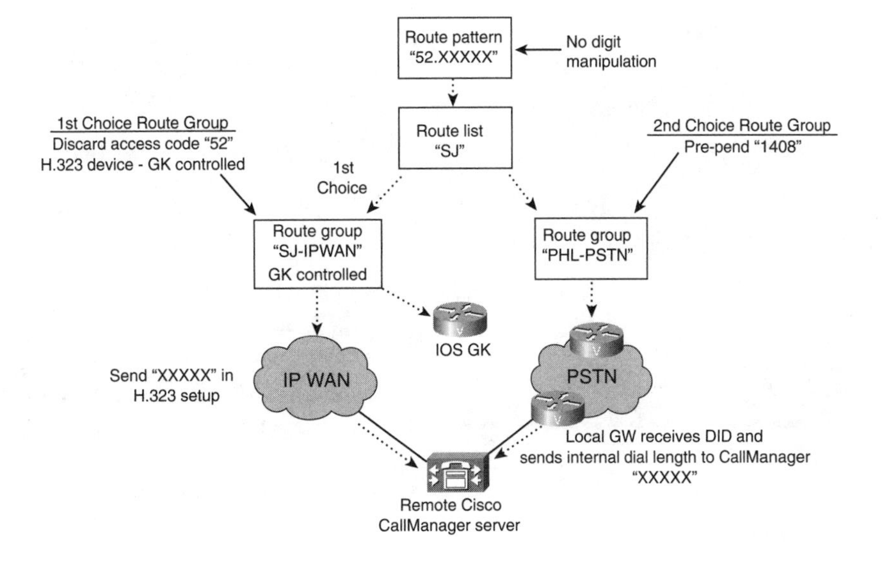

Bandwidth consumed by calls between devices, such as Cicso IP Phones and gateways, can be controlled by specifying the codec usage when setting up regions. Devices are placed in regions that have a codec specified for all intra-region calls; a particular code can likewise be specified for inter-region calls. Regions are assigned to devices using a device pool. The supported codecs defined in regions are G.711, G.729, and G.723 (G.723 is only supported on Cisco IP Phone models 12SP+ and 30VIP) and also on Cisco IP SoftPhone. Figure 12-12 illustrates the use of regions for distributed call processing environments where often only two regions need be assigned.

DSP Resources

This section briefly considers DSP resources in distributed call processing environments. If each site requires hardware conferencing and transcoding, they need to have DSP resources in a multiple site WAN deployment with distributed call processing. Conferencing and media termination point services can be enabled by the IP Voice Media Streaming application.

Figure 12-12 *Codec Selection Based on Regions*

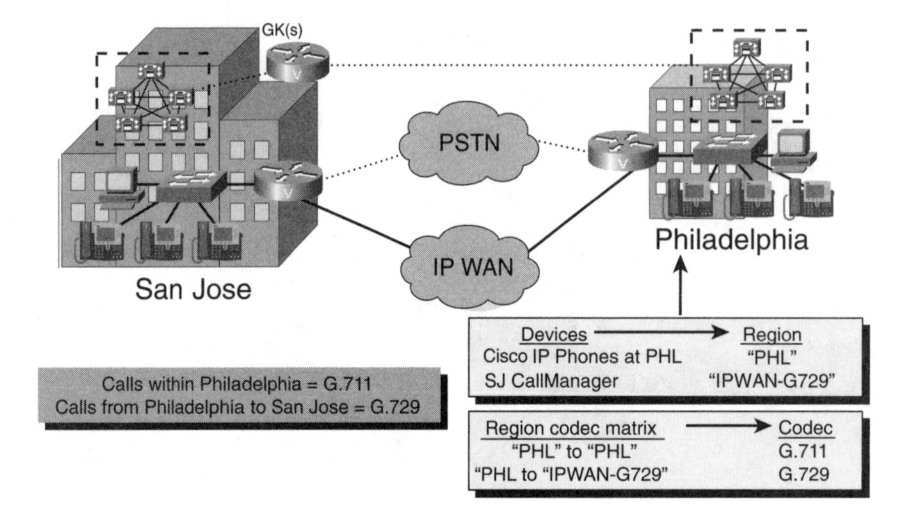

The main purpose of transcoding DSP resources is to perform conversion between different codec types in an RTP stream in the event of a codec mismatch. For example, a compressed G.729 media RTP stream across the IP WAN might need to terminate on a device that only supports G.711, such as many voice mail systems. The transcoding DSP resource would terminate the G.729 media stream and convert it to G.711. This enables the media stream to remain compressed across the WAN. These DSP resources are only used when needed. Figure 12-13 depicts the function of DSP resources across the IP WAN in the following steps:

Step 1 Caller 555-1212 in Region B dials Region A voice mail.

Step 2 CallManager B sees that the destination is Region A, which requires an LBR codec.

Step 3 CallManager A sees an LBR incoming call for a G.711-only device.

Step 4 The media stream is directed to the terminating side DSP farm.

Step 5 The media stream is converted to the proper codec and redirected to the endpoint (voice mail).

The number of resources allocated is based on the requirements for transcoding to G.711 for applications such as IP/IVR, IP/ICD, Personal Assistant, and software conferencing.

Figure 12-13 *DSP Resources for Transcoding*

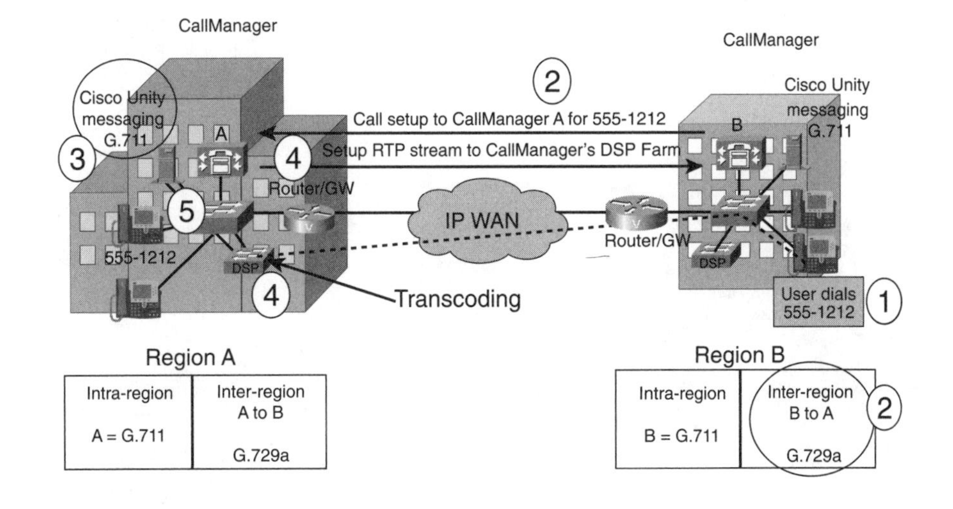

Centralized Call Processing

In a multiple site WAN deployment that uses centralized call processing, Cisco CallManagers are centrally located at the hub or aggregation site with no local call processing at the branch location, as shown in Figure 12-14.

The central site includes a Cisco CallManager server or cluster with all phones registered to a single Cisco CallManager server. The voice messaging system and DSP resources are at the central site. Cisco IP Phones register to an active Cisco CallManager at the central site. CAC imposes a limit on the number of calls per site (location). If the IP WAN link goes down, there is no service unless there is a dial backup configured in the dial plan or the remote router is configured with SRST.

Where centralized call processing is used, CAC is provided using the locations constructed as shown in Figure 12-15. Under this scheme, locations are created with a geographical correspondence, such as a branch office. For example, a location could be designated as Branch 1, Mountain View Office; a postal code could also be used. The location should correlate to a geographical location that is serviced by a wide-area link. A maximum bandwidth to be used by inter-location voice calls is then specified for the location. Devices within that location are then designated as belonging to that location.

Figure 12-14 *Centralized Call Processing Deployment Model*

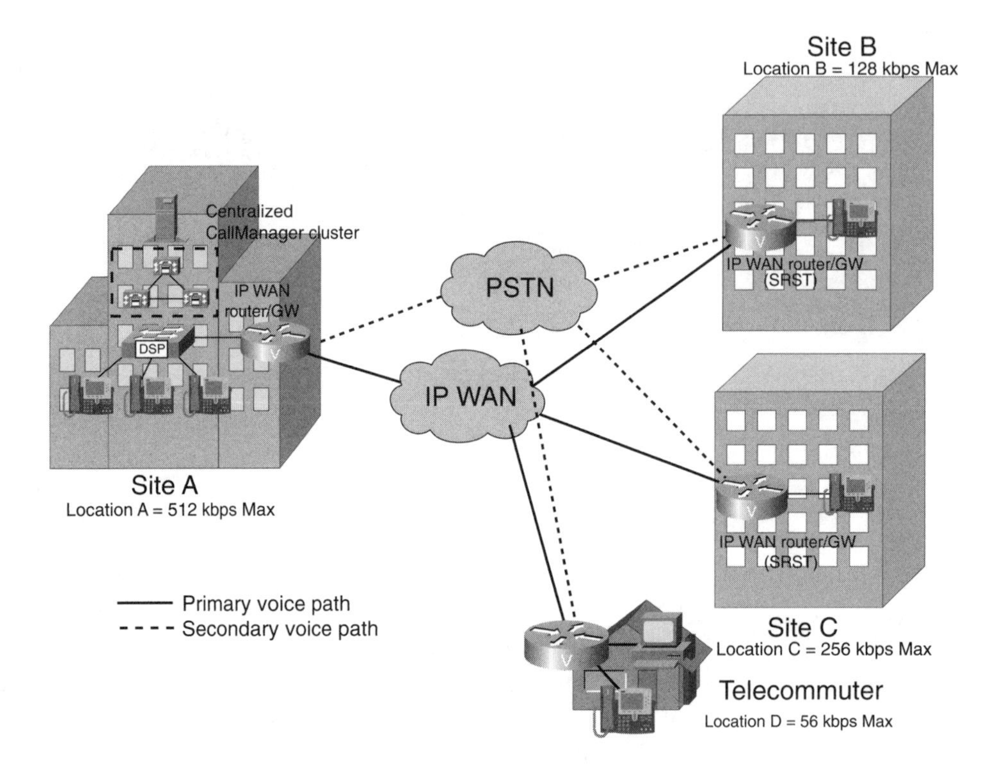

Figure 12-15 *CAC by Locations*

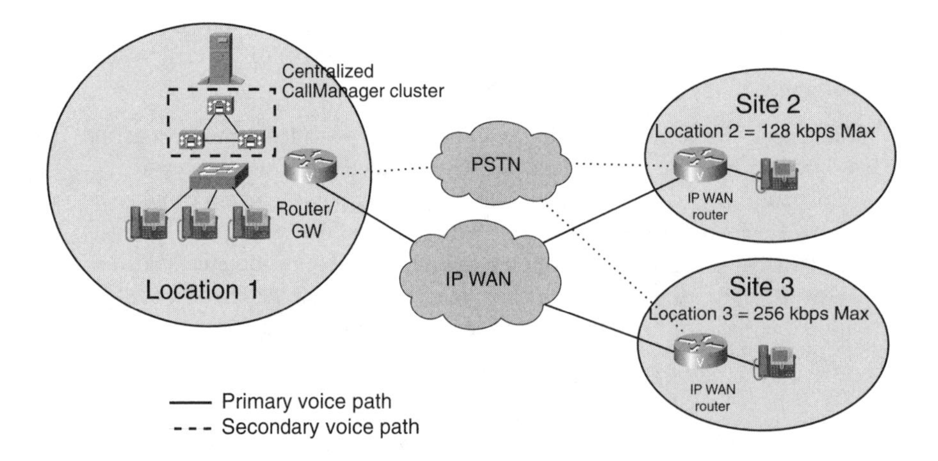

The centralized Cisco CallManager keeps track of the current amount of bandwidth consumed by inter-location voice calls from a given location. If a new call attempted across the IP WAN exceeded the configured setting, a busy signal would be issued to the caller as well as a configurable visual display, such as "All Trunks Busy," on devices with this capability. If the caller gets a busy signal, the caller must hang up the phone and dial the access code for the location's PSTN gateway to facilitate an outgoing call.

The following caveats should be considered when deploying locations-based CAC:

- Mobility of devices between locations is not possible because Cisco CallManager decrements the specified location of the device, not the physical location.
- Calls are admitted based upon the availability of 24 kbps of bandwidth. G.729 calls consume 24 kbps, and G.711 calls consume 80 kbps. Thus, if mixed codecs are used over the WAN, all calls should be assumed to consume 80 kbps and the bandwidth allocated accordingly. Where possible, a single codec should be configured for the WAN. In this case, the bandwidth allocated should be done so in *n* x 24 kbps increments for G.729 or *n* x 80 kbps increments for G.711.
- The locations-based CAC mechanism works across different servers. This means that the Cisco CallManager cluster in a centralized call processing deployment can now contain up to four active Cisco CallManagers to support a maximum of 10,000 Cisco IP Phones or 20,000 total device units (when Cisco CallManager runs on a larger supported server). Devices such as gateways, conferencing resources, voice mail, and other applications "consume" device units according to their relative "weight."
- Cisco CallManager deployments of centralized call processing are limited to hub-and-spoke topologies.
- Where more than one circuit or virtual circuit exists to a spoke location, the bandwidth should be dimensioned according to the dedicated resources allocated on the smaller link.

To use locations-based admission control in a centralized call processing WAN deployment, the remote branch offices (locations) can be registered to a Cisco CallManager in a cluster at the central site. Locations cannot be used between clusters at the central site connected to separate remote sites (locations).

Call Types

Intra-location calls are generally made between Cisco IP Phones and other devices such as analog phones connected to gateway devices based on MGCP or the Skinny gateway protocol. As within a cluster, all devices register with a single Cisco CallManager so that the availability of all devices and WAN bandwidth is known. When a call is attempted, the outcome is one of the following:

- The call succeeds.

- A busy tone is issued due to the remote device being active.
- A busy tone is issued due to insufficient WAN resources; a message might also appear on the device.

Figure 12-16 shows the three types of calls that need to be accommodated using a centralized call processing deployment model.

Figure 12-16 *Three Types of Calls*

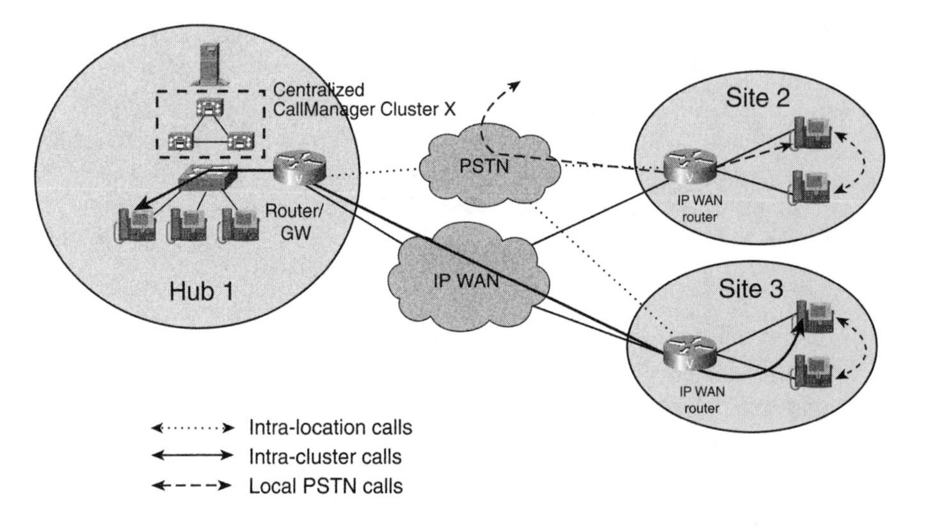

No configuration of a dial plan is required for intra-cluster calls in the majority of cases.

Intercluster calls are made using H.323 and, with Cisco CallManager, intercluster calls can use alternative routing, including PSTN fallback. Between clusters connected over a WAN, a gatekeeper is required for CAC.

Each site can dial a single number to access the local PSTN. The same code can be used for PSTN access, and a local gateway is selected based on the partition and calling search space (CSS).

Using Figure 12-16, the partitions detailed in Table 12-2 would be configured to allow users to have access to either all intra-cluster locations or all intra-cluster locations and a local gateway.

Table 12-2 *Intra-Cluster and Local Gateway Access Required Partitions*

Partition Name	Designated Directory Numbers or Route Patterns Assigned to the Partition
Cluster-X Users	All Cisco IP Phone directory numbers within the cluster.
Cluster-X Hub PSTN Access	Route pattern for PSTN access pointing to gateway(s) located at the hub location.
Cluster-X Site 2 PSTN Access	Route pattern for PSTN access pointing to gateway(s) located at the Site 2 location.
Cluster-X Site 3 PSTN Access	Route pattern for PSTN access pointing to gateway(s) located at the Site 3 location.

The next thing that has to be defined is the CSS. Table 12-3 defines the CSSs for the partitions defined in Table 12-2.

Table 12-3 represents perhaps the simplest example of the required configuration for multiple-site WAN local-call processing. The dial plan consists essentially of a single pattern for PSTN calls, typically a 9. The gateway traversed depends entirely upon the calling device's partition and selected CSS as detailed in Table 12-3.

Table 12-3 *Intra-Cluster and Local Gateway Access Required Partitions*

Calling Search Space	Partitions	Devices Assigned to Calling Search Space
Cluster-X Internal Only	Cluster-X Users	Devices that can only make internal calls.
Cluster-X Hub Unrestricted	Cluster-X Users Cluster-X Hub PSTN Access	Devices that can make internal calls and access the PSTN using a gateway(s) in the hub location.
Cluster-X Site 2 Unrestricted	Cluster-X Users Cluster-X Site 2 PSTN Access	Devices that can make internal calls and access the PSTN using a gateway(s) in the Site 2 location.
Cluster-X Site 3 Unrestricted	Cluster-X Users Cluster-X Site 3 PSTN Access	Devices that can make internal calls and access the PSTN using a gateway(s) in the Site 3 location.

Survivable Remote Site Telephony (SRST)

As enterprises extend their IP telephony deployments from central sites out to remote offices, one of the factors considered vital in deployment is the ability to cost-effectively provide backup redundancy functions at the remote branch office. However, the size and number of these small-office sites precludes most enterprises from deploying dedicated call processing servers, unified messaging servers, or multiple WAN links to each site to achieve the high availability required. The Cisco CallManager IP Telephony solution combined with the newly released SRST feature in Cisco IOS Software enables companies to extend high-availability IP telephony to their small branch offices with a cost-effective solution that is extremely simple to deploy, administer, and maintain. Another term used with SRST is Cisco CallManager fallback mode. Figure 12-17 shows the general process a Cisco IP Phone uses when in the Cisco CallManager fallback mode.

Figure 12-17 *SRST*

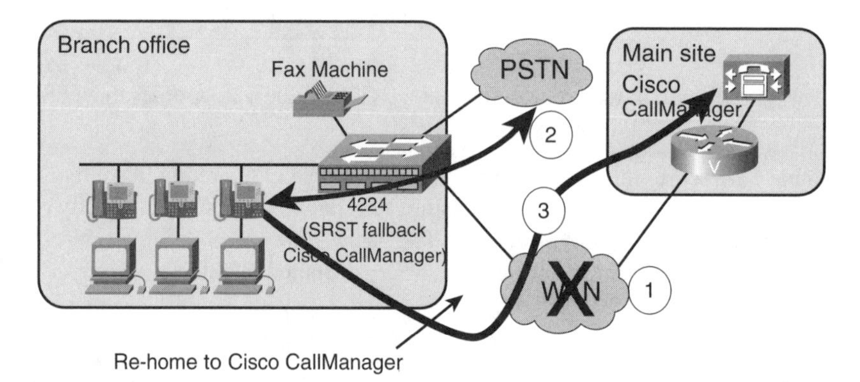

Figure 12-17 shows the following:

1. The IP WAN is down.
2. The Cisco IP Phone registers with the default gateway, which is running SRST.
3. The IP WAN is back up and the Cisco IP Phone registers with the Cisco CallManager at the central site.

Key Features and Benefits

IP telephony is currently undergoing explosive growth, driven by access to value-added features and applications only IP telephony can provide to the end user. Additionally, the cost benefits of running voice and data on a single network is fueling the rapid acceptance of this technology. An architecture in which a Cisco CallManager cluster and application servers are located at the central site provide telephony services for all sites of a corporation has the following benefits:

- Centralized configuration and management.
- Access at every site to all Cisco CallManager features, such as next-generation call centers, unified messaging services, embedded directory services, mobility, and Cisco IP SoftPhones, during normal operations.

- IT staff not required at each remote site.
- Cost-effective operations through converged voice and data network.
- Remote maintenance and troubleshooting.

When remotely placing Cisco IP Phones from a Cisco CallManager cluster, call processing redundancy must be provided to ensure service during a WAN failure. This is especially critical when emergency calls, such as those to 911 in the United States, need to be placed during a WAN outage.

Cisco has developed SRST for Cisco 2600 or Cisco 3600 Series Branch Office Access Routers and the Catalyst 4224 Access Gateway Switch, which comprises network intelligence integrated into Cisco IOS Software. Now SRST is supported on the following platforms:

- Cisco IAD2420 router

NOTE Although the Cisco IAD2420 supports the SRST feature, it is not recommended as a solution for the enterprise branch office.

- Cisco 2600 series
- Cisco 3600 series
- Cisco MC3810 concentrators

NOTE In Cisco IOS Software Release 12.2(2)XG, the only platform supported is the Cisco MC3810.

This service can act as the call processing engine for Cisco IP Phones located in the branch office during the WAN outage

SRST is a capability embedded in Cisco IOS Software that runs on the local branch office access router. Cisco IP Phones automatically detect a failure in the network and initiates a process to intelligently auto-configure the router to provide call processing backup redundancy. The router provides call processing for the duration of the failure, ensuring that the phone capabilities remain active and operational. Upon restoration of the WAN and connectivity to the network, the Cisco IP Phones automatically shift call processing functions to the primary Cisco CallManager cluster. Configuration for this capability is performed once in the local branch office access router. IT staff is not needed at the remote sites to enable and disable this functionality as a result of the intelligence and simplicity of the SRST feature.

The following are the features of SRST:

- Support for re-homing of Cisco IP Phones to use call processing on the local router upon Cisco CallManager failure
- Support for IP and POTS phones on the router
- Extension to extension dialing
- Extension to PSTN dialing
- Primary line on phone
- DID
- Calling party ID (Caller ID/ANI) display
- Calling party name display
- Last number redial
- Call transfer without consultation
- Call hold and retrieve on a shared line

The following lists the supported types of phone calls using SRST:

- Cisco IP Phone to Cisco IP Phone
- Cisco IP Phone to any router voice port
- Cisco IP Phone to VoIP H.323/SIP OnNet
- Cisco IP Phone to VoFR/VoATM OnNet

The following lists the phone features with SRST:

- Multiple lines per Cisco IP Phone
- Multiple line appearance (shared lines)
- Call hold and retrieve
- Caller ID
- Call transfer of local calls:
    - Supports local Cisco IP Phone calls
    - Supports FXS/FXO/BRI/PRI calls
    - Supports VoFR/VoATM call transfers
    - Does not support H.323 or SIP transfer at the time of writing this book
    - Does not support E&M or analog DID transfer

Table 12-4 shows the Cisco IOS Software release information to support SRST.

Table 12-4 *Cisco IOS Software Release Information*

Platform	Cisco IOS Software Images
Cisco 2600 and 3600	12.1(5)YD through 12.1(3)YD and 12.2(7)T
Cisco IAD 2400	12.1.5YD
Catalyst 4224 Access Gateway Switch	12.1.5YE
MC3810	12.2(2)XG

Cisco IP Phone Support

Up to 24 Cisco IP Phones are supported on the smaller platforms, and up to 48 Cisco IP Phones are supported on the larger platforms.

While 24 or 48 Cisco IP Phones can be configured per system, up to two lines per phone are supported as a default, providing, for example, up to 48 lines on a Cisco 2650 and 96 lines on a Cisco 3640.

Table 12-5 shows the number of phones and lines supported per platform.

Table 12-5 *IP Phone Support per Platform*

Platform	Number of Phones Supported	Number of Lines Available
Cisco 175x[1], IAD2400, 2600, 3620, Catalyst 4224	Up to 24 phones	48
Cisco 3640 Catalyst 4000 AGM[1]	Up to 48 phones	96
Cisco 3660[2]	Up to 144 phones	288

1 SRST will become orderable at a later date on these platforms.

2 Starting with 12.1.5YD1

Minimum System Requirements

Table 12-6 outlines the minimum system requirements to support SRST.

Table 12-6 *Minimum System Requirements for SRST*

Platform	Minimum Requirement	Comment
Cisco 2600 3600	Cisco IOS Software Plus Image	
Cisco 2600 and 3620 and 3640	16 MB Flash and 64 MB DRAM	96 MB DRAM is preferred.
Cisco 3660	16 MB Flash and 64 MB DRAM for up to 48 phones, 96 MB for up to 96 phones, and 128 MB for up to 144 phones	
Cisco IAD2400 Catalyst 4224	Standard shipping configuration	
Cisco CallManager	3.0(5) or higher	
New firmware images (loads) for Cisco IP Phones	Cisco 7940 and 7960P003D302.bin Cisco 7910P004D302.bin	All Cisco CallManager phone loads support SRST.

Cisco Catalyst 4224 Access Gateway Switch

The Catalyst 4224 Access Gateway switch is an integrated Layer 2/Layer 3 switch and gateway combined into a single box solution for small enterprise remote offices. The Catalyst 4224 supports up to 24 remote IP telephony/data users configured as part of a centralized Cisco CallManager deployment model. The Catalyst 4224 also provides basic backup call processing during IP WAN network outages via IOS SRST.

Figure 12-18 shows a typical setup of the Catalyst 4224 Access Gateway switch in a CIPT network. The following summarizes what is provided by a Catalyst 4224:

- IP IOS routing
- Voice gateway
- 24 ports 10/100 Ethernet inline power
- SRST
- Single Box Micro Branch Solution
- Centralized or local Cisco CallManager
- ISDN dial backup capable
- Single T1 branch access

Figure 12-18 *Cisco Catalyst 4224 Access Gateway Switch*

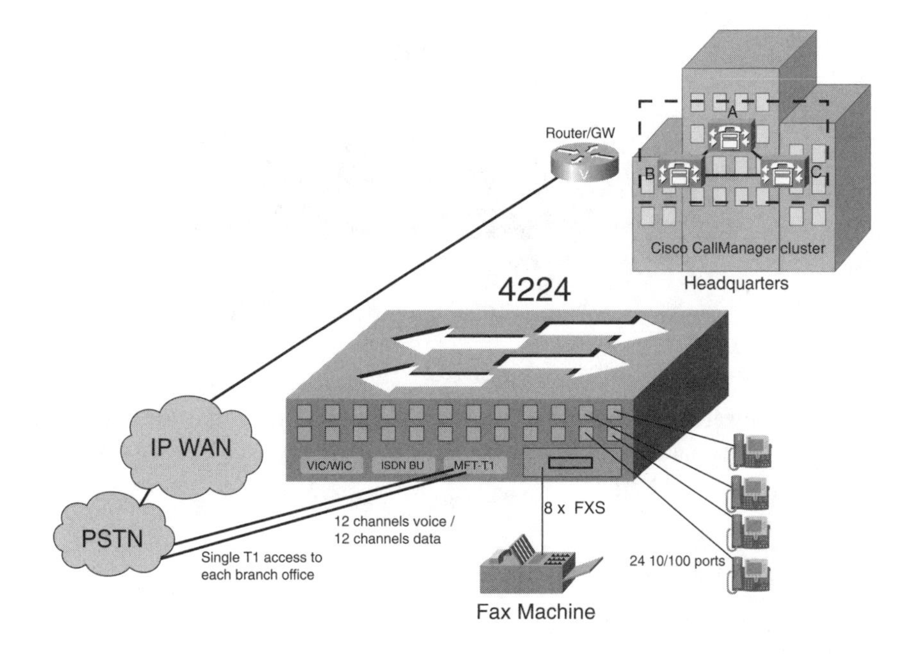

From a hardware perspective, the Catalyst 4224 has five main components:

- TDM subsystem that comprises 2 VIC/WIC slots, 1 VIC slot, built in 8 port FXS module and a TDM switch.
- Layer 3 CPU subsystem that provides routing functionality and interfaces with the TDM and switch subsystems.
- DSP subsystem interfaces to the TDM subsystem and CPU subsystem for converting voice streams to IP packets.
- Layer 2 switch subsystem, a 24 port 10/100 QoS enabled switch, which interfaces to the CPU and Power subsystems.
- Power subsystem that provides power to the Catalyst 4224 and inline power to the Cisco IP Phones connected to the 10/100 ports.

Figure 12-19 shows the front view of the Catalyst 4224 Access Gateway Switch.

The Catalyst 4224 has DSPs installed on the motherboard. These DSPs are used for converting between voice and packet. They also compress and decompress packets based on the codec configured. Transcoding and hardware conferencing services are provided at the central Cisco CallManager location, not on the Catalyst 4224. There is no DSP farm capability as there is on the Catalyst 400X Access Gateway Module.

Figure 12-19 *Cisco Catalyst 4224 Front View*

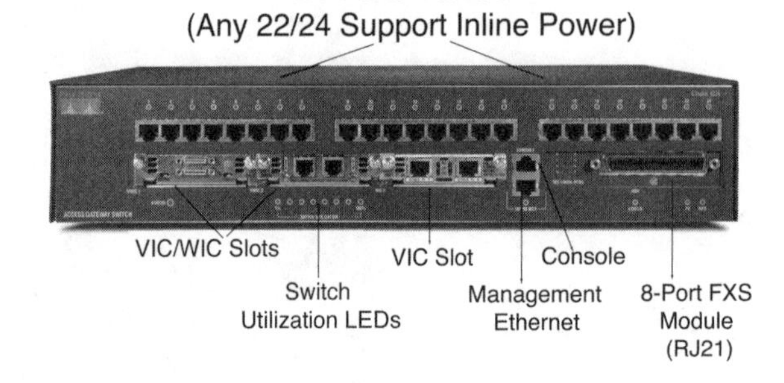

From a software perspective, the Catalyst 4224 is configured from the Cisco IOS Software command-line interface (CLI) using a 12.1.4T-based image. The following Cisco IOS Software images are available:

- IP Plus (standard)
- IP Plus with Firewall
- IP Plus with IPsec 56
- IP Plus with 3DES
- IP Plus with Firewall and IPsec 56
- IP Plus with Firewall and 3DES
- Optional feature license required to use SRST

There is also support for an external redundant power supply, which connects in the back. The Catalyst 4224 ships with 64 MB of DRAM and 32 MB of inline SIMM Flash. The Catalyst 4224 also has a built in hardware encryption engine. The 10/100 Management port is to be used solely for management and not as a 25th Fast Ethernet port.

Voice support includes T1-CAS, T1/E1-PRI (voice only), E1-R2, and ISDN-BRI (voice and data) as well as analog interfaces as Table 12-7 shows.

Table 12-7 *Supported VIC/WICs*

VICs	VWICs	WICs
VIC-2FXS	V WIC-1MFT-T1	WIC-1DSU-56K4
VIC-2FXO	V WIC-2MFT-T1-DI[2]	WIC-2A/S
VIC-2FXO-EU	V WIC-2MFT-T1	
VIC-2FXO-M1	V WIC-1MFT-E1	
VIC-2FXO-M2	V WIC-2MFT-E1-DI[2]	
VIC-2FXO-M3	V WIC-2MFT-E1	
VIC-2BRI-S/T-TE[1]		

1 Supports voice or data

2 No DI support to external PBX

Analog DID and E&M interfaces are not supported. Also the WIC-1B-S/T is not supported and there are no plans to do so.

As previously illustrated in Figure 12-19, slots 1 and 2 can support the full range of VICs, WICs, and VWICs because these slots have connections directly into the TDM subsystem, (VICs and VWICs) as well as the Layer 3 CPU (WICs). Slot 3 supports the VICs listed above as well as the VWICs. WICs are not supported in slot 3 because there is no SCC connection to the Layer 3 CPU.

The following codecs are supported:

- G.711 A-law 64 kbps
- G.711 μ-law 64 kbps
- G.729 abr 8 Annex–A and B 8 kbps
- G.729 ar8 G729 Annex–A 8 kbps
- G 729 r8 G729 8 kbps

The MFT VWICs provide a single physical interface into a carrier's network and essentially provide Channel Bank functionality. The channels may be allocated as either voice or data channels with the WAN carrier provisioning within the WAN cloud as appropriate.

The Catalyst 4224 brings scalability to enterprise networks. The 4224 can participate in all three supported deployment models: isolated, distributed, and centralized.

As Figure 12-20 illustrates, the Catalyst 4224 supports a variety of designs.

Figure 12-20 *Cisco Catalyst 4224 Enterprise Scalability*

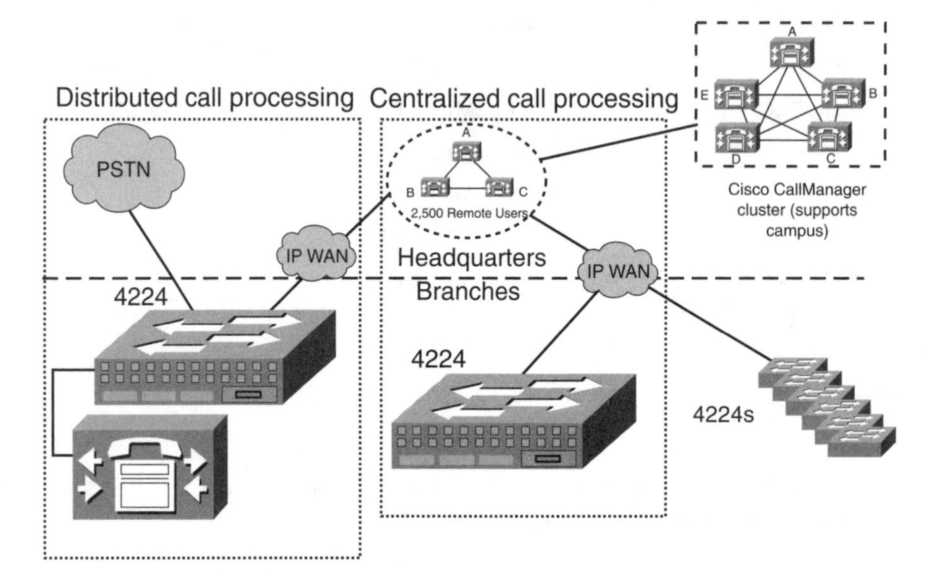

Configuring SRST

Most of the magic (software capability) for SRST happens in the phone load itself. The MAC address and phone number is all stored in the phone itself. When everything is working, the phone (under the Network Settings) displays its default router as a "Standby." If the WAN connectivity dies, the load on the phone tells it to try to register with this default gateway. The default gateway setting will now display "Active."

Table 12-8 shows the commands and optional commands used to configure SRST on a supported platform.

Table 12-8 *Configure Cisco CallManager Fallback*

Command	What It Does
call-manager-fallback	Enables fallback capability and puts you in a submenu.
ip source-address *ip address* [**port** *port #*]	Enables router to receive Skinny messages on this particular port. The default port is 2000.
max-ephones	Defaults to 0. Maximum phones that will be allowed to register. The 3640 supports up to 48 phones, and 3660 supports up to 14 phones; the IAD24xx, and 26xx, and the 3620 platforms support up to 24 phones.

Table 12-8 *Configure Cisco CallManager Fallback (Continued)*

Command	What It Does
max-dn	Defaults to 0. Maximum number of directory numbers that can be configured. The 3640 supports up to 96 phones; the 3660 supports up to 288 DNs, the IAD24xx, and 26xx; and the 3620 platforms up to 48 phones.
Optional Command	**What It Does**
keepalive *seconds*	Skinny KeepAlive timer. Default setting is 30 seconds.
default-destination *number*	Causes incoming calls that don't have a called number (for example, from an FXO port) to be routed to the specified Cisco IP Phone extension.
dialplan-pattern *tag pattern* [**extension-length** *number*]	Creates a global prefix that can be used to expand the abbreviated extension numbers (automatically obtained from the Cisco IP Phones) to expand into fully qualified E.164 numbers.
transfer-pattern *number*	By default you can transfer from Cisco IP Phone to Cisco IP Phone. If you want to be able to transfer to an H.323 endpoint, you need to have the number to which you're transferring defined as a transfer-pattern. Transfer is currently only a blind transfer.
access-code {**bri** | **e&m** | **fxo** | **pri**} *number*	Configure trunk access codes for each type of line so that the Cisco IP Phones can access the trunk lines only during Cisco CallManager fallback mode when the SRST feature is enabled.
huntstop	Used to set the huntstop attribute for the dial peers associated with the Cisco IP Phone phone lines.
voicemail *number*	Used to configure a number that is dialed when the **messages** button or soft key is pressed.
reset {*MAC addr* | **all**}	Very useful command used to reset individual or all Cisco IP Phones.

Example 12-2 shows a sample configuration for SRST.

Example 12-2 *Configuring the SRST*

```
call-manager-fallback
 ip source-address 10.1.0.1 port 2000 strict-match
 max-ephones 24
 max-dn 24 dialplan-pattern 1 408734.... extension-length 4
 transfer-pattern 510650....
 voicemail 11111
```

Summary

When interconnecting two or more isolated campus LAN CIPT deployments, consider the following:

- QoS tools ensure voice quality in two ways: by giving voice priority over data and by preventing voice from oversubscribing a given WAN link.
- The maximum bandwidth setting for a zone should take into account the limitation that the WAN link may not be filled with more than 75 percent voice.
- Each remote Cisco CallManager must configure across the IP WAN as a gatekeeper-controlled H.323 intercluster trunk.

Locations construct is used for WAN deployments that use centralized call processing and is limited to a hub-and-spoke topology.

SRST is configured on supported IOS platforms and provides limited call processing to remote branch offices in a centralized deployment model.

Post-Test

How well do you think you know this information? The post-test is designed to help you gauge your knowledge about this chapter. Of the 10 questions, if you answer one to three questions correctly, we recommend that you re-read this chapter. If you answer four to seven questions correctly, we recommend that you review those sections that you need to know more about. If you answer 8 to 10 questions correctly, you probably understand this information well enough to move on to the next chapter. You can find the answers to these questions in Appendix B, "Answers to Chapter Pre-Test and Post-Test Questions."

1 For deployments of the distributed call processing model, what is used to provide CAC?

____

____

____

2 True or False? In a CIPT environment the gatekeeper is designed to work with other Cisco IOS H.323 gateways.

____

____

____

3 For each codec listed below, what bandwidth will the Cisco CallManager deliver to the gatekeeper?

- G.711 ____________
- G.729 ____________
- G.723 ____________

4 What keeps track of the current amount of bandwidth consumed by inter-location voice calls from a given location?

____

____

____

5 What Cisco CallManager route plan feature in a centralized call processing deployment enables for the same access code to be used for PSTN access through a local gateway at multiple sites within the same Cisco CallManager cluster?

____

____

____

6 What is the feature that automatically detects a failure in the network, and using Cisco Simple Network Automated Provisioning capability, initiates a process to intelligently auto-configure the router to provide call processing backup redundancy for the Cisco IP Phones in that office?

____

____

____

7 List at least three of the four supported types of phone calls using SRST:

____

____

____

8 What are the two ways QoS tools ensure voice quality?

____

____

____

9 List at least two of the four caveats that should be considered when deploying locations-based CAC.

____

____

____

10 The maximum bandwidth setting for a zone should take into account the limitation that the WAN link may not be filled with more than what percentage for voice?

____

____

____

This chapter covers the following topics:

- System Guidelines
- Route Plan Guidelines
- Service Guidelines
- Feature Guidelines
- Device Guidelines
- User Guidelines
- Applications Guidelines

CHAPTER 13

Best Practices for Cisco IP Telephony Deployment

This chapter provides some guidelines and rules for a Cisco IP Telephony network solution. Some of the tips and guidelines in this chapter were covered throughout the book and, for some, this will be the first mention of such guidelines. The best practices described in this chapter have been arranged by topic; for example, if you are implementing a route plan, you should review the route plan section and ensure the best practices outlined in that section are followed.

You can find detailed information about maximizing your CIPT solution in the Cisco Press book, *Cisco CallManager Fundamentals* (ISBN: 1-58705-008-0). Chapter 2 of Cisco *CallManager Fundamentals* is of particular interest for those seeking detailed information on NANP and international call routing and dial plans.

System Guidelines

This section provides best practices for menu items under the System menu in Cisco CallManager Administration (CCMAdmin). This section includes cluster considerations, auto-registration, and some database replication issues.

The Publisher Should Be—At Most—a Backup Cisco CallManager

If at all possible the Publisher should be only a database server. The following guidelines apply based on the number of users and other circumstances:

- In clusters that support up to 10,000 users, the Publisher server should be used only as the database server.
- In clusters that support up to 5000 users, the Publisher should be used as the database server and run the TFTP service.

- In a cluster that supports up to 2500 users, the Publisher server is used as the database server and can also run the TFTP service.
- If circumstances dictate, the Publisher server can act as the database server, run the TFTP service, and also be used as the backup Cisco CallManager server for devices. When is it okay to use the Publisher server as a backup Cisco CallManager server? I have seen documentation indicating this is acceptable for deployments of up to 5000 users. As a best practice, however, you should limit the Publisher's use as a backup Cisco CallManager server for deployments of 2500 users or less.
- Proper attention to device weights is important. If the deployment uses CTI ports or multiple trunks, the number of phones registered to a particular CallManager is reduced.

Verify Cisco CallManager Servers in a Cluster Are Running the Same Software Version

All Cisco CallManagers in a cluster should be running the same software version. Verify that all the Cisco CallManagers in the cluster and applications upgraded properly.

Click **Help** > **Component versions** in Cisco CallManager Administration to verify all applications are the same version.

Determine if an Excessive Number of Devices Exist on a System Based on Cisco CallManager Memory

Table 13-1 shows the weights associated with each device type in a CIPT solution. Device weights is a way to measure the amount of load a type of device has on Cisco CallManager when that type of device is registered in a Cisco CallManager cluster. Refer to the following URL for the current weights per device:

www.cisco.com/univercd/cc/td/doc/product/voice/ip_tele/network/dgclustr.htm#xtocid103062

Table 13-1 *Weights by Device Type*

Device type	Weight per Session/Voice Channel	Session/DS-0 per Device	Cumulative Device Weight
Cisco IP Phone	1	1	1
Analog gateway ports (MGCP)	3	Varies	3 per DS-0
Analog gateway ports (Skinny)	1	Varies	1 per DS-0
T1 gateway	3	24	72 per T1
E1 gateway	3	30	90 per E1
H.323 client	3	Varies	3 per call

Table 13-1 *Weights by Device Type (Continued)*

Device type	Weight per Session/Voice Channel	Session/DS-0 per Device	Cumulative Device Weight
H.323 gateway	3	Varies	3 per call
Music on hold stream	10	20	200[4]
Transcoding resource	3	Varies	3 per session
Software MTP	3	24	72[1]
Conference resource (hardware)	3	Varies	3 per session
Conference resource (software)	3	24	72[1]
CTI route point	2	Varies	Varies[2]
CTI client port	2	1	2
CTI server port	2	1	2
CTI 3rd party control[3]	3	1	3
CTI agent phone[3]	6	1	6
Messaging (voice mail)	3	Varies	3 per session
Intercluster trunk	3	Varies	3 per call

1 When installed on the same server as Cisco CallManager, the maximum number of sessions is 24.

2 CTI route point's cumulative weight is dependent on the associated CTI ports used by the application.

3 Includes the associated Cisco IP Phone.

4 When MOH is installed co-resident, the maximum number of streams is 20.

The total number of device units that a single Cisco CallManager can control depends on the server platform. Table 13-2 gives details of the maximum number of devices per platform.

Table 13-2 *Maximum Number of Devices per Server*

Platform Server Characteristics	Maximum Device Units per Server	Maximum Cisco IP Phones per Server
MCS-7835-1000 PIII 1000 MHz, 1 GB RAM	5000	2500
MCS-7835 PIII 733 MHz, 1 GB RAM	5000	2500
MCS-7830 PIII 500 MHz, 1 GB RAM	3000	1500

continues

Table 13-2 *Maximum Number of Devices per Server (Continued)*

Platform Server Characteristics	Maximum Device Units per Server	Maximum Cisco IP Phones per Server
MCS-7830 PIII 500 MHz, 512 MB RAM	1000	500
MCS-7825-800 PIII 800 MHz, 512 MB RAM	1000	500
MCS-7822 PIII 550 MHz, 512 MB RAM	1000	500
MCS-7820 PIII 500 MHz, 512 MB RAM	1000	500

Determine if a Database Change Is Waiting for a Cisco CallManager Restart

Although the majority of changes to the Cisco CallManager database do not require a restart of Cisco CallManager before the change takes effect, there are still a few actions that require a restart. For example, when you add a new subscriber to the cluster, all other members of the cluster must be restarted for the SDL links to the new subscriber to be formed.

Cisco CallManager restarts affect devices in a cluster, so if a restart is necessary, perform it during off-hours when possible. The majority of configuration changes you make will not require a restart. However, check the online help in Cisco CallManager Administration (**Help > Contents and Index** or **Help > For This Page**) to determine whether a restart is required.

Distribute Devices in a Cluster Among Cisco CallManager Servers

Use device pools to evenly distribute devices among Cisco CallManagers in a cluster. Device pools should be named for logical groups; for example, use building numbers, sites, office codes, function, or locations of devices. Using logical nomenclature for device pools makes it easier to find devices.

Also, device pool names could reflect the specific region or Cisco CallManager group for that device pool. This also helps to quickly identify which devices are contained in a specific device pool. For example, you might name a device pool **DefaultDallasLobby** for lobby phones that belong to the default Cisco CallManager group and the Dallas region.

Create a Rogue-Phone Calling Search Space if Auto-Registration Is Enabled

Use a calling search space (CSS) to restrict operations for rogue or unadministered phones. A rogue or unadministered phone is any phone on the network that registers with Cisco CallManager without the administrator knowing about it. Create a CSS for those auto-registered phones so that they will have nowhere to call. To control this even further, create a rogue-phone CSS as a private line automatic ring-down (PLAR) so that when the phone goes off-hook, it automatically dials security or some other phone enforcement tool.

For the rogue-phone CSS to work, you can't have any callable devices in the **<none>** partition. If a device is in the **<none>** partition, then anyone on the system has access to that device's number because every CSS implicitly has at least the **<none>** partition in it's configuration. See Chapter 2, "Navigation and System Setup," and Chapter 3, "Cisco CallManager Administration Route Plan Menu," for details. For configuration of PLAR, see Chapter 3.

Consider Account and Password Information for Increased Security

Ensure that SQL and NT accounts have passwords, and be sure that all the servers in the cluster have the same password for authentication and replication.

Changing passwords is very involved. Follow all the steps provided in the Administrative Accounts and Passwords section in the *System Guide* of the Cisco CallManager Administration and System Guides, Release 3.1(1) documentation at the following site:

> www.cisco.com/univercd/cc/td/doc/product/voice/c_callmg/3_1/sys_ad/adm_sys/ccmsys/a10accts.htm

Because most attempts at cracking into a server utilize the Administrator user name, you should consider disabling the Guest account on the server, removing any users in the Guest Group, and changing the name of the Administrator to something other than "Administrator."

Determine Whether Unnecessary Services Are Running

Terminal Services, while useful and convenient to use, can provide a potential means for a security breach. Terminal Services relies on Microsoft Windows 2000 authentication and valid administrator privileges, but you may still consider this a security risk. Depending on your level of concern about crackers gaining access to a machine via Terminal Services, you may consider disabling the service altogether or changing it to a Manual start. You can do this in the Services area of the Control Panel.

On Subscriber servers, it is preferable for IIS and the WWW service to be disabled. In Chapter 8, "Installation, Backups, and Upgrades," refer to the post-installation section for those services that should be disabled.

Only One Server Needs to Run TFTP in a Cluster

You need to run TFTP on only one server in the cluster. In fact during the installation of Cisco CallManager, if you already have a TFTP server installed, you should deselect this component and not install this service on other servers in the cluster.

In releases of Cisco CallManager subsequent to Release 3.1(1), you can configure an array of IP addresses into DHCP option 150 for TFTP server redundancy, so it may be desirable to have multiple TFTP servers running in the cluster. Also, it's a good idea to have at least one other machine running TFTP so that if the primary TFTP server fails, you can easily change option 150 to point to the new IP address. This will enable you to continue to add users and make other changes while the primary TFTP is down.

Set Trace Levels Appropriately for Cluster Size

If trace levels are set at too fine a detail or too many trace bits are enabled, tracing may cause performance degradation on Cisco CallManager (which will affect the quality of calls). Consider the size of your cluster and set the trace level accordingly to minimize this negative impact. Cisco recommends that you disable tracing and then enable it only for a specific device or specific period of time using Cisco CallManager Serviceability.

Check the CDR Database Size and Purge When Necessary

Performance problems can result when the CDR database size is over 1 GB. Occasionally, you should check the file size by clicking **Start** > **Programs** > **Microsoft SQL Server 7.0** > **Enterprise Manager** > **Databases** > **CDR**. Look at the file size. If it exceeds or is approaching one gigabyte, use ART (also known as CAR, CDR Analysis and Reporting) to purge the CDR database.

This task can be automated by scheduling automatic database purge in ART/CAR (see the online help for ART or CAR for instructions) or by setting the **Max CDR Records** service parameter in CCMAdmin to limit the number of records entered in the CDR database (**Service > Service Parameters >** *select a server > click on the Cisco Database Layer Monitor Service in the list box at left*).

Route Plan Guidelines

This section discusses the best practices that apply to the Cisco CallManager route plan.

Prevent Redundant Route Plans and Conflicting Route Patterns That Inhibit Outside Dial Tone

Ensure Different Call Park Numbers Are Used for Each Server in a Cluster

Ensure the Voice Mail Pilot Number Is Unique in a Cluster

To address all three of these issues, check the Route Pattern page in CCMAdmin. To get a better picture, view the route plan report in a file and print it. You can print the route plan by clicking **Route Plan > Route Plan Report** in CCMAdmin. On the Route Plan Report page, select to view file, save to disk, and then open the file and print it. Figure 13-1 shows a portion of the Route Plan report.

Figure 13-1 *Route Plan Report Page in CCMAdmin*

System Route Plan Service Feature Device User Application Help

Cisco CallManager Administration
For Cisco IP Telephony Solutions

Cisco Systems

Route Plan Report

View In File

Listing of all Call Park, Call Pickup, Conference, Route Pattern and Translation Pattern in the system:

Records 1-48 of 48

All ('All' may take a long time to load)

Pattern/ Directory Number	Type	Partition	Route Detail
12345	Call Park		
5000	Call Park	ciscocm1	
1234	Call Park	cssHouston	
2323	Conference		
1111	Conference	cssHouston	
1212	Conference	cssHouston	
22222	Conference	cssHouston	
34343	Conference	cssHouston	
44444	Conference	cssHouston	
	Translation Pattern		
77723 where (LOCAL-AREA-CODE DOES-NOT-EXIST AND LOCAL-DIRECT-DIAL EXISTS) OR (AREA-CODE EXISTS)	Translation Pattern		
55566	Translation Pattern	Partition1	
55567	Translation Pattern	Partition2	
1234566	Call Pickup		
55632	Call Pickup		
1234566	Call Pickup	ciscocm1	
55632	Call Pickup	ciscocm1	

Review the print out carefully to be certain there are no pattern overlaps or conflicts. Remember that a directory number is also a route pattern; so for example, check the call park extensions or ranges and voice mail pilot numbers to ensure there is no overlap among servers in a cluster. Be certain that translation patterns, route patterns, directory numbers, Meet-Me conference numbers, and other patterns do not conflict.

For NANP, Discard Digits Based on the PSTN's Expectations

The PSTN expects to receive digits in a certain format. Depending on the local area dialing rules, the PSTN will expect to receive seven digits (office code + extension), 10 digits (local area code + office code + extension) or 11 digits (1 + area code + office code + extension). These are the most commonly used formats that the PSTN expects to receive digits. Be certain that if you are using an access code (for example, dialing "9" for an outside line) in a route pattern, the access code is discarded before sending the digits to the PSTN.

Also, a common problem you may encounter is when the ISDN numbering plan or type are not configured the way the provider wants them to be. This is just as important as sending 11 versus 10 digits.

Send Calling Line Identification (CLID) Correctly

Use calling party number transformations so that CLID is sent correctly. For example, your phone number is 5-digit extension, such as 31234, but the PSTN expects to receive a 10-digit number. You can use various mechanisms to transform the calling party number, including the external phone number mask on line, the calling party transform mask on a route pattern or route list, or on the caller ID directory number on the gateway. Using the external phone number mask on the line gives you the greatest amount of flexibility.

See the online help in CCMAdmin or Chapter 2 in the Cisco Press book, *Cisco CallManager Fundamentals* for detailed information.

For NANP: Create a Route Filter to Allow Seven-Digit Local Dialing Where Necessary

NOTE This best practice applies only if you are using the North American Numbering Plan (NANP).

If you are in an area where a local number can be successfully connected by dialing seven digits, create a route filter that accepts the seven digits with the @ wildcard. Also, Cisco CallManager provides a seven-digit-dialing route filter that filters LOCAL-AREA-CODE DOES-NOT-EXIST.

Prevent Excessive Forwarding

Excessive hop count (phone forwarded too many times) results in a *forward loop*. For example, a call to phone A is forwarded to phone B. Phone B has designated all calls be forwarded to phone A. When the call is forwarded from phone A to phone B, it is forwarded to phone A again, resulting in a forward loop. The number of times a call can be forwarded is controlled by the **MaxForwardsToDn** service parameter (**Service > Service Parameters >** *select a server > click on the Cisco CallManager Service in the list box at left*). The default setting is 12; Cisco CallManager will attempt to forward the call to the same number only 12 times. After 12 attempts, Cisco CallManager drops the call.

Restrict the Calling Search Space/Partition Fields to 1024 Characters

One CSS cannot have a number of partitions whose character length is equal to or greater than 1024. In other words, a CSS with 100 partitions with a 15-character name would exceed the field size for that one CSS.

The character limit also includes one extra character per partition for the colon that separates each partition name. Also, the CSS on a phone is concatenated with the CSS on a line resulting in possibly a longer CSS if either one contains partitions the other doesn't.

Service Guidelines

This section discusses the best practices that apply to service parameters in Cisco CallManager Administration (CCMAdmin).

Verify Consistent Service Parameters on Each Cisco CallManager in the Cluster

Service parameters (**Service > Service Parameters**) are configured on a per-Cisco CallManager basis. Although this is a tedious task, you must ensure that service parameters are consistently applied throughout the entire cluster.

Enterprise parameters are applied to all nodes in a cluster.

Update New Server Names After Installation

The URL for the directory is configured in the Enterprise Parameters Configuration page (**System > Enterprise Parameters**). When a new server is installed, it will have the following URL by default:

http://<*Server_name*>/CCMUser/xmldirectory.asp

If you are not using DNS or have not configured DNS, you need to edit this parameter and change the <*Server_name*> to the IP address of the server.

Verify Transcoder Availability for Devices with Multiple Codecs

If a cluster has multiple codecs, be sure that devices have access to a transcoding resource. Ensure that a transcoding resource is part of the media resource group and media resource group list that is assigned to devices.

Ensure that Transcoders and Voice Mail 'Talk' G.711 to Each Other

Place the voice mail server (G.711) in the same region that the transcoder is in and ensure that G.711 is the codec being used within that region.

Build Media Resource Groups/Lists

Build media resource groups and media resource group lists to ensure that all devices have access to the media resources you configured. If new media resources are added to a cluster, Cisco CallManager will build a "default MRG/MRGL" that any devices not assigned a MRGL will access.

Feature Guidelines

This section discusses the best practices that apply to the Feature menu in CCMAdmin. This section includes a suggestion for call pickup considerations.

Prevent Excessive Directory Numbers with Shared Line Appearances in a Pickup Group

Too many directory numbers with shared line appearances in a call pickup group results in lots of phones ringing. Be certain that the call pickup groups are efficiently organized so that only the necessary users, located in close geographic proximity to each other, are included in a given call pickup group. Phones with directory numbers belonging to a call pickup group need to use a distinctive ring and be within hearing of other phones in that group.

Device Guidelines

This section discusses the best practices that apply to the Device menu in CCMAdmin. This section includes tips for gateways, CTI ports, Cisco IP Phones, gateways, directory numbers, and voice mail.

Be Aware of Devices That Do Not Use Default Loads

The Firmware Load Information page (**Device > Firmware Load Information**) displays devices that are not using the default device loads. Go to **Device** > **Firmware Load Information** to view this page, as shown in Figure 13-2.

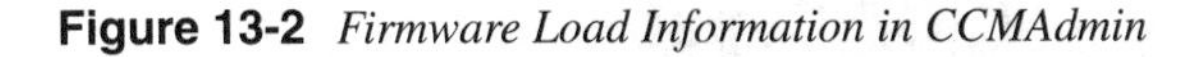

Figure 13-2 *Firmware Load Information in CCMAdmin*

System Route Plan Service Feature Device User Application Help

Cisco CallManager Administration
For Cisco IP Telephony Solutions

Cisco Systems

Firmware Load Information

Device Defaults Configuration

Device Type	Device Not Using Default Load
12 S	None
12 SP	None
12 SP+	None
14-Button Line Expansion Module	None
30 SP+	None
30 VIP	None
Analog Access	None
Analog Access WS-X6624	None
Conference Bridge WS-X6608	None
Digital Access	None
Digital Access WS-X6608	None
Digital Access+	None
IP Phone 7910	3 devices found (View detail...)
IP Phone 7935	None
IP Phone 7940	None
IP Phone 7960	None
Media Termination Point WS-X6608	None

It is best to have all devices use the same firmware load, except in cases where certain devices are used solely for test purposes and are not part of the production environment.

Configure All Ports of a WS-X6608-E/T1, Even If Not All Ports Are Used

Because all the E1/T1 ports on the module share the same XA processor, a port that continuously resets has the capability to reset a registered port that is next to it. For example if port 2 continuously resets, it can reset ports 1 and 3 even if ports 1 and 3 are registered, thus creating a domino effect.

Configure and register all eight ports, even if they are not being used.

Do Not Configure a Gateway in Its Own Calling Search Space

This is usually something to be concerned with when integrating with a legacy PBX or another Cisco CallManager cluster. The CSS of the gateway should not include the route pattern assigned to the gateway. That way, if there is an incoming call from another system (that is, PBX or Cisco CallManager cluster) and the cluster does not have that number, it will not route it back through the gateway to the system it just came from, resulting in a looped call.

Configure H.323 Gateways That Are MGCP Capable to Support Call Preservation

Call preservation is the capability of a gateway or device to maintain the voice stream of a call when the Cisco CallManager that it is registered to is no longer available (that is, the network failed or a Cisco CallManager service stopped). If you have H.323 gateways that are capable of using MGCP, consider configuring the gateways using MGCP because MGCP supports call preservation, whereas, H.323 does not.

MGCP does not currently support Q.931 gateways. If you are using Q.931 gateways, gateway call preservation is not possible using the practice described here.

Do Not Configure an Excessive Number of Cisco IP SoftPhones or CTI Ports

Watch the count of CTI ports per Cisco CallManager and within the cluster. Refer to *Cisco IP Telephony Design Guide*:

> www.cisco.com/univercd/cc/td/doc/product/voice/ip_tele/network/dgclustr.htm#xtocid125962

In addition, refer to Table 13-1 and Table 13-2 for specifications.

Register Cisco IP Phones to a Specific Device Pool, Instead of the Default Device Pool

The default device pool is designed for use with deployments of 2500 users or less and laboratory environments for quick set up and registration of phones. If you have a larger system, ensure that all phones are registered to a device pool other that the default device pool. In a large system, having all phones registered to a device pool other than the default will help to accurately monitor and troubleshoot your CIPT solution.

To identify phones that are using the default device pool, click **Device** > **Phone** and search for phones using device pool "default."

Ensure that Phones Have a Calling Search Space

All phones should be assigned a CSS. You can identify phones that do not have a CSS assigned by searching the list of phones in CCMAdmin (**Device > Phone**). Search for phones using the filter "calling search space" and leaving the other criteria blank. All phones that have **<None>** for their CSS will be displayed. Click on each phone that was returned in the search results and assign a CSS on the Phone Configuration page.

Check the Directory Number Configuration

Proper directory number configuration is critical. You need to check for partition assignments and forwarding designations. Directory number settings can be found on the Phone Configuration page (**Device > Phone >** *find and select a phone*). You can also use the Bulk Administration Tool (BAT) to search for and update some or all of the following issues.

Table 13-3 highlights those best practices associated with directory number configuration that you can address at the Directory Number Configuration page.

Table 13-3 *Addressing Directory Number Configuration Problems*

Issue	What to Check for
Directory numbers without partitions	Check to make sure each phone's directory numbers are assigned to a partition.
Phones without call forward busy (CFB) designation	Do all phones have a directory number to which calls are forwarded if the phone is busy? Usually the voice mail directory number is entered in this field.
Phones without a CSS for call forward no answer (CFNA)	Does the CSS on CFNA on all the phones have restrictions on it? Depending on your company needs, you might want to consider using a CSS that has restrictions to prevent a user from receiving forwarded calls that were originally placed to a toll-free number at the workplace. For example, user A could forward calls to their home number, go home, and then instruct friends and family to reach user A by dialing a company-sponsored toll-free number. User A's phone would forward calls to User A's house and bypass any cost to the user's friends or family.
For devices that have voice mail, ensure that they have CFB and CFNA to the voice mail pilot number	Ensure CFB and CFNA fields have been configured with the directory number of the voice mail system. This ensures unanswered calls are routed to a voice mail system where callers have the choice of leaving a message for the user.

Forward the Voice Mail Pilot Number and Subsequent Voice Mail Ports

Verify that the voice mail pilot number has a CFB and CFNA designation to the next voice mail port in the system. Depending on the size of your system, you may have several voice mail ports. You want the pilot number to forward to a voice mail port, and that voice mail port to forward to another voice mail port, and so on. This ensures that calls destined for voice mail are routed to an available port.

For example, the voice mail pilot number is 2000 and there are five additional voice mail ports, with a directory number range from 2001 through 2005. Directory number 2000 will have directory number 2001 as its CFB and CFNA designations. Likewise, directory number 2001 will have directory number 2002 as its CFB and CFNA designations, and so forth. The number of voice mail ports needed depends on the size of your system.

You can configure the CFB and CFNA designations for voice mail ports by clicking **Device > Cisco Voice Mail Ports** and clicking on the voice mail ports in the left box to view or configure forwarding designations.

Do Not Run Cisco Messaging Interface (CMI) on More than One Cisco CallManager Server in a Cluster

Running more than one CMI service in a Cisco CallManager cluster can result in voice mail integration corruption. Only one CMI should be running in a cluster. If any other servers are running this service, disable CMI via Cisco CallManager Serviceability or using the **Windows 2000 Administrative Tools > Services**. Stopping the CMI service via Cisco CallManager Serviceability only stops the service for this session; using Windows 2000 Services, you can change the startup properties so that it does not restart next time you boot the server.

Ensure a Transcoder Is Available If Voice Mail Is Used

Some voice mail systems will use only G.711 because of voice quality issues and the CPU intensive nature of supporting codecs other than G.711. If this is the case, be sure that devices that need to connect to the voice mail system have a transcoder in the MRGL that is assigned to that device.

Cisco Unity supports G.729 and G.711, so a transcoding resource is generally not needed.

User Guidelines

This section discusses the best practice that applies to the User menu in CCMAdmin.

Ensure Cisco IP SoftPhone Users Have Been Assigned a CTI Port

To be able to use Cisco IP SoftPhone, a user must be assigned a CTI port. CTI ports are allocated to users who have the **CTI Enabled** box checked in the User Configuration page in CCMAdmin (**User** > **Global Directory** > *click* **Search** *for a complete list or enter search criteria* > **Search** > *click on the user you want to view configuration information for*).

Application Guidelines

This section discusses the best practices that apply to the applications you may have installed on CCMAdmin, including Cisco WebAttendant, BAT, and the Administrative Reporting Tool (ART, also known as CAR, the CDR Analysis and Reporting tool, available in Cisco CallManager Serviceability).

Ensure Unique WebAttendant Directory Number

When you are configuring the Cisco WebAttendant's directory number, be sure that this directory number is unique throughout the cluster.

Ensure No Forwarding Designation Has Been Configured on WebAttendant Directory Numbers

On the Cisco IP Phone that will be used in conjunction with Cisco WebAttendant, be certain that the directory numbers on that phone are not configured with CFB or CFNA directory numbers. The configuration of the hunt group will do this work.

Prevent Loops in Hunt Groups

Do not configure the last member of a Cisco WebAttendant hunt group with the first member of the same hunt group. Doing so creates a loop that can cause Cisco WebAttendant to not work as designed. The last member of the hunt group should always be the voice mail pilot number or the pilot number of the Auto Attendant application.

Best Cisco WebAttendant Configuration Practices

To ensure optimum Cisco WebAttendant performance, disable call waiting on all Cisco WebAttendant lines. Also, you cannot use shared line appearances on Cisco WebAttendant directory numbers.

Enable Service Parameters for Administrative Reporting Tool (ART)/CDR Analysis and Reporting (CAR)

To use the reporting features in ART/CAR (the name changed depending on which release you are using), you must enable two Cisco CallManager service parameters in CCMAdmin: Call Diagnostics Enabled and CDREnabled. To do so, click **Service > Service Parameters** > *select a server* > *click on the Cisco CallManager Service in the list box at left* > *enable the specified parameters*. The Service Parameters Configuration page displays, as shown in Figure 13-3.

Figure 13-3 *Service Parameters Configuration in CCMAdmin*

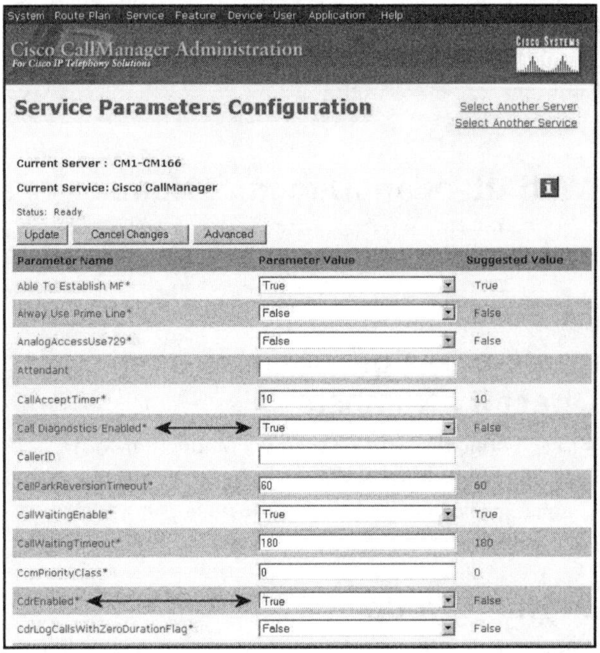

System Route Plan Service Feature Device User Application Help

Cisco CallManager Administration
For Cisco IP Telephony Solutions

Service Parameters Configuration

Select Another Server Select Another Service

Current Server : CM1-CM166
Current Service: Cisco CallManager
Status: Ready

Update | Cancel Changes | Advanced

Parameter Name	Parameter Value	Suggested Value
Able To Establish MF*	True	True
Alway Use Prime Line*	False	False
AnalogAccessUse729*	False	False
Attendant		
CallAcceptTimer*	10	10
Call Diagnostics Enabled*	True	False
CallerID		
CallParkReversionTimeout*	60	60
CallWaitingEnable*	True	True
CallWaitingTimeout*	180	180
CcmPriorityClass*	0	0
CdrEnabled*	True	False
CdrLogCallsWithZeroDurationFlag*	False	False

Use the Bulk Administration Tool (BAT)

Simply using BAT is a best practice. BAT was designed to relieve the monotonous and time-consuming process involved when adding, updating, deleting, or searching for devices and users. For detailed information on how BAT can help you save time and make managing your system faster and easier, review Chapter 7, "Understanding and Using the Bulk Administration Tool (BAT)," and the BAT documentation at

http://www.cisco.com/univercd/cc/td/doc/product/voice/sw_ap_to/admin/bulk_adm/index.htm

Summary

This chapter supplied some guidelines to follow when deploying your CIPT solution. These tips are an accumulation of numerous installations and pages of feedback from deployment teams around the world that can help make for a successful deployment of the CIPT network.

Use this chapter as a final checklist for all deployments. Your support and monitoring team will thank you.

Applications

Upon completing this chapter you will be able to do the following tasks:

- Given a Cisco CallManager server and a second PC, install and configure Cisco WebAttendant.
- Given a Cisco CallManager server, install and configure Cisco IP SoftPhone.
- Given a Cisco CallManager server and a Cisco Unity server, integrate Cisco Unity with Cisco CallManager.

CHAPTER 14

Applications

Cisco and third-party companies are developing numerous applications to integrate with Cisco CallManager and the discussion of some of these applications is beyond the scope of this book. This chapter focuses on three of the primary applications available from Cisco that integrate with Cisco CallManager—Cisco WebAttendant, Cisco IP SoftPhone, and Cisco Unity. This chapter covers the following topics:

- Abbreviations
- Cisco WebAttendant
- Cisco IP SoftPhone
- Cisco Unity

Pre-Test

Do you already know this information? The pre-test is designed to help you gauge your knowledge about this chapter. Of the 10 questions, if you answer one to three questions correctly, we recommend that you read through this chapter. If you answer four to seven questions correctly, we recommend that you read this chapter, reading those sections that you need to know more about. If you answer 8 to 10 questions correctly, you probably understand this information well enough to skip this chapter. You can find the answers to these questions in Appendix B, "Answers to Chapter Pre-Test and Post-Test Questions."

1 Which Cisco service must be running on the Cisco CallManager server for Cisco WebAttendant to work?

2 What is the purpose of the pilot point directory number as it relates to Cisco WebAttendant?

3 What should the last member of a Cisco WebAttendant hunt group be?

4 As an administrator, what must you enable for a user to be able to use Cisco IP SoftPhone?

5 What information does the user need to install the Cisco IP SoftPhone application on their PC?

6 What does Cisco Unity use as the message store and directory service to unify your system administration?

7 What should be configured first in Cisco CallManager to prepare for integration with Cisco Unity?

____

____

____

List three of the four service parameters in Cisco CallManager that need to be configured for integration with Cisco Unity.

8

____

____

____

9

____

____

____

10

____

____

____

Abbreviations

This section defines the abbreviations used in this chapter. For more information about terms and abbreviations used in this chapter refer to the *IP Telephony Network Glossary* at the following URL:

www.cisco.com/univercd/cc/td/doc/product/voice/evbugl4.htm

Table 14-1 provides the abbreviation and the complete term for abbreviations used frequently in the chapter.

Table 14-1 *Abbreviation with Definition*

Abbreviations	Definitions
BLF	busy lamp field
CCM	Cisco CallManager
CCMAdmin	Cisco CallManager Administration
CSS	calling search space
CTI	computer telephony interface
DirN or DN	directory number
DSS	direct station select
GUI	graphical user interface
HTML	Hypertext Markup Language
LAN	local-area network
LDAP	Lightweight Directory Access Protocol
MAC	media access control
MWI	message waiting indicator
PBX	private branch exchange
PSTN	public switched telephone network
RAID	Redundant Array of Inexpensive Disks
TAPI	Telephony Application Programming Interface
TCDSRV	Telephony Call Dispatcher Service
WAN	wide-area network
XFER	Transfer

Cisco WebAttendant

Cisco WebAttendant is a Web-based client/server application that supports the traditional role of a manual attendant console. By associating with a Cisco IP Phone, Cisco WebAttendant enables a receptionist or operator to quickly answer and direct calls to enterprise users. An integrated directory service provides traditional BLF and DSS functions for any line in the system. The application is Web-based and, therefore, portable to Windows 98, NT, and 2000 platforms.

Cisco WebAttendant is the first of many new IP telephony applications that integrate Old World telephony functions with New World applications and services such as LDAP directory and HTML services. A primary benefit of Cisco WebAttendant over traditional attendant console systems is its capability to monitor the state of every line in the system and to efficiently dispatch calls. The absence of a hardware-based line monitor device offers a much more affordable and distributable manual attendant solution than traditional consoles.

Cisco WebAttendant more efficiently automates both the user operations and the administrative operations of a manual attendant function. Cisco WebAttendant uses a Web-based interface for call handling and line-state monitoring. In the next section, Figure 14-1 shows the Cisco WebAttendant user interface. The lower half of the display provides a directory of all users in the system. The state of each user's primary line appearance is presented with each user record entry. The benefit over Old World consoles and line extenders is that each user's line is monitored, as opposed to monitoring only a select few in an Old World system.

Drag-and-drop capabilities and access to corporate LDAP directories combine to offer key advantages over Old World manual-attendant stations. In a system with hundreds or thousands of users, the Cisco WebAttendant operator accepts calls and performs directory lookup by selecting the field title in the directory section and typing the first few characters of the user's extension, last name, first name, or department. A directory search that matches the query is returned. The operator can view the status of the user's line (busy or available) and advise the caller of the line state. The operator can then transfer the call to the user either through the Transfer (**XFER**) button or by dragging and dropping the call from the selected loop to the desired user's record. The primary benefit of this user interface is quicker transaction time and subsequent customer satisfaction improvement.

Cisco WebAttendant is scalable. Call distribution groups can be assigned to any pilot number, which can in turn be assigned to one or more Cisco WebAttendant loops. These loops represent answerable lines in a multiple attendant system. Calls are dispatched to one or more online attendants' loops, thereby enabling scalability and distribution among multiple operators. Multiple Cisco WebAttendants may be configured to monitor the same lines, affording scale to multiple operators when conditions require.

Cisco WebAttendant Prerequisites

The following prerequisites must be met before you can begin using Cisco WebAttendant:

- All phones, users, and resources (such as conference rooms with phones) must be added in CCMAdmin.
- The TCDSRV must be activated and running on Cisco CallManager.
- Pilot points, hunt groups, and Cisco WebAttendant users must be configured in CCMAdmin.
- Directory numbers assigned to a Cisco IP Phone associated with Cisco WebAttendant must be unique throughout the system; they cannot appear on any other device in the system.

- The phone button template for the Cisco IP Phone that is associated with Cisco WebAttendant must have **Hold**, **Transfer**, and **Answer/Release** buttons defined. (These are provided by default on Cisco IP Phone models 7960 and 7940.) Up to eight lines can be defined for use with each Cisco WebAttendant.

Cisco WebAttendant client requirements:

- Microsoft Internet Explorer 5.0 or higher Web browser with ActiveX enabled.
- Platform and operating system: PC running Microsoft Windows 95, Windows 98, Windows NT 4.0 (Service Pack 3 or higher), and Windows 2000 workstation or server.
- Display adapter color palette setting: Minimum of 256 colors; select 16-bit color or higher for optimal display.

Cisco WebAttendant User Interface

Figure 14-1 shows the Cisco WebAttendant user interface. The following are the features of this interface.

Figure 14-1 *Cisco WebAttendant User Interface*

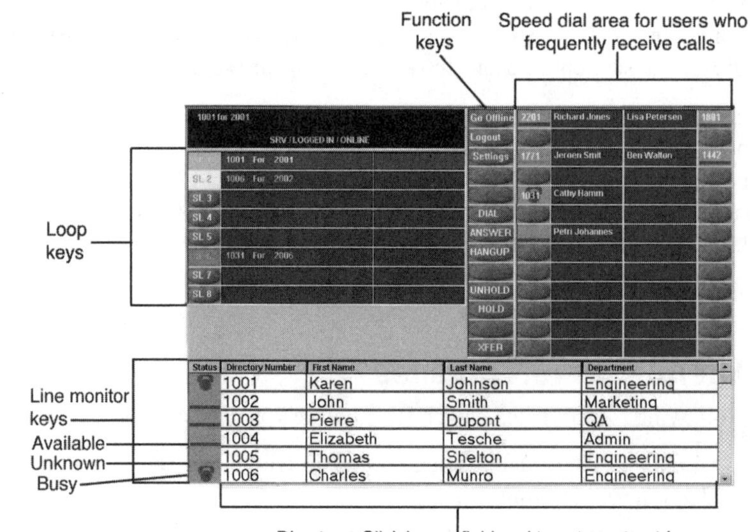

- Loop keys (simultaneous management of lines)
- Line monitor keys—26
  — Line states—Busy, available, and unknown
  — User label per line monitor key for easy reference to user

- Function keys
  - — Log on, log off
  - — System supplementary features—Hold, resume, transfer, call waiting
  - — Answer/Hang Up
- Directory
  - — Line state—One record for every line appearance in the Cisco CallManager cluster
  - — Query—Fields are searchable by last name, first name, extension, department
- Per-call drag-and-drop transfer and hold—Drag a call from the loop key to the line monitor key or directory record for transfer; drag a call from the loop key to the hold key for hold
- Area for users who receive frequent calls or whose lines the receptionist needs to monitor

Cisco WebAttendant also provides headset capabilities of Cisco IP Phones and keyboard shortcuts as an alternative to mouse operation.

The pane in the top left corner is the display pane showing the state of the client when connected to TCDSRV with the text "SRV," the login state by the text "LOGGED IN," and the online state by the text "ONLINE." When online, the state text appears in green, when offline but logged in, the state text appears in yellow, and when not logged in, the state text appears in red. The upper left corner of the display pane shows the current call state such as "3001 for 0," meaning directory number "3001" is calling directory number "0." The display pane also displays digits currently in the dial string buffer. The *dial string buffer* contains digits the attendant has entered using the numeric keypad.

Online/Offline refers to the Cisco WebAttendant user's ability to receive (online) or not receive (offline) calls at the attendant's phone. Logged In/Logged Out refers to whether the Cisco WebAttendant user is able to control an attendant phone. Only users who have logged in can control the attendant phone.

The attendant lines pane is located just under the display pane with lines labeled SL1 through SL8, assuming the attendant's phone has eight lines. The currently selected line shows a lighter shade of blue when selected. The SL button reverses color when the state of the line is "OFFHOOK" or will be red when "ONHOLD."

Blink rates with colors indicate various call states. A slow blink rate consists of an on/off frequency of ≈ 1 Hz with a duty cycle of 50 percent. A fast blink rate consists of an on/off frequency of ≈ 16 Hz with a duty cycle of 50 percent. The blink rate events are as follows:

- A slow blink rate in yellow indicates a ringing attendant line.
- A slow blink rate in red indicates an attendant line with a call on hold.

- A fast blink rate in red indicates an attendant line with a call on hold with another active call or call with dial tone.

The bottommost grid is the directory pane. The directory pane shows the system directory, including users' directory numbers, first names, last names, and departments. A status column shows the line state for each directory number in the directory. A blue underline indicates an IDLE state, meaning the directory number is ready to accept a call; a red line indicates an UNKNOWN state, meaning the state of the directory number cannot be determined (it may not be registered with Cisco CallManager); and a telephone icon indicates an OFFHOOK state, meaning the directory number is already is use.

The attendant can search for a user by placing the cursor in one of the columns and typing a few characters. The contents of the column sort by the characters typed. For example, if an incoming caller requested Mr. Jones, the attendant would place the cursor in the Last Name column and start typing "jon" and watch the names as they sort. Last names starting with the letters entered would sort to the top of the list, making it easy for the attendant to locate the requested party.

The function keys are a column of feature buttons to the right of the display pane. The feature button pane consists of the following buttons:

- **Go Online/Offline**—This button toggles between the online and offline state. The online state refers to the state in which the Cisco WebAttendant user is able to receive calls at the attendant's phone.
- **Login/Logout**—This toggle button changes the login state of the Cisco WebAttendant user. A Cisco WebAttendant user cannot control an attendant phone without being logged in.
- **Settings**—This button opens a configuration dialog for the Cisco WebAttendant client.
- **DIAL**—This button initiates a call from the currently selected attendant line (also known as the button name SL) to the directory number currently in the dial string buffer.
- **ANSWER**—This button answers an incoming call to the currently selected SL, the same functionality as answering a call on the attendant's phone.
- **HANGUP**—This button disconnects a call on the currently selected SL, the same functionality as disconnecting a call on the attendant's phone.
- **UNHOLD**—This button retrieves the call on the currently selected SL, removing it from the HOLD state. This is the same functionality as pressing the **Resume** soft key on the attendant's phone.
- **HOLD**—This button places a call on the currently selected SL on hold. This is the same functionality as pressing the **Hold** soft key on the attendant's phone.
- **XFER**—This button transfers the call on the currently selected SL to the directory number entered by the attendant (the number currently in the dial string buffer).

The speed dial pane is located to the right of the feature button pane. The attendant configures the speed dials so that they contain directory numbers and labels for the most active users. By clicking any speed dial button, a call is placed from the currently selected SL to the directory number associated with that speed dial.

Cisco WebAttendant Administrator Features and System Capabilities

The administrator features of Cisco WebAttendant are as follows:

- Remote system/device installation and configuration through an Internet Explorer version 5.0 or higher Web browser.
- Simultaneous line monitoring by multiple operators. Any operator can view the state of any line from his or her Cisco WebAttendant.
- Call distribution from a single pilot number to multiple directory numbers or user-line pairs.
- Simultaneous monitoring of incoming calls from multiple operator positions.
- Creation of up to 32 pilot numbers/distribution groups.

The system capabilities of Cisco WebAttendant are as follows:

- Scalability
    - 32 hunt groups (pilot numbers) per Cisco CallManager cluster.
    - 16 hunt group members per hunt group.
    - 8 call loops per WebAttendant. Any loop may be assignable as a hunt group member.
- 96 Cisco WebAttendant clients per cluster
- Maximum of 16 * 32 = 512 simultaneous calls on as many as 96 configured WebAttendants
- Availability—Provision for multiple operators on same loop/pilot and monitoring same line; if an operator station fails or is offline, calls are distributed to all other operators with the same loops
- Manageability—System device configuration and operation through a Web interface; no operator applications to install at each operator's PC
- Affordability—No line extender hardware devices required.

Cisco WebAttendant TCDSRV

Figure 14-2 depicts the WebAttendant and TCDSRV within the Cisco CallManager system and shows the interconnections between TCDSRV and Cisco CallManager using CTI. TCDSRV is the call control server for the Cisco WebAttendant client and communicates with Cisco CallManager via CTI, just as TAPI is a call control server for TAPI applications and communicates with Cisco CallManager via CTI.

Figure 14-2 *Cisco WebAttendant Architecture Overview*

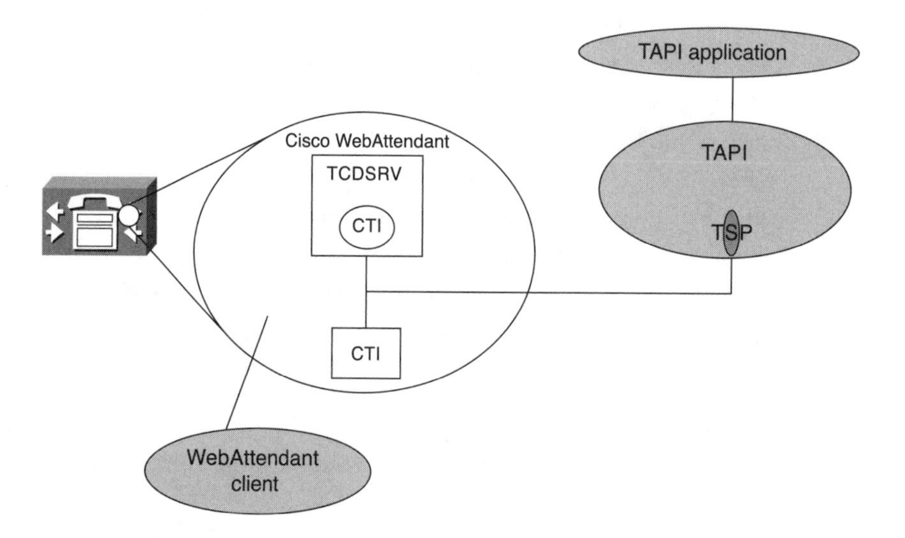

TCDSRV provides centralized services for WebAttendant clients and pilot points. For Cisco WebAttendant clients, TCDSRV provides call control functionality, line state information for any accessible line within Cisco CallManager(s) domain, and caching of directory information. For pilot points, TCDSRV provides automatic redirection to directory numbers listed in hunt groups and N+1 failover on Cisco CallManager failure.

Following is a list of TCDSRV features and requirements:

- **Pilot Point**—A pilot point is a CTI route point with an associated hunt group. A call placed to the pilot point directory number is accepted by TCDSRV, which then redirects the call to the first available directory number in its associated hunt group.
- **Call Control**—A full range of call control functions consisting of call answer, call origination, call redirect, call hold, call resume (unhold), and call disconnect (hang up). These operations occur through third-party call control of Cisco IP Phones.
- **Line State Information**—Line state information consists of line available, line on-hook, and line off-hook.

- **Caching of Directory Information**—TCDSRV caches directory information for fast lookup and quick response times for Cisco WebAttendant clients. The cache is automatically updated each day.
- **Pilot Point N+1 Failover/Recovery**—TCDSRV also provides N+1 failover for pilot points in the event of Cisco CallManager failure. A web of TCDSRV services communicates with one another to implement a nearby neighbor or least distance failover, handing off pilot point servicing to a nearby neighbor TCDSRV. Failover recovery returns pilot point servicing to the original TCDSRV upon restoration of the failed Cisco CallManager server. For example, given a configuration of two CallManagers referred to as CCM1 and CCM2 and two corresponding TCDSRVs referred to as TCD1 and TCD2, the following scenario shows how the failover and recovery works:
  - — TCD1 is servicing route point 2000.
  - — TCD1 detects loss of communications with CCM1.
  - — TCD1 is aware of the presence of TCD2 and requests TCD2 to take over servicing of route point 2000.
  - — TCD2 accepts and opens route point 2000.
  - — Calls destined for route point 2000 are now handled by CCM2.
  - — TCD1 now detects that CCM1 is back.
  - — TCD1 requests TCD2 to relinquish control of all route points previously failed over.
  - — TCD2 closes route point 2000 and acknowledges request.
  - — TCD1 reopens route point 2000 and restores original service.

Installing and Configuring Cisco WebAttendant

This section provides instructions for installing and configuring Cisco WebAttendant. Following is an overview of the steps you must perform:

Step 1 Verify TCD is running on the intended Cisco CallManager.

Step 2 Configure a phone for use with WebAttendant.

Step 3 Establish a shared database.

Step 4 Add a Cisco WebAttendant user.

Step 5 Add and configure pilot points.

Step 6 Add and configure hunt groups.

Step 7 Install and configure the WebAttendant client.

Cisco WebAttendant Configuration Step 1: Verify TCD is Running on the Intended Cisco CallManager

The Cisco TCDSRV starts running automatically when Cisco CallManager is started. The following procedure describes how to verify that the Cisco TCD service is running and how to start Cisco TCD if it is stopped.

NOTE If you add new Cisco WebAttendant users or modify the user information or password for an existing user, you must wait approximately six minutes for the changes to take effect.

Step 1 In CCMAdmin, choose **Application > Cisco CallManager Serviceability**.

Step 2 Choose a Cisco CallManager server from the server list on the left side of the pane. The pane refreshes.

The Service Name column lists all services that are configured on this server.

Step 3 Look at the Service Status column for the Cisco Telephony Call Dispatcher:

— If an arrow icon appears, the Cisco TCD service is running.

— If a square icon appears, the Cisco TCD service is stopped.

Step 4 If the Cisco TCD service is not running, click the **Start** button in the Service Control column.

Cisco WebAttendant Configuration Step 2: Configure a Phone for Use with Cisco WebAttendant

Review the "Adding and Configuring a Cisco IP Phone" section in Chapter 6, "Cisco IP Telephony Devices," for instructions on configuring a phone to use in conjunction with Cisco WebAttendant.

Cisco WebAttendant Configuration Step 3: Establish a Shared Database

You must create a share for C:\Program files\Cisco\users called "wausers." By default, the client uses cached directory information from the Cisco CallManager Directory user database. The Cisco WebAttendant client displays user and line information in the Directory section of its user interface.

If you choose the default Cisco TCD database setting during Cisco WebAttendant client configuration, you must perform the following procedure to ensure that the Cisco WebAttendant client can display the directory information from the Cisco CallManager directory database.

NOTE When you check the **Full Control** check box, you automatically choose all available permission selections.

Step 19 Click **Apply**.

Step 20 Click **OK**.

Step 21 Perform this procedure on every Cisco CallManager in the cluster.

NOTE To ensure that the changes made to the Shared As properties are visible to Cisco WebAttendant clients, Cisco WebAttendant users should exit the client, log out of Windows, and then log back in to Windows.

Cisco CallManager automatically makes directory database information available to Cisco WebAttendant clients and updates the information every 24 hours with the latest changes.

Cisco WebAttendant Configuration Step 4: Add a Cisco WebAttendant User

Use the following procedure to add a Cisco WebAttendant user:

Step 1 Open CCMAdmin and click **Service** > **Cisco WebAttendant**.

Step 2 Click the **Cisco WebAttendant User Configuration** link in the upper right corner of the page.

Step 3 Enter the appropriate configuration settings as described in the items that follow and illustrated in Figure 14-3:

— **User ID**—Type the login name for the new Cisco WebAttendant user, up to 50 alphanumeric characters.

— **Password**—Type a password, up to 50 alphanumeric characters.

— **Confirm**—Retype the same password.

— **Station Type**—Select the user type:

Attendant—Users who plan to use Cisco WebAttendant to manage call traffic for more than one Cisco IP Phone (receptionists, administrative assistants, or secretaries).

User—Users who plan to use Cisco WebAttendant to control the phone on their desk.

NOTE If you are running Cisco CallManager in a cluster environment, perform this proce every Cisco CallManager server in the cluster.

Perform the following steps to set up the wauser shared directory:

Step 1 Log in to the Cisco CallManager server.

Step 2 Use Windows Explorer to browse to the following folder: C:\Program Files\Cisco\Users.

Step 3 Right-click the **Users** folder and choose **Sharing**.

Step 4 Click the **Share this Folder** radio button.

Step 5 Change the default share name from "Users" to "wausers." Make sure the share name, which is not case-sensitive, is wausers, so that directory information can display.

Step 6 Click the **Permissions** button. Click the **Share Permissions** tab if it does not automatically appear.

Step 7 Choose **Everyone** in the Name pane, if it is not already chosen.

Step 8 Check the **Full Control**, **Change**, and **Read** check boxes; then click **OK**.

Step 9 Click the **Security** tab.

Step 10 Uncheck the **Allow inheritable permissions from parent to propagate to their object** check box.

Step 11 A security dialog box appears. Click the **Remove** button in the dialog box.

Step 12 If the Name pane has any entries in it, choose them one-by-one and click **Remove** after each selection.

Step 13 Click **Add**.

Step 14 In the **Look In** field, choose the name of the machine you are currently using.

Step 15 Choose **Everyone** in the Name pane; then click **Add**.

Step 16 Click **OK**.

Step 17 The Users Properties Security window appears. Make sure that only **Everyone** appears in the Name pane.

Step 18 In the Permissions pane, check the **Full Control** check box.

Figure 14-3 *Cisco WebAttendant User Configuration Page*

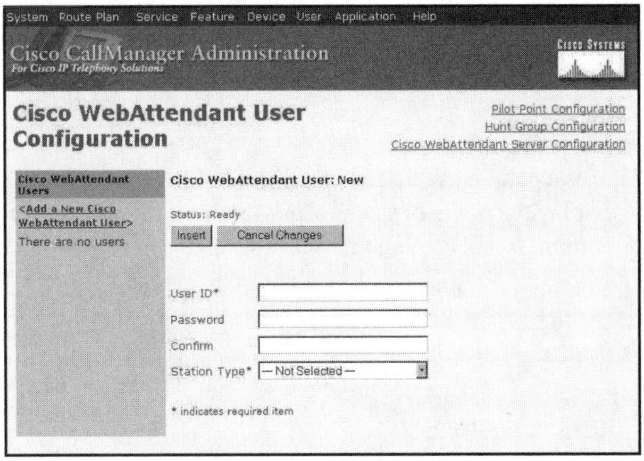

Step 4 Click **Insert** to add the new user. The Cisco WebAttendant User Configuration page refreshes and the new User ID appears in the list on the left side of the page.

Step 5 Repeat Steps 3 and 4 to add additional users.

Step 6 Restart the Cisco TCD service for changes to take effect.

Cisco WebAttendant Configuration Step 5: Add and Configure Pilot Points

Figure 14-4 is the configuration page for adding a pilot point.

Figure 14-4 *Pilot Point Configuration Page*

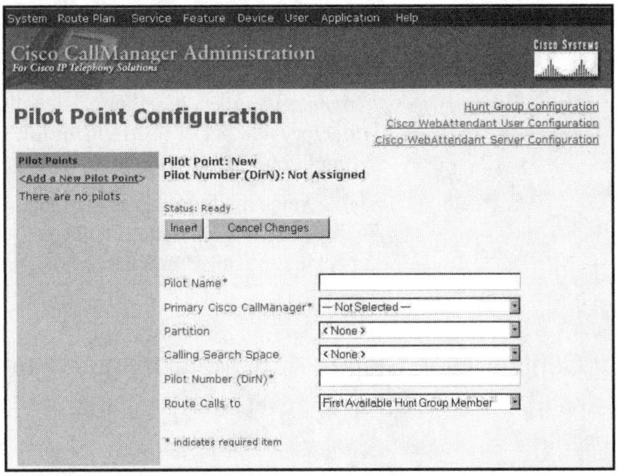

Step 1 Open CCMAdmin, click **Service** > **Cisco WebAttendant**, and then click the link to Pilot Point Configuration in the upper right corner of the page.

Step 2 Enter the appropriate settings. See Table 14-2 for definition of pilot point settings.

Step 3 Click **Insert.**

The pilot point is created and the Pilot Point Configuration page refreshes to display the name of the new pilot point in the list on the left. The new pilot point is selected and its settings appear.

Table 14-2 *Pilot Point Configuration Settings*

Field	Description
Pilot Name	Specify a descriptive name for the pilot point, up to 50 alphanumeric characters including spaces.
Primary Cisco CallManager	From the drop-down list box, choose a name or IP address of the Cisco CallManager whose Cisco TCD service will service this pilot point. When selecting the primary Cisco CallManager, take call processing and device load balancing into account.
Partition	Choose **<None>** from the drop-down list box. Cisco WebAttendant pilot points do not belong to partitions.
Calling Search Space	To designate which partitions the pilot point searches when attempting to route a call, choose a CSS from the drop-down list.
Pilot Number (DirN)	Enter a directory number in this field to designate a directory number for this pilot point. Be certain this number is unique throughout the system (that is, it cannot be a shared line appearance).
Route Calls To	To route incoming calls to the first available member of a hunt group, choose the First Available Hunt Group Member option. To order members based on the length of time that each directory number or line remains idle, choose the Longest Idle Hunt Group Member option. If the voice mail number is the longest idle member of the group, Cisco TCD will route the call to voice mail without checking the other members of the group first.

Cisco WebAttendant Configuration Step 6: Add and Configure Hunt Groups

Figure 14-5 shows the Hunt Group Configuration page.

Figure 14-5 *Hunt Group Configuration Page*

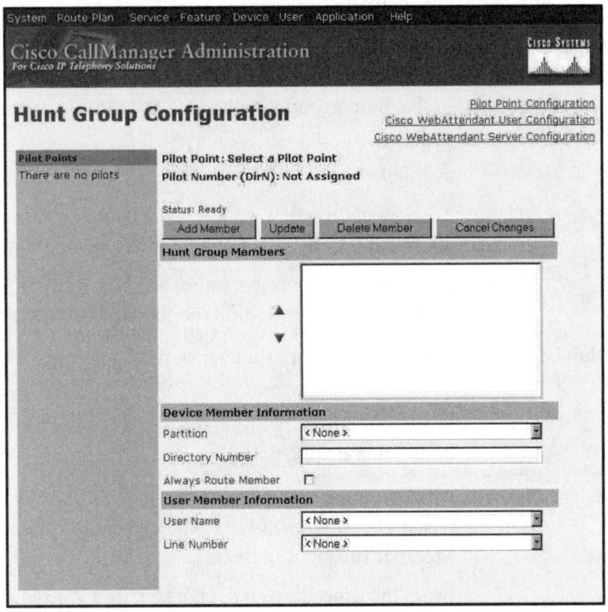

Step 1 Open CCMAdmin and click **Service** > **Cisco WebAttendant** > *select a pilot point* > *click the link to Hunt Group Configuration in the upper right corner of the page.*

Step 2 Click the **New** button. Using the setting definitions in Table 14-3, specify a directory number or specify a user name and line number and then click the **Update** button.

Step 3 When several numbers are listed, numbers can be reordered by clicking on the up or down arrows to the left of the number list followed by clicking the **Update** button.

Table 14-3 *Hunt Group Configuration Settings*

Field	Description
Partition	If a hunt group member is a directory number, complete the Partition and Directory Number fields in the Device Member Information section. This field designates the route partition to which the directory number belongs: If the directory number for this hunt group member is in a partition, you must choose a partition from the drop-down list. If the directory number is not in a partition, choose **<None>**.

continues

Table 14-3 *Hunt Group Configuration Settings (Continued)*

Field	Description
Partition (Cont.)	Always Route Member, an optional check box, applies only to directory numbers. If this box is checked, TCD always routes the call to this hunt group member, whether it is busy or not. If this box is not checked, TCD does not check whether the line is available before routing the call. To manage overflow conditions, check this box for voice mail or auto attendant numbers that handle multiple, simultaneous calls. For linked hunt groups, only check the **Always Route Member** box when you are configuring the final member of each hunt group.
Directory Number	Enter the directory number of the hunt group member device in this field. When the directory number is not in the specified partition, an error dialog box displays.
User Name	If the hunt group member is a user and line number, complete only the Cisco WebAttendant User Name and Line Number fields in the User Member Information section. From the drop-down list, choose Cisco WebAttendant users that will serve as hunt group members. Only Cisco WebAttendant user names added using Cisco WebAttendant User Configuration appear in this list.
Line Number	From the drop-down list, choose the appropriate line numbers for the hunt group.

Cisco WebAttendant Configuration Step 7: Install and Configure the Cisco WebAttendant Client

Step 1 Check to make sure that you have added to the Cisco CallManager database the Cisco WebAttendant user and the phone you wish to associate with Cisco WebAttendant.

Step 2 Write down the MAC address of the phone that is to be associated with the Cisco WebAttendant client you are installing. The MAC address is a 12-character hexadecimal number located on a label on the underside of the Cisco IP Phone.

Step 3 Log in to the PC on which you want to install the Cisco WebAttendant client.

Step 4 Open Internet Explorer (version 5.0 or higher) and browse to CCMAdmin.

Step 5 Choose **Application** > **Install Plug Ins**.

Step 6 Click on the icon for the Cisco WebAttendant client.

Step 7 The Cisco WebAttendant installation wizard runs.

Step 8 Click **Next** at the initial screen, and then click **Yes** to accept the License Agreement.

Step 9 Click **Next** to install the Cisco WebAttendant client to the default location or use the **Browse** button to specify a new location and then click **Next**.

Step 10 Select a program folder and click **Next**.

Step 11 Enter the following information on the Customer Information screen:

— Login ID: Enter the Cisco WebAttendant user ID for the attendant.

— Password: Enter the Cisco WebAttendant password for user ID specified above.

Step 12 Click **Next**.

Step 13 Enter the following information:

— IP Address: IP address or host name of the primary Cisco CallManager for TCD (usually the Cisco CallManager that the Cisco WebAttendant's phone is registered with).

— MAC ID: MAC address (using uppercase letters) of the Cisco IP Phone that will be used with Cisco WebAttendant. (See Step 2 for a description of the MAC address.) You must use uppercase letters when entering the MAC address.

Step 14 Click **Next**.

Step 15 After the installation program finishes installing files, select whether you want to restart the computer now or later. Click **Finish**.

Step 16 Restart the computer.

Step 17 Once the application is installed, you can configure or update any client settings that you did not configure during the installation process.

The Settings dialog shown in Figure 14-6 contains entries for setting up Cisco WebAttendant to be associated with a particular attendant Cisco IP Phone.

Figure 14-6 *Cisco WebAttendant Client Settings Dialog Box*

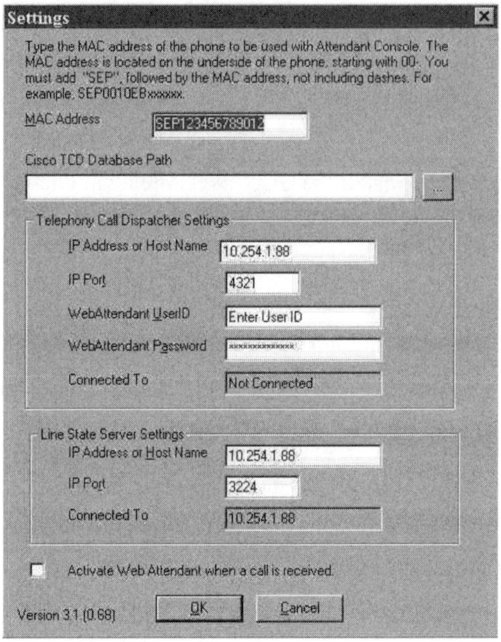

The first entry is the MAC address of the phone to be used as the attendant's phone. The next entry is a path to an optional directory database. When left blank, the directory will use the LDAP directory currently specified in the CallManager setup. Under the Telephony Call Dispatcher Settings, the IP Address or Host Name is the address of the Cisco CallManager this client will connect to. The IP Port defaults at 4321 and should not be changed, and the WebAttendant User ID and WebAttendant Password comprise the Cisco WebAttendant user setup in Cisco CallManager under Cisco WebAttendant users. Under the Line State Server Settings, the IP Address or Host Name is the IP address of the Cisco CallManager server this client will use. The IP Port defaults to 3224 and should not be changed.

The Connected To fields for Telephony Call Dispatcher and Line State Server are read-only fields that indicate which Cisco CallManager the WebAttendant client is connected to. They can be different and are used as a tool for administrative debugging.

Cisco IP SoftPhone

The Cisco IP SoftPhone is a virtual telephone that operates from a laptop or desktop PC. It controls your hardware IP telephone and features an intuitive user interface and context-sensitive controls. Because the Cisco IP SoftPhone integrates with Microsoft's NetMeeting, advanced multimedia collaboration tools are right at your fingertips with a single click.

The Cisco IP SoftPhone takes advantage of LDAP services that are part of Cisco AVVID. Calling a user is now as simple as looking up names in a directory and dragging and dropping that information into the Cisco IP SoftPhone. And with your personal directory/phone book, you can always find your contact list and connection information, even if you are not connected to a main directory server.

Cisco IP SoftPhone Features and Benefits

The Cisco IP SoftPhone provides standard telephone features as well as advanced benefits available only with the integration of Cisco AVVID networks and PC applications:

- Collaboration integration. The Cisco IP SoftPhone integrates with Microsoft NetMeeting.
- Use Cisco IP SoftPhone to connect as a software-only virtual Cisco IP Phone or work in tandem with a Cisco IP Phone.
- Users have a self-configurable display of dial pad, directory, call history, and other features.
- Drag-and-drop support for speed and ease of use.
- Cisco IP SoftPhone plays user-recorded sound files to callers.
- Record call history. The Cisco IP SoftPhone offers the capability to automatically maintain a record of your calls; called number, call timestamp, and recorded call length.
- Integrated help and contextual function keys.
- Cisco IP SoftPhone provides all the features of a desktop business telephone. It is fully integrated with the Cisco IP Phone, and both devices reflect the same, current call state.
- Users make or receive calls within a converged enterprise network or legacy technology networks (PSTN or PBX).
- Features include caller ID display, transfer (regular or blind), hold, conference, redial, voice mail integration, ringer volume, microphone volume, mute, and do not disturb.
- Call deflection. Users route their calls automatically to voice mail or another destination.
- Public and private (phone book) directory integration.
- Call initiation is performed via directory name completion or drag and drop operation.
- Users dial from a keyboard or onscreen dial pad.
- PC speakers or headset volume controls are included.

Cisco IP SoftPhone Technical Specifications

The following are the technical specifications for the Cisco IP SoftPhone:

- Requires Windows 95, Windows 98, Windows NT 4.0 (SP4 or higher), or Windows 2000
- Basic System Requirements
  - Pentium 166-MHz MMX processor (Pentium 266 MMX recommended for standalone phone mode)
  - 32 to 64 MB of RAM (depending on feature activation)
  - Up to 40 MB free disk space (depending on installation options)
  - Windows-compatible full-duplex sound card (for standalone phone mode)
- Standards Supported
  - TAPI compliant
  - T.120 (via NetMeeting integration)
  - H.323
  - G.711, G.723.1, and G.729a codec support
- Installable from CD-ROM or over a network
- InstallFromTheWeb package

Installing and Configuring a Cisco IP SoftPhone

Figure 14-7 shows the Phone Configuration page for adding a Cisco IP SoftPhone.

An overview of the configuration steps is as follows:

Step 1 Add a CTI port for the phone, when applicable.

Step 2 Add or enable users and associate them to CTI ports.

Step 3 Install and configure Cisco IP SoftPhone.

Step 4 Start Cisco IP SoftPhone and add directories.

NOTE Cisco IP SoftPhone can be used to control a hardware Cisco IP Phone or used in conjunction with a hardware Cisco IP Phone for virtual phone functionality. If a user plans to use Cisco IP SoftPhone only to control the hardware Cisco IP Phone, that user does not need to be assigned a CTI port.

Figure 14-7 *Phone Configuration Page (CTI Port)*

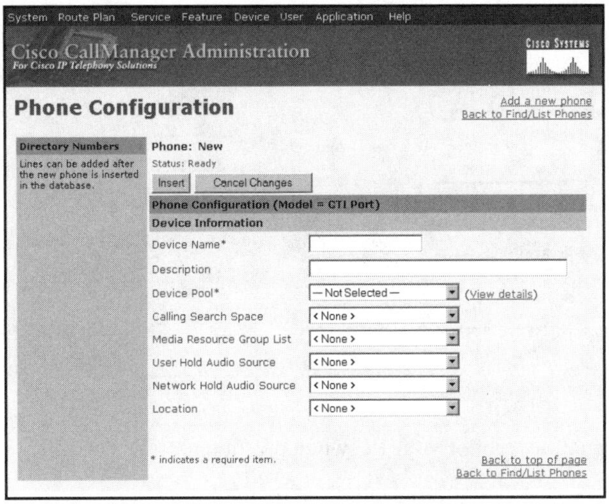

Add a CTI Port for the Phone if Applicable

Configure a phone in CCMAdmin. When configuring a Cisco IP SoftPhone in CCMAdmin, select CTI port as the phone type. Enter the device name, device pool, and the directory number. The directory number can be the primary line of the controlled phone. After the phone is configured, you must associate a user.

Add or Enable Users and Associate Them to CTI Ports

When a user is created or a user is going to be a user of a CTI type application such as Cisco IP SoftPhone, check the box for **Enable CTI Application Use**. If this box is not checked for the user, the user will not see lines when he or she launches the Cisco IP SoftPhone application. After selecting a user, assign the Cisco IP SoftPhone device previously configured.

After the Cisco IP SoftPhone device is assigned to a user, the user then needs to install the Cisco IP SoftPhone application on their associated PC.

Install and Configure Cisco IP SoftPhone

Run the install package on the PC on which you want to install Cisco IP SoftPhone. Enter the user name and password, which should match the user logon user name and password as shown in Figure 14-8. The IP address will be the Primary Cisco CallManager IP address, which you can obtain from a Cisco IP Phone 7960 by pressing **settings**, **3** and then **21**.

Figure 14-8 *Cisco IP SoftPhone Installation*

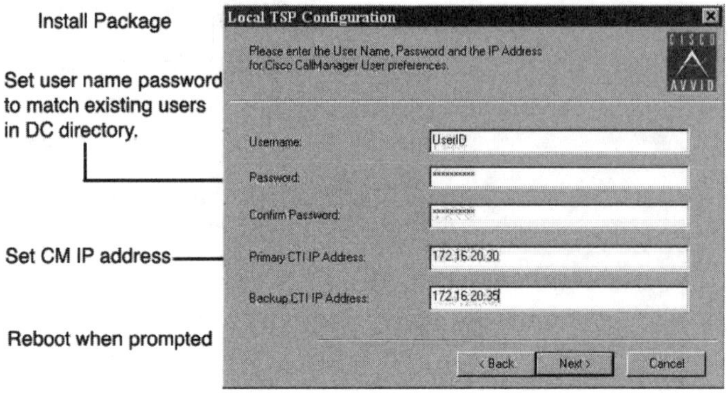

Install Package

Set user name password to match existing users in DC directory.

Set CM IP address

Reboot when prompted

Local TSP Configuration

Please enter the User Name, Password and the IP Address for Cisco CallManager User preferences.

Username: UserID

Password:

Confirm Password:

Primary CTI IP Address: 172.16.20.30

Backup CTI IP Address: 172.16.20.35

< Back | Next > | Cancel

After installation, reboot your PC when prompted.

If changes are needed after installation, do the following:

Step 1 From the Windows Control Panel, open the Telephony applet (Windows 95/98/NT) or Phone and Modem Options applet (Windows 2000).

Step 2 Click the **Telephony Drivers** tab (Windows 95/98/NT) or **Advanced** tab (Windows 2000).

Step 3 Select **Cisco IP PBX Service Provider** or **CiscoTSP001.tsp** (depending on the version of Cisco IP SoftPhone you are using) in the selection box and click **Configure**.

The IP address for the Cisco CallManager you are using with Cisco IP SoftPhone will be displayed in the CallManager Location section of the Cisco IP PBX Service Provider window. If this is not the same IP address that your Cisco IP Phone is using, you will need to edit it to match your Cisco IP Phone's IP address.

The user name and password configured for Cisco IP SoftPhone must be exactly the same as the user name and password assigned to the user in Cisco CallManager. In the Security section, type the user name and password assigned to you for Cisco CallManager.

Start Cisco IP SoftPhone and Add Directories

Start the Cisco IP SoftPhone by going to **Start** > **Programs** > **Cisco IP SoftPhone**.

NOTE The first time you start Cisco IP SoftPhone, the Microsoft NetMeeting configuration program displays, followed by the Cisco IP SoftPhone Line Selection window. In the NetMeeting configuration window, follow the instructions on the screen to tune audio parameters and specify user information for your system. The collaboration features of Cisco IP SoftPhone will not work unless you configure settings for NetMeeting.

Select a line/device to control as shown in Figure 14-9. When you select a line that corresponds with a CTI port, the media for that call terminates on the Cisco IP SoftPhone so that you can use the Cisco IP SoftPhone as a standalone phone. When you select a line that corresponds with a hardware phone (it will show up as SEP<*mac address*>), the Cisco IP SoftPhone will not terminate the media stream. Instead, the Cisco IP SoftPhone just controls the hardware phone. After you have selected a line/device, the Cisco IP SoftPhone will be operational, as shown in Figure 14-10.

Figure 14-9 *Selecting Lines/Devices for Cisco IP SoftPhone*

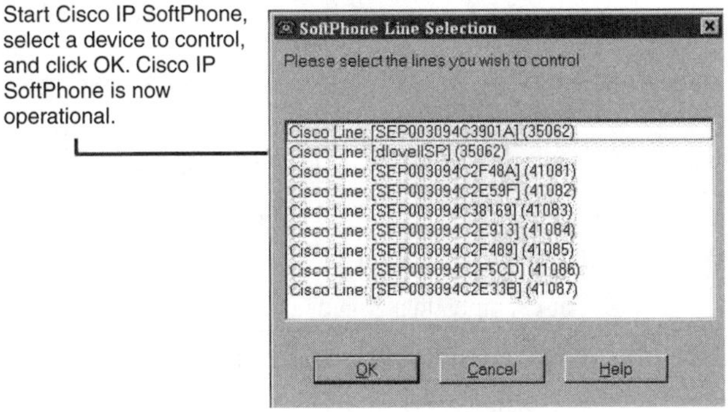

To create and use the Cisco IP SoftPhone directory, do the following:

Step 1 Click the Settings toolbar icon.

Step 2 Select the **Directories** tab.

Step 3 Select **Add**. The Directory Service window opens.

Step 4 Configure the directory settings as described in the Table 14-4.

Step 5 Click **OK**. The Directories tab reappears showing the directory name you just added.

Figure 14-10 *Cisco IP SoftPhone*

Table 14-4 *Cisco IP SoftPhone Directory Settings*

Setting	Description
Display Name	Enter a name for the LDAP directory. For example, SoftPhone-CM.
Server Name	Enter the name for the LDAP server. For example, ldap.company.com.
Port Number	Enter the port number used by the directory. For example, 8404.
Search Base	Enter the base or root of the directory service in which to search for names. For example, ou=users, o=company.com.

Cisco Unity

Cisco Unity is a powerful unified communications server that provides advanced, convergence-based communication services—such as voice mail and unified messaging—your company needs on a platform that offers the utmost in reliability, scalability, and performance. As an integral part of the Cisco AVVID environment, Cisco Unity complements the full range of Cisco IP-based voice solutions by providing advanced capabilities that unify data and voice.

Cisco Unity's server architecture is unified with your data network, minimizing installation, administration, and maintenance costs. Built on a platform that can scale to meet your organization's needs as it grows, Cisco Unity also uses streaming media and an intuitive browser-style system administration interface that makes life easier for the people who install and support your system, ultimately lowering your organization's total cost of ownership.

Digital Networking Capability

Cisco Unity's optional digital networking module enables the system to connect to other Cisco Unity servers at the same site via the LAN or remote sites using a WAN or the Internet. Digital networking makes communicating with coworkers at remote locations fast and efficient by giving staff the ability to send subscriber-to-subscriber messages anywhere in the world.

Cisco Unity Features

- Delivers advanced voice mail and powerful unified messaging in a unified environment.
- Designed for convergence, Cisco Unity provides optimum scalability, reliability, and performance.
- Leverages your communications infrastructure investment by integrating with Cisco CallManager and leading legacy telephone systems—even simultaneously—paving the way for a smooth transition to IP telephony.
- Easy-to-use browser-based system administration interface enables maintenance from any PC on the network, saving time, expense, and effort.
- True unified architecture enables IT staff to set one backup procedure, one message storage policy, and one security policy.
- Superior component-based server architecture provides a solid and flexible foundation for future growth.
- Intuitive browser-based system administration console and tools simplify installation, maintenance, and daily use.
- Browser-based personal administrator enables IT staff to enable end users to manage more of their own accounts, saving time and decentralizing routine administration.
- Innovative use of streaming media provides efficient audio delivery.
- Fault-tolerant system tools include robust security, file replication, event logging, and optional software RAID levels 0–5.
- International product offering fully localized versions in multiple languages—including Dutch, four dialects of English (Australian, New Zealand, U.K., and U.S.), French, German, Norwegian, and Spanish—and, depending on the language, feature everything from system prompts and subscriber conversations to the browser-based administration consoles and product documentation in the customer's language of choice.

Integrating Cisco CallManager With Cisco Unity

The purpose of this section is to assist with the initial integration of Cisco Unity and Cisco CallManager. Cisco Unity and Cisco CallManager communicate with each other via TSP that is installed on the Unity server. Please be sure to consult the Qualified Product Combinations table contained in the readme file before installing.

NOTE The screen captures used in this document were taken on a Cisco Unity server running on Windows 2000. Some screens and names may appear slightly different under Windows NT or may not appear at all. In addition, this document assumes that you have already installed Cisco Unity and Cisco CallManager.

Cisco CallManager Configuration

Use the following instructions to configure Cisco CallManager for Cisco Unity.

Step 1 In CCMAdmin, click **Device** > **Cisco Voice Mail Port** or **Device** > **Cisco Voice Mail Port Wizard**.

Step 2 Enter the appropriate information in each of the fields as defined in Table 14-5.

Table 14-5 *Cisco Voice Mail Port Wizard Field Descriptions*

Field	Description
Port Name	Enter a name to identify the Cisco voice mail port. You must add a device for each port on Cisco voice mail. If there are 24 ports, you must define 24 devices. Note: For Cisco uOne systems, make sure the name matches the information in the uOne .ini files, such as CiscoUM-VI1 or CiscoUM-VI2. Use the following naming convention for the ports: CiscoUM-VI<*consecutive number for each port*>.
Description	Enter the purpose of the device.
Device Pool	Choose the default value device pool.
Calling Search Space	Choose the appropriate CSS. A CSS comprises a collection of partitions that are searched for numbers called from this device.
Location	Choose the default value **<None>**. The location specifies the total bandwidth available for calls to and from this device. A location setting of <*None*> means that the locations feature does not keep track of the bandwidth consumed by this device.

Table 14-5 *Cisco Voice Mail Port Wizard Field Descriptions (Continued)*

Field	Description
Directory Number	Enter the number associated with this voice mail port. Make sure this field is unique in combination with the Partition field.
Partition	Choose the partition to which the directory number belongs. Choose **<None>** if partitions are not used.
Calling Search Space	Choose the appropriate CSS. A CSS comprises a collection of partitions that are searched for numbers called from this directory number.
Display	Enter the text that you want to appear on the calling party phone when a call is placed to this line.
Forward All	Leave this field blank.
Forward Busy	Enter the voice mail directory number where calls are forwarded if this port is busy (for example, the next sequential voice mail port number). For this number, use the next sequential Cisco voice mail port or, if it is the last port, an operator number. Make the Forward Busy and Forward No Answer fields have the same value. If your voice mail ports are in a partition, the CFB for each voice mail port must have a CSS that enables them to call the partition in which the subsequent port's DN exists. So for example, if you have four voice mail ports (1000 through 1003) and they are in the Internal partition, then you would configure CFB for port 1000 to go to 1001. You must also ensure, however, that the CSS for CFB is set so that it can access the partition that 1001 is in.
Forward No Answer	Enter the voice mail directory number where calls are forwarded if this port does not answer the call (for example, the next sequential port). Make this number the next sequential Cisco voice mail port or, if it is the last port, an operator number. Make sure the Forward Busy and Forward No Answer fields have the same value. If your voice mail ports are in a partition, the CFNA for each voice mail port must have a CSS that enables them to call the partition in which the subsequent port's DN exists. So for example, if you have four voice mail ports (1000 through 1003) and they are in the Internal partition, then you would configure CFNA for port 1000 to go to 1001. You must also ensure, however, that the CSS for CFNA is set so that it can access the partition that 1001 is in.

Step 3 Click **Insert** to add the newly configured port.

Step 4 Repeat Steps 1–3 for all of the ports defined in Step 12, incrementing the port number for each port or use the Cisco Voice Mail Port Wizard, which creates a range of ports automatically.

Once completed the configuration should look similar to Figure 14-11.

Figure 14-11 *Cisco Voice Mail Ports as Shown in CCMAdmin*

Step 5 Click **Service** > **Service Parameters**.

Step 6 In the Server list on the Service Parameters Configuration page, click the server you are configuring for use with Cisco Unity, then click **Next**.

Step 7 In the Services list, click **Cisco CallManager**. The list of parameters appears.

Step 8 Configure or update the parameters shown in Table 14-6, as appropriate.

Table 14-6 *Cisco CallManager Service Parameter Settings*

Service Parameter to Configure	Setting
AdvancedCallForwardHopFlag	Set to **True**. Used with service parameter **VoiceMailMaximumHopCount**
MessageWaitingOffDN	The unique extension that turns MWIs off. This must match the value entered in Step 12 in the **Cisco Unity Configuration** section. Also, make sure that this number does not conflict with a route pattern.
MessageWaitingOnDN	The unique extension that turns MWIs on.
VoiceMail	The extension that users call to access Cisco Unity. This is typically the extension number (directory number, in CCMAdmin) of the first voice mail port. If you specify an extension for voice mail, users can call Cisco Unity by pressing the **messages** button on their Cisco IP Phones.

Table 14-6 *Cisco CallManager Service Parameter Settings (Continued)*

Service Parameter to Configure	Setting
VoiceMailMaximumHopCount	Used together with service parameter **AdvancedCall ForwardHopFlag**, the maximum number of voice mail ports skipped to find the next available voice mail port. Specify a value that is twice the number of Cisco CallManager ports that are connected to the Cisco Unity server. For example, if you have a 48-port Cisco Unity system, enter **96** for this setting.

Step 9 Click **Update** to save the setting(s). If this is a CallManager cluster, updates to the service parameters must be performed on all servers in the cluster.

Step 10 Stop and start all Cisco CallManagers and the Cisco Unity server.

Cisco Unity Configuration

The following steps provide an overview of the configuration steps for Cisco Unity with Cisco CallManager:

Step 1 Review and gather system installation requirements.

Step 2 Add voice mail ports in Cisco CallManager.

Step 3 Specify voice mail service parameters.

Step 4 Install, configure, and test the TSP.

Step 5 Test the integration.

You can find detailed explanations for each of these steps at the following link:

> www.cisco.com/univercd/cc/td/doc/product/voice/c_unity/unity30/integuid/callma31/itcicmip.htm#xtocid223599

Following are detailed steps for configuration. You can download the TSPs at the following link:

> www.cisco.com/cgi-bin/tablebuild.pl/unity

NOTE The TSP required to install and run with Cisco CallManager may change with each new version of Cisco CallManager that is installed. Before loading a new version of Cisco CallManager or Cisco Unity, always check the URL listed above for updated TSPs and their associated release notes.

Step 1 Double-click on the downloaded TSP file. Click **Unzip**. Leave the default **Unzip to folder** value the same unless you have a need to change it.

Step 2 Click **Start**, then **Run**, and enter the path to which the TSP was unzipped. The default is C:\AVCiscoTSP Install. Click **OK**.

Step 3 Double-click the **setup** icon. Status will appear.

Step 4 Click **OK**.

Step 5 Click **Next**.

Step 6 Click **Next** to choose the default install path, or click **Change** if you wish to install into a different path. The default is recommended.

Step 7 Click **Next**.

Step 8 Click **Finish**.

NOTE Before clicking **Yes** in Step 9, make sure to save all data and close all windows you may have open.

Step 9 Click **Yes** to reboot the server.

Step 10 After the machine reboots and you have logged in to the console, you will be prompted with the AV-Cisco Service Provider window. Click **Add**.

Step 11 Enter the IP address of your primary CallManager and then click **OK**.

Step 12 Enter the appropriate information in each of the fields as follows:

- In the **Number of voice ports** field, enter the number of ports you have purchased. The numbers of voice ports will automatically propagate on the **CallManager Device List** located on the right side of the screen.
- In the **Device name prefix** field, enter the device name prefix you want to use. The name of the voice ports will automatically propagate on the **CallManager Device List** located on the right of the screen.

NOTE When configuring uOne ports, you will use the prefix name, so make note of the value entered.

— Enter the message waiting on and off directory numbers under **MessageWaitingOnDN** and **MessageWaitingOffDN**, respectively.

NOTE These are the same values you used when configuring MWI on Cisco CallManager.

Step 13 Click **Add IP Address** to add the IP address of a redundant Cisco CallManager, if applicable.

Step 14 Click **OK**.

Step 15 Check the **Automatically reconnect to the Primary CCM on Failover** box if you would like to automatically reconnect to the primary Cisco CallManager on failover. This is recommended.

Step 16 Click **OK**.

NOTE Before clicking **OK** in Step 17, make sure to save all data and close all windows you may have open.

Step 17 Click **OK** to reboot the computer.

Testing the Integration

Use the following procedures to test the integration of Cisco CallManager and Cisco Unity.

Step 1 Click **Start > Settings > Control Panel** and double-click **Phone and Modem Options**.

Step 2 Select **AV-Cisco Service Provider** and click **Configure**.

Step 3 Click **Test**.

NOTE If the test does not complete successfully, begin troubleshooting with the *MWI Troubleshooting Guide* at the following link:

http://avforums.isomedia.com/cgi-bin/showthreaded.pl?Cat=&Board=ipswitch&Number=1313&page=0&view=expanded&sb=5

Configure MWI Ports for Multiple Clusters

If the Cisco Unity server services multiple Cisco CallManager clusters, perform the following procedure to enable MWIs to be activated on extensions in each cluster.

Step 1 In Cisco Unity Administrator, dedicate at least one port to send MWIs to each cluster.

Step 2 In CCMAdmin, click **Device > Cisco Voice Mail Port**.

Step 3 In the Cisco Voice Mail Ports list, click the name of the port before the first dedicated MWI port.

Step 4 Under Call Forwarding Information, change the extensions to the number of the first voice mail port or of the operator.

Step 5 Click **Update**.

Step 6 In the Cisco Voice Mail Ports list, click the name of a dedicated MWI port.

Step 7 Disable call forwarding for this port.

Step 8 Click **Update**.

Step 9 Repeat Steps 6 through 8 for all remaining dedicated MWI ports.

Summary

Cisco WebAttendant is an application that supports the traditional role of a manual attendant console. Associated with an Cisco IP Phone, Cisco WebAttendant enables the receptionist or operator to quickly answer and dispatch calls to enterprise users.

Cisco IP SoftPhone is a virtual telephone operating from your laptop or desktop computer. It controls your hardware IP telephone and features an intuitive user interface and context-sensitive controls.

Cisco Unity delivers advanced voice mail and powerful unified messaging in a unified environment. Cisco Unity provides an easy-to-use, browser-based system administration interface that enables maintenance from any PC on the network, saving time, expense, and effort.

Post-Test

Do you already know this information? The pre-test is designed to help you gauge your knowledge about this chapter. Of the 10 questions, if you answer one to three questions correctly, we recommend that you read this chapter. If you answer four to seven questions correctly, we recommend that you read this chapter, reading those sections that you need to know more about. If you answer 8 to 10 questions correctly, you probably understand this information well enough to skip this chapter. You can find the answers to these questions in Appendix B, "Answers to Chapter Pre-Test and Post-Test Questions."

1 Which Cisco service must be running on the Cisco CallManager server for Cisco WebAttendant to work?

____

____

____

2 What is the purpose of the pilot point directory number as it relates to Cisco WebAttendant?

____

____

____

3 What should the last member of a Cisco WebAttendant hunt group be?

____

____

____

4 As an administrator, what must you enable for a user to be able to use Cisco IP SoftPhone?

____

____

____

5 What information does the user need to install the Cisco IP SoftPhone application on their PC?

____

____

____

6 What does Cisco Unity use as the message store and directory service to unify your system administration?

7 What should be configured first in Cisco CallManager to prepare for integration with Cisco Unity?

List three of the four service parameters in Cisco CallManager that need to be configured for integration with Cisco Unity.

8

9

10

Appendixes

Cisco CallManager Architecture

The internal architecture of Cisco CallManager is based on layers. The layers communicate to identify device registration, call control, and device de-registration. This appendix provides a brief background into the Cisco CallManager architecture. For more information, refer to the book, *Cisco CallManager Fundamentals: A Cisco AVVID Solution* (ISBN: 1-58705-008-0). This chapter covers the following topics:

- Abbreviations
- Architecture Layer Overview
- Link Layer
- Device Layer
- Media Control Layer
- Call Control Layer

Abbreviations

This section defines the abbreviations used in this appendix. For more information about terms and abbreviations used in this chapter refer to the *IP Telephony Network Glossary* at the following URL:

www.cisco.com/univercd/cc/td/doc/product/voice/evbugl4.htm

Table A-1 provides the abbreviation and the complete term.

Table A-1 *Abbreviations and Complete Terms Used in This Appendix*

Abbreviation	Complete Term
Cc	Call Control
Cdcc	Call Control Child Process
DA	digit analysis
DLCX	Delete Connection
IE	Information Element
ISDN	Integrated Services Digital Network
ITU	International Telecommunications Union
LAN	local-area network
MDCX	Modify Connection
MGCP	Media Gateway Control Protocol
PRI	Primary Rate Interface
PSTN	public switched telephone network
QoS	quality of service
RAS	Registration, Admission, and Status
RIP	Request in Progress
RQNT	Notification Request
RSIP	Restart in Progress
RTCP	RTP Control Protocol
RTP	real-time transport protocol
SCCP	Skinny Client Control Protocol
SDL	signal distribution layer
SSAPI	Supplementary Service Application Programming Interface
UDP	User Datagram Protocol

Architecture Layer Overview

Figure A-1 shows the layered architecture of Cisco CallManager as well as the method the layers use to communicate with each other. The list that follows describes each of the six layers in greater detail.

Figure A-1 *Cisco CallManager Architecture*

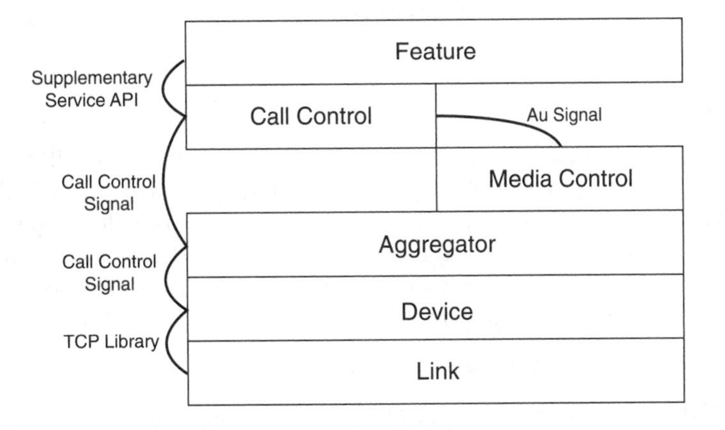

- **Feature layer**—The top layer that handles all the other features that the phones enable other than normal call operations.
- **Call Control layer**—Communicates to three other layers; the Feature layer, Media layer, and Aggregator layer using the following methods:
  - **Feature layer**—Using SSAPI
  - **Media layer**—Using Au signals
  - **Aggregator layer**—Cc signals
- **Media Control layer**—Closely interacts with the Call Control layer to provide media-related support to the calls to provide connectivity between devices.
- **Aggregator layer**—Communicates with the Call Control layer on the Aggregator layer's upper side and uses the Device layer to determine device registration.
- **Device layer**—Interacts with the upper layers and database of Cisco CallManager on one side and uses the Link layer for transmitting the signals to the stations on the other side.
- **Link layer**—Closely interacts with the Call Control layer to provide link-related support to the calls to provide connectivity between devices.

The sections that follow discuss the layers of the Cisco CallManager from the bottom up.

Link Layer

Transport Control Protocol (TCP) and User Datagram Protocol(UDP) sockets are used by the Link layer to communicate the information across the network. Appropriate libraries for implementing the same are included.

The Link layer communicates with the different devices across a Cisco AVVID solution and provides link connectivity to the devices.

Device Layer

The Device layer controls all the devices that register with Cisco CallManager. The Device layer hosts three protocols for supporting various types of devices: MGCP (for gateways), SCCP (for phones), and H.323. The Device layer interacts with the upper layers and database of Cisco CallManager on one side and uses the Link layer for transmitting the signals to the devices on the other side. Figure A-2 highlights the Device layer of the Cisco CallManager architecture.

The following are entities of the Device layer:

- **H.323 interface**—Provides capabilities of connecting H.323 devices, such as Microsoft NetMeeting, or Cisco IOS Software H.323 gateways, such as 2600 or 3600, to Cisco CallManager. The H.323 interface is responsible for H.323 device initialization, device registration, device deletion, and other call-related processing.
- **Station interface**—Processes all the call signaling messages between Cisco CallManager and the stations (phones). The SCCP station interface is responsible for device registration, device deletion, and other call-related processing.
- **PRI**—Processes all the call signaling messages between the Cisco CallManager and PRI gateways. Typically, PRI establishes communication between the gateways and Cisco CallManager to control, maintain, and clear the calls through the PSTN.
- **MGCP**—Processes all the call signaling messages between Cisco CallManager and MGCP gateways. MGCP is responsible for gateway registration, gateway deletion, and other external call-related processing.

Device Definition

Two classes of devices exist in a Cisco CallManager environment:

- **Terminals**—Terminals may be further subdivided into two classes:
  - Phones (clients)
  - Applications (servers)
- **Gateways**—Gateways are further subdivided into two classes:
  - Voice-only
  - Voice/router

Figure A-3 illustrates the functionality of each device category.

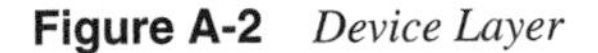

Figure A-2 *Device Layer*

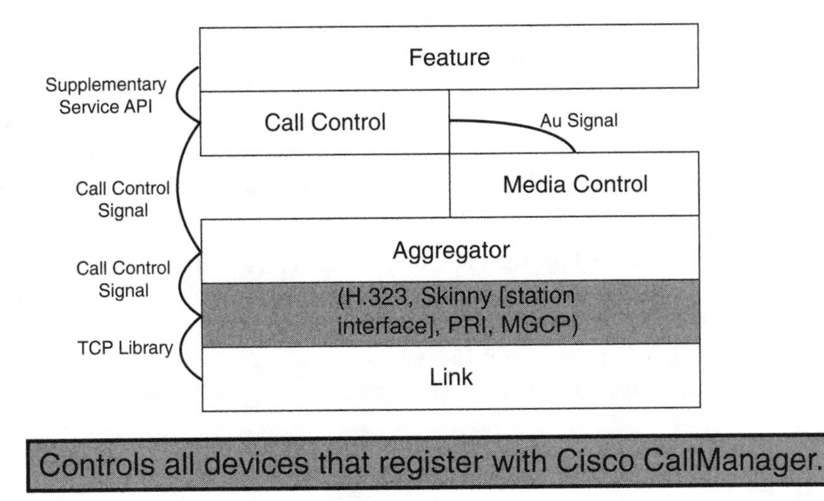

Figure A-3 *Device Definition*

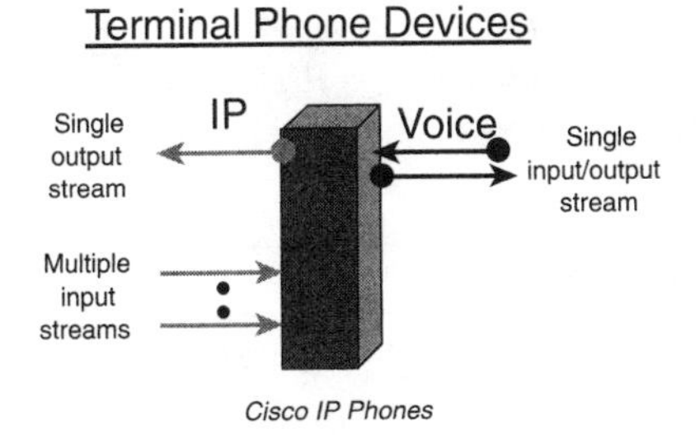

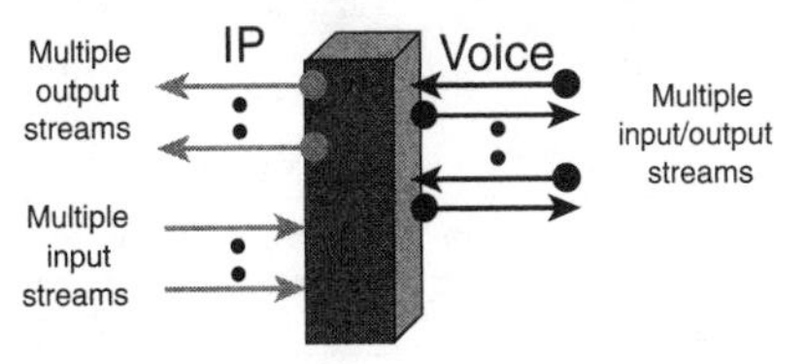

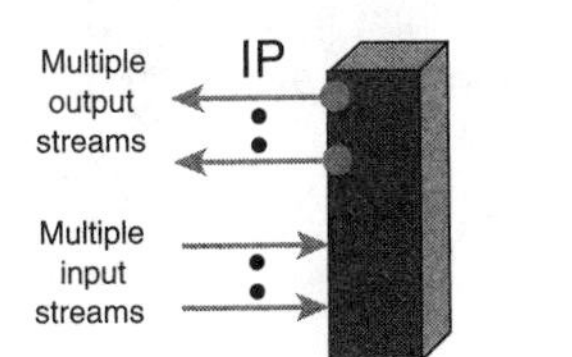

Software Conference Bridge, Software MTP, DSP Service Cards

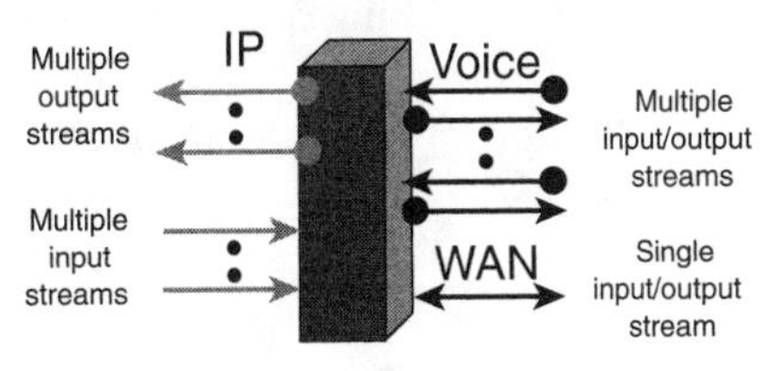

26xx, 36xx, 53xx, Cisco ICS 7750

H.323 Interface

The H.323 standard provides a foundation for audio, video, and data communications across IP-based networks, including the Internet. H.323 is an umbrella recommendation from the ITU that sets standards for multimedia communications over LANs that do not provide a guaranteed QoS. These networks dominate today's corporate desktops and include packet-switched TCP/IP and IPX over Ethernet, Fast Ethernet, and Token Ring network technologies. The H.323 protocol suite is based on several protocols, as illustrated in Figure A-4.

The H.323 standards are important building blocks for a broad new range of collaborative, LAN-based applications for multimedia communications. The standards include parts of H.225.0—RAS; Q.931; H.245 RTP/RTCP; audio/video codecs, such as the audio codecs (G.711, G.723.1, G.728, and so on); and video codecs (H.261 and H.263) that compress and decompress media streams. H.225 manages call setup and termination.

Figure A-4 *H.323 Protocol Suite*

<table>
<tr><td>Audio/Video Applications</td><td colspan="4">Terminal Control and Management</td><td>Data Applications</td></tr>
<tr><td>G.XXX
H.XXX</td><td rowspan="2">RTCP</td><td rowspan="2">H.225.0
Terminal to Gatekeeper Signaling
(RAS)</td><td rowspan="2">H.225.0
Call Signaling</td><td rowspan="2">H.245
Media Control</td><td rowspan="2">T.124

T.125</td></tr>
<tr><td>RTP</td></tr>
<tr><td colspan="3">Unreliable Transport (UDP)</td><td colspan="2">Reliable Transport (TCP)</td><td rowspan="3">T.123</td></tr>
<tr><td colspan="5">Network Layer (IP)</td></tr>
<tr><td colspan="5">Link Layer</td></tr>
<tr><td colspan="6">Physical Layer</td></tr>
</table>

Media streams are transported on RTP/RTCP. RTP carries the actual media and RTCP carries status and control information. The signaling is transported reliably over TCP. The following protocols deal with signaling:

- **RAS**—Manages registration, admission, and status.
- **H.245**—Negotiates channel usage and capabilities.
- **H.235**—Manages security and authentication. This is part of the standard but unsupported by Cisco CallManager.

Registration, Admission, and Status (RAS)

The RAS channel is used to carry messages used in the gatekeeper discovery and endpoint registration processes which associate an endpoint's alias address with its call signaling channel transport address. The RAS channel is an unreliable channel. Because the RAS messages are transmitted on an unreliable channel, H.225.0 recommends timeouts and retry counts for various messages. An endpoint or gatekeeper that cannot respond to a request within the specified timeout may use the RIP message to indicate that it is still processing the request.

H.225 Protocol

H.225 is a standard that covers narrow-band visual telephone services defined in H.200/AV.120-Series recommendations. The H.225 standard specifically deals with those situations where the transmission path includes one or more packet-based networks, each of which is configured and managed to provide non-guaranteed QoS. H.225.0 describes how audio, video, data, and control information on a packet-based network can be managed to provide conversational services in H.323 equipment. The structure of H.225 follows the Q.931 standards.

H.245 Protocol

H.245 is line transmission of non-telephone signals. It includes receiving and transmitting capabilities as well as mode preference from the receiving end, logical channel signaling, and control and indication. H.245 handles the negotiation of media streams between H.323 endpoints. It enables the endpoints on a call to agree on a method for encoding voice traffic (codec) and allows the endpoints to exchange an IP address and port for the media to actually stream to.

H.323 Implementation in Cisco CallManager

The H.323 recommendation for the stations provides mechanisms for establishing, controlling, and clearing of information flows, including audio information between two H.323-compliant terminals. To implement a full H.323 compliant terminal as shown in Figure A-5 requires high expenditure for computer power and memory size.

An H.323 *Proxy* can be implemented in a relatively high-powered server and can communicate to a simplified Skinny client efficiently using SCCP.

By implementing the station telephone set as a Skinny client over IP and using a proxy for H.225 and H.245 signaling, a relatively inexpensive Cisco IP Phone can be constructed to coexist in an H.323 environment. By coupling with an H.323 proxy, the Skinny client can interoperate with H.323-compliant terminals to establish, control, and clear audio calls.

Figure A-5 *H.323 Implementation*

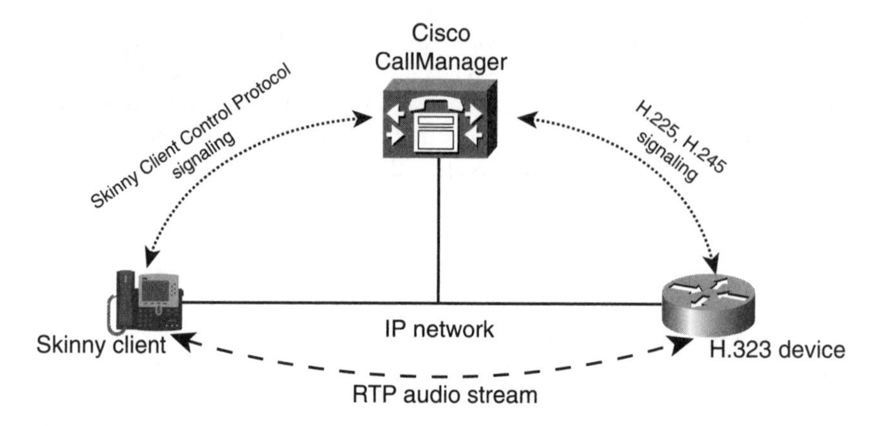

Exchange Messages Between an H.323-based Client and a Cisco IP Phone

Figure A-6 shows a sample exchange of messages between an H.323-based client and a Skinny client. Note that this is a sample only and may not match exactly the sequence of messages.

Station Interface

The station interface is responsible for device registration, device deletion, and other call-related processing. Typically, the station interface communicates with devices such as the Cisco IP Phone using SCCP. In addition, the station interface interacts with the Call Control layer of Cisco CallManager via the line control (Aggregator layer) to establish, control, and clear the audio calls.

SCCP Support

The Skinny client uses the following protocols:

- TCP/IP to/from one or more Cisco CallManagers to transmit and receive stimulus.
- RTP/UDP/IP to/from a similar Skinny client or H.323 terminal for audio.

Registration of a Station Device

The message sequence in Figure A-7 illustrates the operations during the registration of a Cisco IP Phone.

Figure A-6 *Call Flow for H.323 Client to Skinny Client*

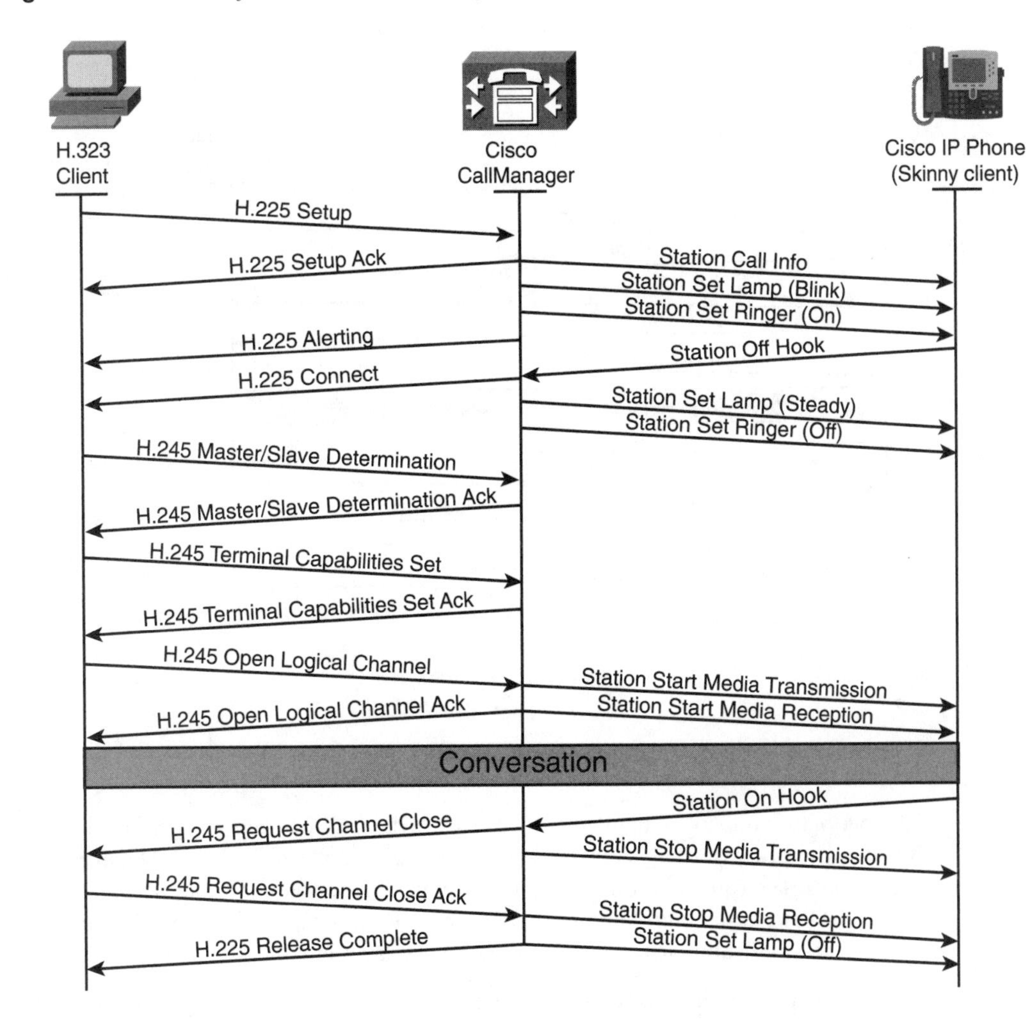

Normal Call Processing

Figure A-8 displays the messages between two Cisco IP Phones and Cisco CallManager during normal call processing.

Figure A-7 *Skinny Station (Client) Registration Messages*

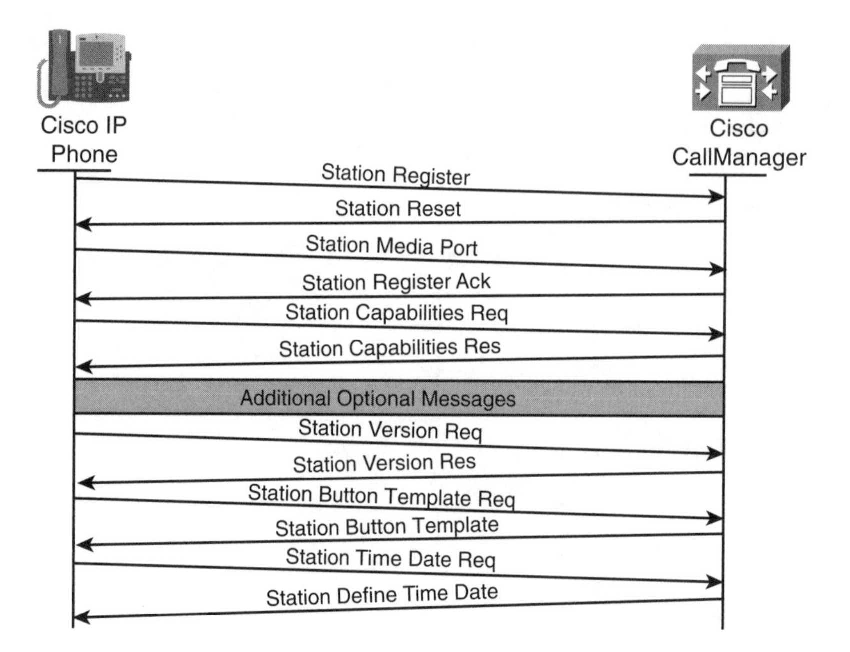

PRI

PRI is one of the protocols supported by the Device layer of Cisco CallManager for gateways. This interface processes all the call signaling messages between Cisco CallManager and the gateways. Typically, PRI establishes communication between the gateways and Cisco CallManager to control, maintain, and clear the calls through the PSTN. The following sections summarize the role of the Device layer and the PRI protocol.

PRI consists of 23 B channels and one 64-kbps D channel for a T1 line or 30 B channels and a D channel for an E1 line. Thus, PRI provides 1.544 Mbps on a T1 line and up to 2.048 Mbps on an E1 line. PRI uses the Q.931 protocol over the D channel for signaling purposes.

Figure A-8 *Call Flow Messages from Cisco IP Phone to Cisco IP Phone*

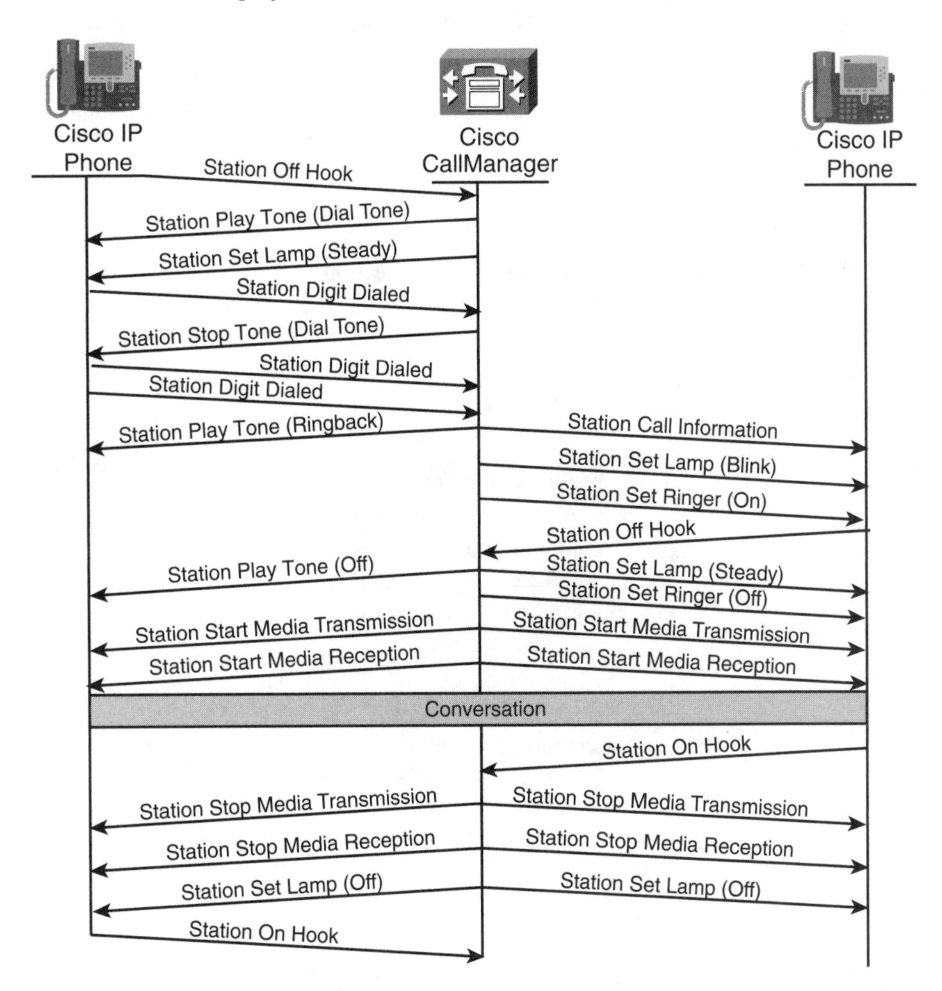

Q.931 Overview

Q.931 is ISDN's connection control protocol, roughly comparable to TCP in the Internet protocol stack. Q.931 doesn't provide flow control or perform retransmission because the underlying layers are assumed to be reliable and the circuit-oriented nature of ISDN allocates bandwidth in fixed increments of 64 kbps. Q.931 does manage connection setup and tear-down.

Figure A-9 shows the general format of a Q.931 message, while Figure A-10 shows a sample call. The general format of a Q.931 message includes the following:

- A single byte *protocol discriminator* (eight for Q.931 messages).
- A *call reference* value to distinguish between different calls being managed over the same D channel.
- A *message type*.
- Various.
- IEs as required by the message type in question.

Figure A-9 *Q.931 Message*

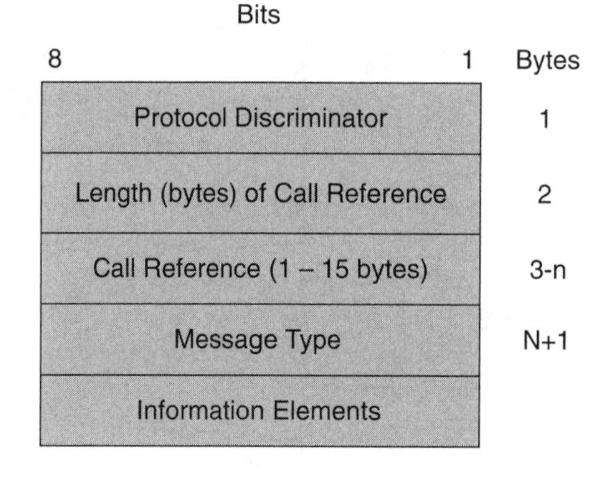

A typical Q.931 message might look like Figure A-10.

MGCP

MGCP is one of the protocols supported by the Device layer of Cisco CallManager. MGCP is used by Cisco CallManager to control VoIP gateways. It processes all the call signaling messages between Cisco CallManager and the MGCP gateways.

MGCP is responsible for gateway registration, gateway deletion, and other external call related processing. MGCP communicates with telephony gateways and interacts with the Call Control layer of Cisco CallManager via the line control (Aggregator layer) to establish, control, and clear the audio calls.

Figure A-10 *Typical Q.931 Message*

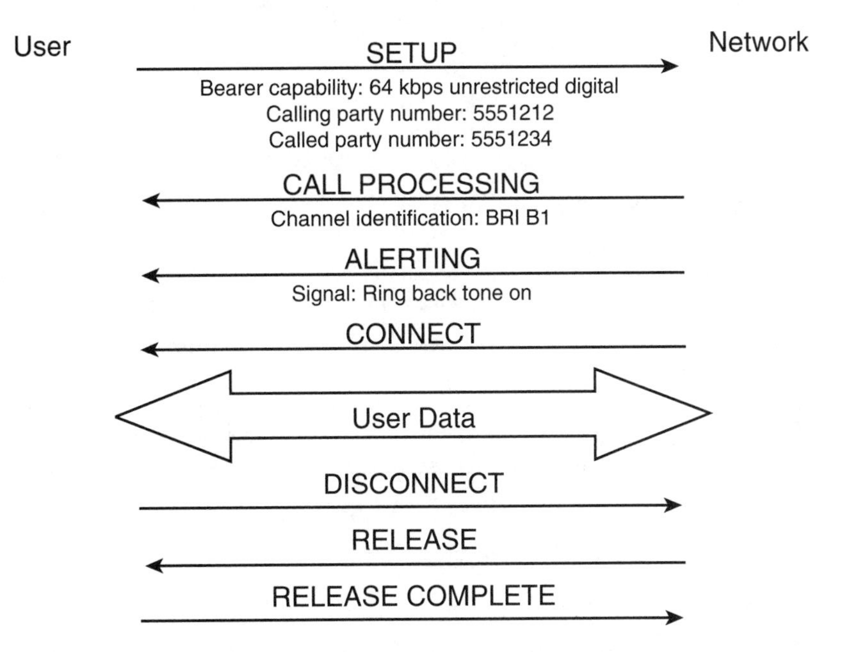

Table A-2 lists and describes the eight commands used with MGCP as displayed in a trace. This information proves useful when reviewing traces for troubleshooting.

Table A-2 *MGCP Commands*

Command	Description
RQNT—NotificationRequest	Cisco CallManager issues a **NotificationRequest** command to a gateway, instructing the gateway to watch for specific events such as hook actions or DTMF tones on a specified endpoint; it is also used to request the gateway to apply a specific signal to endpoint (for example, dial tone, ringback, and so forth).
NTFY—Notify	The gateway uses the **Notify** command to inform Cisco CallManager when the requested events occur.
CRCX—CreateConnection	Cisco CallManager uses the **CreateConnection** command to create a connection that terminates in an endpoint inside the gateway.
MDCX—ModifyConnection	Cisco CallManager uses the **ModifyConnection** command to change the parameters associated to a previously established connection.

continues

Table A-2 *MGCP Commands (Continued)*

Command	Description
DLCX—DeleteConnection	Cisco CallManager uses the **DeleteConnection** command to delete an existing connection. The **DeleteConnection** command may also be used by a gateway to indicate that a connection can no longer be sustained.
AUEP—AuditEndpoint	Cisco CallManager uses the **AuditEndpoint** command to audit the status of an endpoint associated with it.
AUCX—Audit Connection	Cisco CallManager uses the **AuditConnection** command to audit the status of any connection associated with it.
RSIP—Restart in Progress	The gateway uses the **RestartInProgress** command to notify Cisco CallManager that the gateway or a group of endpoints managed by the gateway is being taken out of service or is being placed back in service. There are three types of restart: • Restart—Endpoint in service • Graceful—Wait until call clearing • Forced—Endpoint out of service

Media Control Layer

The Media Control layer is responsible for interconnecting the IP media to the clients. IP media is transported over the RTP protocol, which is a UDP protocol. Control over RTP source and receiver addresses is typically done via MGCP/SCCP (client/server) and SIP/H.323 (peer/peer).

The media functions performed between devices consist of providing:

- Simple connections
- MDCXs
- Multiple connections

Simple Connections

Simple connections are half- or full-duplex connections between two devices. Different codecs are used depending upon the situation. Cisco CallManager manages connection setup. Media connections are virtual connections directly between devices.

MDCXs

The MDCX function is typically used for such features as hold, transfer, and so on to effect transfers of media streams from one device to another.

Multiple Connections

There are two categories of multiple connection functions with several types within each category:

- **Uni-directional**—Used for features such as music on hold and paging. Multiple uni-directional connections consist of two types:
  - Unicast
  - Multicast
- **Bi-directional**—Used for such features as Ad Hoc conferencing and Meet-Me conferencing.

Call Control Layer

The Call Control layer consists of the following major components:

- **Call Control (Cc)**—Only one instance of Cc exists in each Cisco CallManager. Call Control is responsible for initialization/tear-down of other processes. The Cc spawns the Cdcc on a per-call basis. The Line control child process (Line Cdpc) communicates with the Cc for various requests using the SDL messages. DA also communicates with the Cc using SDL messages.
- **Call Control Child Process (Cdcc)**—The Cdcc instance exists on a per-call basis and is responsible for setting up and tearing down a call. The number of instances of Cdcc at any given time in a Cisco CallManager depends on the number of active calls at that time. The life span of Cdcc is the life span of an active call. It is spawned by Cc during call setup and torn down during call teardown. The Cdcc communicates with other processes like Matrix Control (Media Control layer), DA, and Line Control.
- **Digit Analysis (DA)**—Responsible for handling digit analysis. DA uses tables to drive its understanding of national dialing plans, arbitrary tags to drive its understanding of digit substrings, and wildcards to drive its understanding of ranges of numbers.

Summary

Cisco CallManager architecture is based on layers; the following identifies and describes those layers:

- **Feature layer**—The top layer that handles all the other features that the phones allow other than normal call operations.
- **Call Control layer**—Communicates to three other layers: the Feature, Media, and Aggregator layers. The Feature layer communicates with it using SSAPI, the Media layer using Au signals, and the Aggregator layer using Call Control signals.
- **Media Control layer**—Closely interacts with the Call Control layer to provide media-related support to the calls to provide connectivity between devices.
- **Aggregator layer**—Interacts with the upper Call Control layer on one side and uses Device layer to determine device registration.
- **Device layer**—Interacts with the upper layers and database of Cisco CallManager on the one side and uses the Link layer for transmitting the signals to the stations on the other side.
- **Link layer**—Closely interacts with the Call Control layer to provide link-related support to the calls to provide connectivity between devices.

Answers to Chapter Pre-Test and Post-Test Questions

Answers to Chapter 2 Pre-Test and Post-Test Questions

1. Why is it recommended to change the server name to an IP address in CCMAdmin user interface?

   Answer: By changing the server name to an IP address, it enables devices to resolve the Cisco CallManager location by IP address rather than by trying to resolve a server name to IP address for dial tone and call processing. One less layer to go through.

2. Assuming DHCP and your network are configured correctly and besides un-checking the box **auto-registration Disabled on this Cisco CallManager**, what must you provide so that auto-registration works properly?

   Answer: You must provide a directory number range, consisting of a starting and ending directory number.

   If you are doing a green field deployment, you will want to use the DID range provided by your local telephone company.

   If you are setting up auto-registration for un-administered phones registering in the cluster, you may want to use a directory number range out to the DID range provided by your local telephone company.

3. Based on configuring one Cisco CallManager group in a Cisco CallManager cluster that has two or three servers, how many users can this cluster support?

   Answer: 2500 users

4 If you expected some growth in your organization, what would be a better way to support 2500 users? How many servers? What would your Cisco CallManager Groups be like? (If it is easier, you may draw your answer.)

Answer: Three servers.

One dedicated Publisher and two Subscribers. The Publisher does not do any call processing.

One of the Subscribers is the Primary Cisco CallManager and the other Subscriber is the Backup Cisco CallManager.

Or each Subscriber is a Primary Cisco CallManager for 1250 users and each is the Backup Cisco CallManager for each other.

5 When would you assign a Date/Time Group other than CMLocal to a device?

Answer: When there are devices registering to the cluster from a different time zone, you want to have that device display the date/time that the device is in. For example, if you have a cluster in San Jose and a device in Dallas registers to that cluster, it would be nice for the device in Dallas to display the Central time rather than the CMLocal time, which is probably Pacific time.

6 What is the main reason for using G.711 between devices as much as possible?

Answer: Although G.711 uses the most bandwidth (excluding wideband), G.711 provides the best audio quality.

7 A device pool is made of characteristics. List the three required characteristics of a device pool.

Answer:

Cisco CallManager Group

Date/Time Group

Region

8 After configuring a device pool and assigning it to a phone, what must you do to the phone so that those changes take effect?

Answer: Reset the phone. You can either restart or reset the phone.

If you Restart, you will restart the phone without shutting it down.

If you Reset, you will shut down the phone and bring it back up.

Remember that restarting or resetting a gateway drops any calls in progress using that gateway. Other devices wait until calls are complete before restarting or resetting.

9 You have two clusters, Cluster One and Cluster Two. Can you set the enterprise parameters in Cluster One and also apply those changes to Cluster Two?

Answer: No. Enterprise parameters are cluster specific. The changes made in Cluster One's enterprise parameters only affect Cluster One.

10 Which deployment model uses Locations?

Answer: Multisite deployment with centralized call processing.

Answers to Chapter 3 Pre-Test and Post-Test Questions

1 Write or draw the process used for configuring a route plan.

Answer:

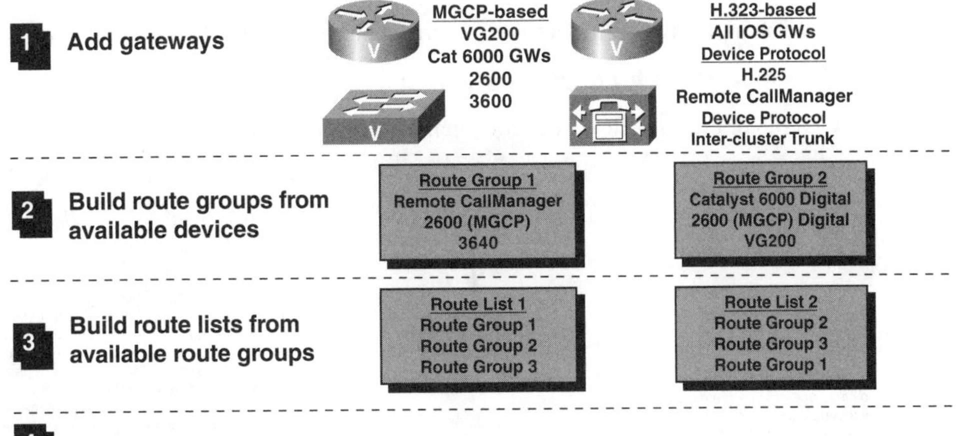

2 Match the most commonly used route pattern wildcards to its definition.

Route Pattern Wildcards						
X	!	#	[x-y]	.	@	[^x-y]

a. One or more digits (0 – 9)

b. Generic range notation

c. Terminates access code

d. Exclusion range notation

e. North American Numbering Plan

f. Terminates inter-digit timeout

g. Single digit (0 – 9)

Answer:

Route Pattern Wildcards

X	!	#	[x-y]	.	@	[^x-y]
g	a	f	b	c	e	d

3 What are the two things a route pattern can be assigned to?

Answer: Route list or gateway devices

4 Write or draw the flow of a call when a user dials digits.

Answer:

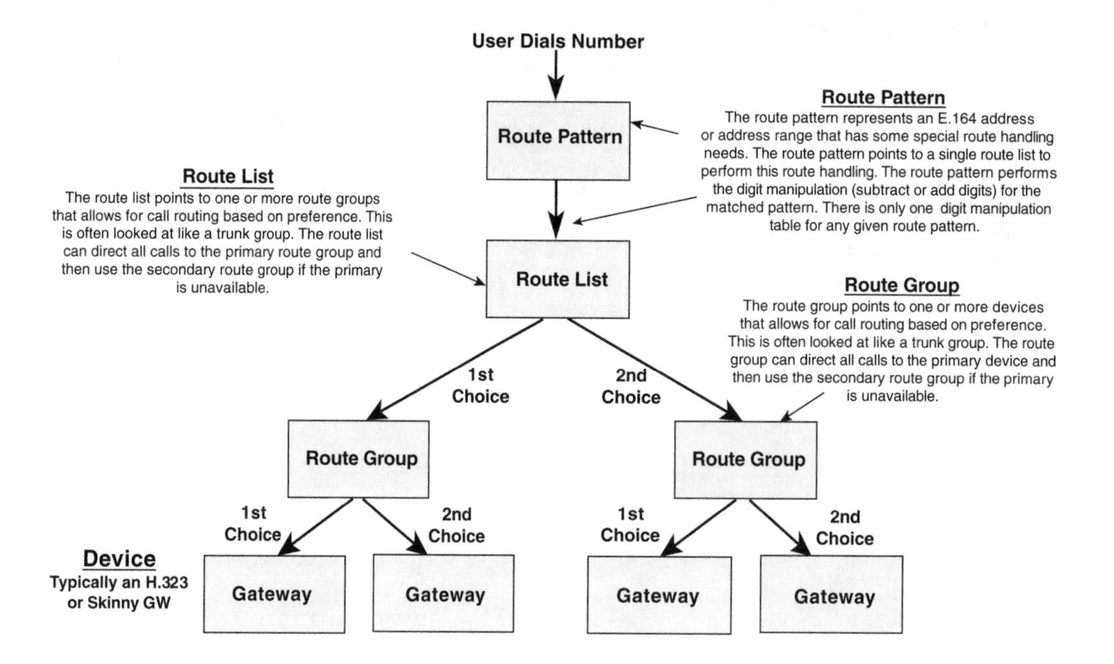
User Dials Number
Route Pattern
Route Pattern
The route pattern represents an E.164 address or address range that has some special route handling needs. The route pattern points to a single route list to perform this route handling. The route pattern performs the digit manipulation (subtract or add digits) for the matched pattern. There is only one digit manipulation table for any given route pattern.
Route List
The route list points to one or more route groups that allows for call routing based on preference. This is often looked at like a trunk group. The route list can direct all calls to the primary route group and then use the secondary route group if the primary is unavailable.
Route List
Route Group
The route group points to one or more devices that allows for call routing based on preference. This is often looked at like a trunk group. The route group can direct all calls to the primary device and then use the secondary route group if the primary is unavailable.
1st Choice
2nd Choice
Route Group
Route Group
1st Choice
2nd Choice
1st Choice
2nd Choice
Device
Typically an H.323 or Skinny GW
Gateway
Gateway
Gateway
Gateway

5 If you are using an "@" in a route pattern, what is automatically recognized as an end-of-dialing character for international calls?

Answer: When @ is used in a routing pattern, the octothorpe (#) is automatically recognized as an end-of-dialing character for international calls. For routing patterns that don't use @, you must include # in the routing pattern to be able to use # character to signal the end-of-dialing.

6 What removes a portion of the dialed digit string before passing the number on to the adjacent system?

Answer: DDIs remove a portion of the dialed digit string before passing the number on to the adjacent system.

7 Which one of the transformation settings changes the dialed number?

Answer: The Called Party Transformation setting changes the dialed number and the Calling Party Transformation setting changes the caller ID number.

8 If dialed digits match a translation pattern, what does Cisco CallManager do with those digits?

Answer: If digits match a pattern and the results of transformations are a translation pattern after the pattern is translated, it will make another trip through digit analysis.

9 What are the prerequisites for using the External Route Plan Wizard?

Answer: The prerequisites are

— **All gateways must be already defined in the system.**

— **The route plan must be for NANP.**

10 List five of the route plan components generated by the external Route Plan Wizard.

Answer: The route plan generated by the external Route Plan Wizard includes the following elements:

— Route filters

— Route groups

— Route lists

— Route patterns

— Partitions

— CSSs

— Calling party digit translations and transformations

— Access code manipulation

11 Match the gateway-type naming convention used by the external Route Plan Wizard to the gateway type.

Gateway types naming convention				
AA	DA	HT	MS	MT

a. Digital Access

b. MGCP Trunk

c. MGCP Station

d. Analog Access

e. H.323 Trunk

Answer:

Gateway types naming convention				
AA	DA	HT	MS	MT
d	a	e	c	b

12 The Route Plan Report can be saved as what type of file?

Answer: The Route Plan Report allows you to save report data into a .csv file that you can import into other applications.

13 What does Cisco CallManager use to apply class of service to devices?

Answer: Partitions and CSSs

14 What are directory numbers and route patterns assigned to?

Answer: Partitions

15 What does a CSS provide for the device it is assigned to?

Answer: Partitions that device can call and, in the end, directory numbers and route patterns that device can call.

16 What are the three specific problems that partitions and CSSs are designed to addresses?

Answer: Partitions and CSSs are designed to address three specific problems:

- **Routing by geographical location**
- **Routing by tenant**
- **Routing by class of user**

Answers to Chapter 4 Pre-Test and Post-Test Questions

1 Name the two types of conferences and briefly describe each type.

Answer: The two types of conferences are Ad Hoc and Meet-Me.

***Ad Hoc Conference*—In an Ad Hoc conference, the user who initiates the conference is the conference controller. Only a conference controller can add participants to a conference. If other participants attempt to join conference, Cisco CallManager ignores the signals. An Ad Hoc conference continues even if the conference controller hangs up, although new participants cannot be added to the conference.**

***Meet-Me Conference*—In a Meet-Me conference, the conference controller provides a bridge or directory number for participants to dial into. Participants can join the conference by dialing the specified directory number. A tone plays each time a participant joins the conference. Participants can leave the conference by hanging up the conference call. The *MeetMe* button or soft key is only used by the conference controller to establish the conference; to join a conference, participants simply dial the directory number supplied by the conference controller. A conference continues even if the conference controller hangs up as long as there are at least two conference participants on the call.**

2 What is the supported codec for ConfBrs?

Answer: Conference devices configured only for software support G.711. Some hardware conference devices, such as the Catalyst 6608, support G.711, G.723, and G.729 codecs, while others, such as the Catalyst 4000, support only G.711 codecs. Any of the G.711-only conference devices can support G.723 and G.729 codecs using a separate transcoding resource.

3 Supplementary services on H.323v1 gateways are enabled through which media resource?

Answer: A *MTP* is a software device that enables supplementary services to calls routed through an H.323v1 gateway. Supplementary services are features, such as hold, transfer, call park, and conferencing, that are otherwise not available when a call is routed to an H.323v1 endpoint.

4 What is the name of the application that provides for software conferencing and MTP?

Answer: The Cisco IP Voice Media Streaming Application automatically adds an MTP device and software ConfBr device for that server.

5 Name the two types of hold that can use MOH.

Answer: The MOH feature provides music for two types of hold: User hold (when the user presses the hold button or soft key) and Network hold (which includes the hold that occurs during transfer, conference initiation, and parking a call).

6 Describe an XCODE. What does it do or provide?

Answer: An XCODE is a device that takes the output stream of one codec and transcodes (converts) it from one compression type to another compression type.

Cisco CallManager uses XCODEs to convert between G.711, G.723, G.729, and GSM codecs. In addition, an XCODE provides MTP capabilities and may be used to enable supplementary services for H.323 endpoints when required.

7 Media resource groups and lists have the same type of relationship and manner of configuration as what two other components in CCMAdmin?

Answer: Media resource groups and lists are configured in a similar manner as route groups and route lists.

8 If you have software ConfBr resources listed before hardware ConfBr resources in a MRG, in what order does the media resource manager allocate ConfBr resources?

Answer: With this arrangement, when a conference is needed, Cisco CallManager allocates the software conference resource first; the hardware conference is not used until all software conference resources are exhausted.

9 Where in Cisco CallManager Administration do you go to assign a bulk number of devices to a MRGL?

Answer: Device Pool. Depending on how your device pools are configured you could add a MRGL to a bulk number of devices by assigning an MRGL to the device pool.

Answers to Chapter 5 Pre-Test and Post-Test Questions

1 What is the benefit of call park?

Answer: The call park feature enables users to place a call on hold at a specified directory number so that it can be retrieved from any other phone in the cluster.

2 After configuring a call park directory number or range of directory numbers, how many calls can be parked at a given call park extension?

Answer: You can park only one call at each call park extension.

3 Describe the two types of call pickup.

Answer: Cisco IP Phones offer two types of call pickup:

Call pickup—**Enables users to pick up incoming calls within their own group. Cisco CallManager automatically dials the appropriate call pickup group number when a user presses the** ***Call Pickup*** **button or** ***PickUp*** **soft key on his or her phone.**

***Group call pickup*—Enables users to pick up incoming calls within their own group or in other groups. Users must press the *Group Call Pickup* button or *GPickUp* soft key and dial the appropriate call pickup group number when using this feature.**

4 Describe what Cisco IP Phone services provide.

Answer: Cisco IP Phone services enable XML applications to display interactive content with text and graphics on Cisco IP Phone models 7960 and 7940.

5 Which two Cisco IP Phone models support phone services?

Answer: Currently only Cisco IP Phone models 7960 and 7940 support Cisco IP Phone services.

6 Name three of the Cisco AVVID IP Telephony components that access user information you configure in CCMAdmin?

Answer: Directory Services, Cisco WebAttendant, and the Cisco IP Phone User Options Web pages access the user information that you enter in CCMAdmin. Other services, such as the IP Auto Attendant and Cisco Personal Assistant, also utilize the user information defined in CCMAdmin.

7 Assuming you are using a Cisco IP Phone model 7940, list four of the seven items a user can select from the Cisco IP Phone User Options Web page.

Answer: At the logon page, enter the user name and password configured when setting up the user. Once logged in, you can configure or select the following settings:

- **Choose the device or device profile to configure**
- **Change the password**
- **Change the PIN**
- **Configure your Cisco Personal Address Book**

When you have selected a device, you are able to configure the following settings in addition to the four settings previously listed:

- **Set or update speed dials**
- **Set or cancel a call forwarding designation**
- **Subscribe or unsubscribe to Cisco IP Phone services if the phone model selected is a Cisco IP Phone 7960 or 7940**

8 What do you need to do in CCMAdmin to ensure that users can access the Cisco IP Phone User Options Web page?

Answer: You must configure user information, including user ID, password, and device association, in Cisco CallManager Administration (*User > Add a New User*).

9 Once phone services have been subscribed to a given phone, how does the user access those services?

Answer: For Cisco CallManager Release 3.1(1), the two phone models that support phone services are the Cisco IP Phone 7940 and 7960. To access phone services, press the *services* button and use the rocker keys, keypad, and soft keys to navigate between and within the services.

10 Name the two Web pages through which a system administrator can subscribe to Cisco IP Phone services.

Answer: System administrators can subscribe to Cisco IP Phone services from the Phone Configuration page in CCMAdmin and the Cisco IP Phones Services page in the Cisco IP Phone User Options Web page.

Answers to Chapter 6 Pre-Test and Post-Test Questions

1 Label the model number for the following Cisco IP Phones:

a.

b.

c.

d.

Answer: You should have labeled the Cisco IP Phone models as follows:

a. Cisco IP Phone 7960

b. Cisco IP Phone 7910

c. Cisco IP Conference Station 7935

d. Cisco IP Phone 7940

2 In a CIPT environment, what are the three supported gateway protocols?

Answer: The three supported gateway protocols are MGCP, H.323, and the Skinny Gateway protocol for legacy gateways.

3 Which gateway protocols are supported by the WS-6608-T/E1 and DT24+ gateways when using Cisco CallManager 3.1 and higher?

Answer: The WS-6608-T/E1and DT24+ gateways support MGCP.

4 Which gateway protocols are supported by the Cisco 2600 and 3600 Series routers?

Answer: The Cisco 2600 and 3600 Series routers support the MGCP and H.323 protocols.

5 In a distributed call processing deployment, what device can be used to provide call admission control between clusters?

Answer: In a distributed call processing deployment, gatekeepers can be used to provide call admission control between clusters.

List the four Cisco provided CTI applications that use a CTI port or CTI route point.

Answer: The following four applications use a CTI port:

6 Cisco IP SoftPhone

7 Cisco IP AutoAttendant

8 Cisco WebAttendant

9 Personal Assistant

10 What are the Cisco CallManager service parameters that need to be configured before integrating Cisco Unity?

Answer: The following Cisco CallManager service parameters need to be configured before integrating Cisco Unity:

MessageWaitingOnDN

MessageWaitingOffDN

VoiceMail

ForwardNoAnswerTimeout

ForwardMaximumHopCount

Answers to Chapter 7 Pre-Test and Post-Test Questions

1 On what server should BAT be installed?

a. Primary

b. Publisher

c. Subscriber

d. Any of the above

Answer: *b.* BAT must be installed on the server running the Publisher database for Cisco CallManager.

2 Where are the BAT Excel template files located?

Answer: BAT Excel template files are located in the C:\CisoWebs\BAT\ExcelTemplate folder. Because you should not have Excel installed on the Publisher database server, you need to copy these template files to your local machine after installation.

3 Describe the purpose of TAPS.

Answer: TAPS enables you to update dummy MAC addresses for phones. Using BAT, you can choose to add phones with dummy MAC addresses rather than entering each phone's MAC address individually. The phones' actual MAC addresses can be updated automatically in the Cisco CallManager database later simply by dialing into a TAPS directory number and following a few voice prompts.

4 Describe the steps you must complete to add phones and users to the Cisco CallManager database using BAT.

a.

b.

c.

d.

Answer: Following are the steps you must complete to add phones and users to the Cisco CallManager database using BAT:

a. Create a BAT phone template, including line details, speed dials, and phone services (if applicable).

b. Create a CSV file containing the device and user information.

c. Use BAT to insert the BAT template and CSV file details to the Cisco CallManager database.

d. (*Optional*) Use TAPS to update the phones if you are using auto-registration and have checked the dummy MAC address option.

5 When creating the CSV file for users or while inserting the users into the Cisco CallManager database using BAT, are password and PIN values required?

Answer: Password and PIN values are required for all users. You can specify these values either in the CSV file or during the bulk insert in BAT, but the values for a password and PIN must be supplied in one location or the other.

6 For values that appear on both the BAT template and the CSV file, which takes precedence?

Answer: Values in the CSV file override values for the same field in the BAT template.

7 Using BAT, can you add, update, or delete Cisco Catalyst 6000 Analog Interface Modules?

Answer: No. Only *ports* on Catalyst 6000 Analog Interface Modules can be added, updated, or deleted using BAT. The modules themselves must be added using CCMAdmin.

8 True or False? Auto-registration does not need to be enabled to use TAPS.

Answer: False. TAPS requires that auto-registration be enabled in Cisco CallManager.

9 How does BAT save time over the usual method of adding, updating, or deleting devices or users in CCMAdmin?

Answer: BAT saves time over the usual method of adding, updating, or deleting devices or users in CCMAdmin because you can create a BAT template for general transactions, such as adding two-line Cisco IP Phone models 7940 and then reuse it for future transactions. Although entering specific information in the CSV file can be time-consuming, you also have the advantage of entering the data in one file without traversing multiple screens in CCMAdmin over and over for each new phone, port, gateway, or user.

10 During BAT installation you have the option of also installing TAPS. Other than BAT and Cisco CallManager, name the required component TAPS needs to function.

Answer: Although you get TAPS for free during BAT installation, you must purchase a Cisco CRA server and configure TAPS on it for TAPS to function properly.

Answers to Chapter 8 Pre-Test and Post-Test Questions

1 When you are installing a cluster of Cisco CallManagers, what must be the first server installed?

a. Subscriber

b. Primary

c. Call Processor

d. Publisher

Answer: d. The Publisher must be the first server installed when you are installing a cluster of Cisco CallManagers.

2 List at least five pieces of information you should have prior to starting an installation.

Answer:

- **Cisco Product Key**
- **User and organization name**
- **Computer name**
- **Workgroup**
- **Domain suffix**
- **TCP/IP properties**
- **DNS information**
- **Database server (Publisher or Subscriber)**
- **Backup server or target**

3 List at least two recommended post-installation tasks.

Answers:

- **Changing passwords for CCMAdmin and SQLSvc accounts.**
- **Configuring DNS.**
- **Configuring the database.**
- **Activating Cisco CallManager services.**

4 If you choose a network directory as the backup destination, how should you configure the directory in Windows 2000?

Answer: If you choose a network directory as the backup destination, the directory must be shared in Windows 2000.

5 List at least five unnecessary services to be set to manual and stopped on all servers in a cluster, and list the two additional services to be set to manual and disabled on Subscriber servers.

Answer:

- **DHCP Client**
- **FTP Publishing Service**
- **Alerter Service**
- **Computer Browser**
- **Distributed File System**

In addition to the services in the preceding list, you should stop and set the following services to manual on the Subscribers:

- **IIS Admin Service**
- **World Wide Web Publishing Service**

6 When you are upgrading a cluster, which server in the cluster must you upgrade first?

Answer: Publisher

This question is worth four points. Describe the four steps used to upgrade a Cisco CallManager cluster of three servers supporting 2500 users.

Answer: The following is the process used to upgrade a Cisco CallManager cluster of three servers supporting 2500 users:

7. Upgrade the Publisher (Server A) and after the upgrade, reboot the server.

8. Upgrade the backup Cisco CallManager (Server C). The backup is upgraded next because no devices are registered to it and service will not be interrupted. When the upgrade is complete, reboot the server.

9. Upgrade the primary Cisco CallManager (Server B). When the upgrade begins, the Cisco CallManager service stops and the devices fail over to the backup Cisco CallManager (Server C). There is a slight interruption in service while the devices register and receive firmware updates. When the upgrade is complete, reboot the server.

10. The final step of the upgrade process is to reboot all of the servers in the cluster. This is done so that the DBConnection values in the registry are synchronized. Start by rebooting the Publisher (Server A). Once the reboot is complete, reboot the primary Cisco CallManager (Server B). When the server comes back online, wait 5–10 minutes to allow the devices to begin the fail-back process. Finally, reboot the backup Cisco CallManager (Server C). The cluster upgrade is now complete.

Answers to Chapter 9 Pre-Test and Post-Test Questions

1 What are some of the basics when considering security for Cisco AVVID?

Answers: In protecting the network and network elements, consider the following topics discussed in this chapter:

- **Maintain physical device security**
- **Use card readers and video surveillance in all data centers and wiring closets**
- **Restrict Telnet access**
- **Use TACACS+/RADIUS for all devices**
- **Change the Administrative password on the Cisco CallManager server**

2 In building a secure Cisco AVVID network, what are some things you should design into the network?

Answers: In building a secure network, design the following in the network:

- **Place all VoIP devices on "voice-only" VLANs. Use separate RFC 1918 addresses for VoIP devices**
- **Place a firewall between the Cisco CallManager cluster and all other devices**
- **Use sensors to monitor for attacks**

3 What are two network factors that affect voice quality?

Answer: Two factors affect voice quality: *lost packets* (jitter) and *delayed packets* (latency).

4 What are the three categories of QoS tools?

Answer: QoS tools can be separated into three categories:

- **Classification**
- **Queuing**
- **Network Provisioning**

5 What are the available power schemes supported by Cisco IP Phones?

Answer: Cisco IP Phones support a variety of power. The available power schemes:

- **Inline power**
- **Wall power**

6 Prior to sending power down the line, which process is used between the Catalyst inline power switch and a Cisco IP Phone?

Answer: The switch and the phone need to perform a discovery process. The switch needs to discover if the device plugged into a port is able to receive power.

7 What are the three phone design IP addressing options?

Answer: There are three phone design IP addressing options:

- **Create a new subnet and use that for Cisco IP Phones in a different IP address space (registered or RFC 1918 address space).**
- **Provide an IP address in the same subnet as the existing data device (PC or workstation).**
- **Start a new subnet in the existing IP address space. (This may require you to re-create the entire IP addressing plan for the organization.)**

8 When configuring the voice VLANs on an interface or a port on a Catalyst switch, which commands are used for the following switches?

a. Catalyst 4000 and 6000: ______________________

b. Catalyst 3524, 2900XL: ______________________

Answers:

8a. Catalyst 4000 and 6000s use the set port auxiliaryvlan CatOS command to create these Cisco IP Phone 802.1Q access trunks in the Catalyst 2948Gs, 2980Gs, 4000s, and 6000s:

```
cat6k-access> (enable) set vlan 10 name 10.1.10.0_data
cat6k-access> (enable) set vlan 110 name 10.1.110.0_voice
cat6k-access> (enable) set vlan 10 5/1-48
cat6k-access> (enable) set port auxiliaryvlan 5/1-48 110

cat4k> (enable) set vlan 11 name 10.1.11.0_data
cat4k> (enable) set vlan 111 name 10.1.111.0_voice
cat4k> (enable) set vlan 11 2/1-48
cat4k> (enable) set port auxiliaryvlan 2/1-48 111
```

8b. In the Catalyst 3500 and 2900 XL series, you can configure this same functionality using a different set of commands:

```
interface FastEthernet0/1
     switchport trunk encapsulation dot1q
     switchport trunk native vlan 12
     switchport mode trunk
     switchport voice vlan 112
     spanning-tree portfast
```

9 On an H.323 gateway, how does the router know which **dial-peer** statement that has the same destination pattern to use first? Or which command in the dial peer configuration is used to identify which dial peer is used first when the dial peers have the same destination pattern?

Answer: How does the router know which dial-peer statement to use first? The preference statement in the dial peer configuration determines which dial-peer statement to use first. The dial-peer statement with the preference set to 0 will be used first, then 1, then 2, and so on.

10 Assume you are running a version of Cisco IOS Software that supports MGCP on a gateway. What is the command that enables the MGCP application on that gateway?

```
router(config)#mgcp
```

Answers to Chapter 10 Pre-Test and Post-Test Questions

1 List the three device requirements for supporting call preservation.

Answer: The three device requirements for supporting call preservation are

- **Active connection maintenance**
- **Disconnect supervision**
- **Switchover algorithm**

2 List the three protocols used by Cisco CallManager to establish calls between devices.

Answer: Cisco CallManager establishes calls between devices via three IP protocols: SCCP, MGCP, and H.323.

3 Cisco CallManager servers in a cluster use what type of links to establish calls between devices registered to different Cisco CallManager servers?

Answer: Cisco CallManager servers in a cluster use Signal Distribution Layer (SDL) links to establish calls between devices registered to different Cisco CallManager servers.

4 What are the three disconnect supervision mechanisms devices must provide for any media connections to be preserved during system failure?

Answer: Devices must provide at least one of the following disconnect supervision mechanisms for any media connections that are preserved during system failure:

- **End user release**
- **Timed**
- **MSF**

5 Define or give an example of the "end user release" disconnect supervision.

Answer: The end user release disconnect supervision is where the device can detect when the user of the connection hangs up.

Complete the missing information marked by questions 6–12 in the following table:

Devices	Switchover Algorithm	Disconnect Supervision
Cisco IP Phones • 12-button series • 30-button series • 79*XX* series	6. **Answer: Graceful**	7. **Answer: End user**
CFB/MTP • Software services • Transcoder	8. **Answer: Immediate**	9. **Answer: MSF**
MGCP Gateways • DT24+ (T1-PRI/CAS) • E1/T1-PRI • T1-CAS • 24 port FXS • IOS platforms (FXO/FXS, T1-PRI/CAS)	10. **Answer: Immediate**	11. **Answer: End user (where Applicable)** 12. **Answer: MSF**

Answers to Chapter 11 Pre-Test and Post-Test Questions

1 List the major components the MRM interfaces with.

Answer: The MRM interfaces with the following major components:

- **Call control**
- **Media control**
- **Media termination point control**
- **Unicast bridge control**
- **MOH control**

2 When does call control interface with the MRM?

Answer: Call control interfaces with the MRM when it needs to locate a resource to set up a conference call or invoke the MOH feature.

3 When does the Media layer interface with the MRM?

Answer: The Media layer interfaces with the MRM when it needs to locate a resource to set up an MTP.

4 What media resource provides the capability to bridge an incoming RTP stream to an outgoing RTP stream on an H.323v1 gateway?

Answer: MTP provides the capability to bridge an incoming RTP stream to an outgoing RTP stream on an H.323v1 gateway.

5 How many full-duplex streams are configurable for software conferences?

Answer: For software conferences, the following limits apply:

- **Up to 128 full-duplex streams are configurable.**
- **With 128 streams, a software conference media resource can handle 128 users in a single conference or**
- **With 128 streams, a software conference media resource can handle up to 42 conferences with three users per conference**

6 How many simplex streams are configurable for MOH media resources?

Answer: For MOH media resources, the following limits apply:

- **Up to 500 simplex streams are configurable.**
- **With 51 configured streams, up to 51 resources are available for MOH application.**

7 How many transcoder resources are available per X-Code device?

Answer: For transcoders, 48 streams register and provide up to 24 X-Code resources.

8 What are the limits for a hardware conferencing resource on a Catalyst 6000?

Answer: For hardware conference on a Catalyst 6000, 32 streams register. Therefore, the hardware conference media resource can handle 32 users in a single conference or up to 10 conferences with three users per conference.

9 List at least two reasons for sharing the media resources within a Cisco CallManager cluster.

Answer: The following are reasons for sharing the media resources within a Cisco CallManager cluster:

- **To enable both hardware and software devices to co-exist within a Cisco CallManager**
- **To enable Cisco CallManager to share and access resources available in the cluster**

- **To enable Cisco CallManager to do load distribution within a group of similar resources**
- **To enable Cisco CallManager to allocate resources based on user preferences**

10 The MOH server is actually a component of which service installed during a Cisco CallManager installation?

Answer: The MOH server is actually a component of the Cisco IP Voice Media Streaming Application (ipvmsapp.exe).

Answers to Chapter 12 Pre-Test and Post-Test Questions

1 For deployments of the distributed call processing model, what is used to provide CAC?

Answer: For deployments of the distributed call processing model, use the H.323 gatekeeper-controlled method to provide CAC.

2 True or False? In a CIPT environment the gatekeeper is designed to work with other Cisco IOS H.323 gateways.

Answer: True. The gatekeeper is designed to work in a CIPT environment with intercluster trunks.

3 For each codec listed below, what bandwidth will Cisco CallManager deliver to the gatekeeper?

- G.711 ____________
- G.729 ____________
- G.723 ____________

Answer: Cisco CallManager will deliver the following bandwidth to the gatekeeper:

- **G.711 (128 kbps)**
- **G.729 (20 kbps)**
- **G.723 (14 kbps)**

4 What keeps track of the current amount of bandwidth consumed by inter-location voice calls from a given location?

Answer: The centralized CallManager keeps track of the current amount of bandwidth consumed by inter-location voice calls from a given location.

5 What Cisco CallManager route plan feature in a centralized call processing deployment enables for the same access code to be used for PSTN access through a local gateway at multiple sites within the same Cisco CallManager cluster?

Answer: The same code can be used for PSTN access and, based upon the partition and CSS, a local gateway is selected.

6 What is the feature that automatically detects a failure in the network, and using Cisco Simple Network Automated Provisioning capability, initiates a process to intelligently auto-configure the router to provide call processing backup redundancy for the Cisco IP Phones in that office?

Answer: SRST automatically detects a failure in the network, and using Cisco Simple Network Automated Provisioning capability, initiates a process to intelligently auto-configure the router to provide call processing backup redundancy for the Cisco IP Phones in that office.

7 List at least three of the four supported types of phone calls using SRST:

Answer: The following lists the supported types of phone calls using SRST:

- **Cisco IP Phone to Cisco IP Phone**
- **Cisco IP Phone to any router voice port**
- **Cisco IP Phone to VoIP H.323/SIP OnNet**
- **Cisco IP Phone to VoFR/VoATM OnNet**

8 What are the two ways QoS tools ensure voice quality?

Answer: QoS tools ensure voice quality in two ways: by giving voice priority over data and by preventing voice from oversubscribing a given WAN link.

9 List at least two of the four caveats that should be considered when deploying locations-based CAC.

Answer: The following caveats should be considered when deploying locations-based CAC:

- **Mobility of devices between locations is not possible because Cisco CallManager decrements the specified location, not the physical location, of the device.**
- **Calls are admitted based on the availability of 24 kbps of bandwidth. G.729 calls consume 24 kbps and G.711 calls consume 80 kbps. Thus, if mixed codecs are used over the WAN, all calls should be assumed to consume 80 kbps and the bandwidth allocated accordingly. Where possible, a single codec should be configured for the WAN. In this case, the bandwidth allocated should be done so in *n* x 24 kbps increments for G.729 or *n* x 80 kbps increments for G.711.**

- **The locations-based CAC mechanism works across different servers. This means that the Cisco CallManager cluster in a centralized call processing deployment can now contain up to four active Cisco CallManagers to support a maximum of 10,000 Cisco IP Phones or 20,000 total device units (when Cisco CallManager runs on a larger supported server). Devices such as gateways, conferencing resources, voice mail, and other applications "consume" device units according to their relative "weight."**
- **Cisco CallManager deployments of centralized call processing are limited to hub and spoke topologies.**
- **Where more than one circuit or virtual circuit exists to a spoke location, the bandwidth should be dimensioned according to the dedicated resources allocated on the smaller link.**

10 The maximum bandwidth setting for a zone should take into account the limitation that the WAN link may not be filled with more than what percentage for voice?

Answer: The maximum bandwidth setting for a zone should take into account the limitation that the WAN link may not be filled with more than 75 percent voice.

Answers to Chapter 14 Pre-Test and Post-Test Questions

1 Which Cisco service must be running on the Cisco CallManager server for Cisco WebAttendant to work?

Answers: The TCDSRV must be activated and running on the Cisco CallManager server for Cisco WebAttendant to work.

2 What is the purpose of the pilot point directory number as it relates to Cisco WebAttendant?

Answer: The pilot point directory number is the directory number used to call Cisco WebAttendant, usually the main office number.

3 What should the last member of a Cisco WebAttendant hunt group be?

Answer: The last member of the hunt group should be the AutoAttendant or voice mail directory number that will be called when the user is offline. You should never have the last member directed to the first member of the group because that creates a continuous loop.

4 As an administrator, what must you enable for a user to be able to use Cisco IP SoftPhone?

Answer: You must check the box to Enable CTI Application Use on the User Information page in CCMAdmin (User > Global Directory > *complete search criteria and click Search > click on the user you want to view configuration information for*) to enable the user to use Cisco IP SoftPhone.

5 What information does the user need to install the Cisco IP SoftPhone application on their PC?

Answer: The user needs the following information:

- **Primary and secondary Cisco CallManager IP addresses. (From a Cisco IP Phone 7960, press the *settings* button, then press 3 for Network Configuration and then 21 for the Cisco CallManager IP address).**
- **Cisco IP SoftPhone installation software and installation/configuration documentation.**
- **Cisco IP Phone User Options logon information (user name and password).**
- **The location of any local customization files that provide dirctory and audio settings if available.**

6 What does Cisco Unity use as the message store and directory service to unify your system administration?

Answer: Cisco Unity uses Microsoft Exchange's message store and directory services to unify your system administration, collecting all messages in a single store and providing you with a single address directory service.

7 What should be configured first in Cisco CallManager to prepare for integration with Cisco Unity?

Answer: To prepare for Cisco Unity integration with Cisco CallManager, you must configure Cisco voice mail ports.

List three of the four service parameters in Cisco CallManager that need to be configured for integration with Cisco Unity.

8

____

____

____

9

____

____

____

10

____

____

____

Answer: The four service parameters that need to be configured in Cisco CallManager are the following:

VoiceMail—The directory number of the first voice mail port.

MessageWaitingOffDN—The directory number that turns off the MWI.

MessageWaitingOnDN—The directory number that turns on the MWI.

VoiceMailMaximumHopCount—The number of call forwards enabled before a call is dropped; this is important to eliminate voice mail forwarding loops.

INDEX

Symbols

Numerics

A

B

C

D

E

F

G

H

I

J-K-L

M

N

O

P

Q

R

S

T

U

V

W-X-Y-Z